PROJECT GEMINI

The Bridge to the Moon - 12 Missions - Five Years

By: TD Barnes

Title of the Book: Project Gemini

Independently published.

Printed in the United States of America
First Edition

Contents

Foreword

This chronology is about Project Gemini - The Bridge to the Moon - 12 Missions - Five Years.

The constellation for which the project was named is commonly pronounced /'dʒɛmɪnaɪ/, the last syllable rhyming with eye. However, staff of the Manned Spacecraft Center, including the astronauts, tended to pronounce the name /'dʒɛmɪni/, rhyming with knee. NASA's public affairs office then started in 1965, declaring "Jeh'-mih-nee" the "official" pronunciation. Gus Grissom, acting as Houston capsule communicator when Ed White performed his spacewalk on Gemini 4, was heard on flight recordings pronouncing the spacecraft's call sign "Jeh-mih-nee 4", and the NASA pronunciation was used in the 2018 film First Man.

Gemini was one of the early pioneering efforts in developing this nation's space capability. The initiation of this program was timed to take advantage of the knowledge gained in our first series of manned space flights - Project Mercury. The Mercury program successfully demonstrated manned orbital flight. Perhaps more important it provided extensive information on how to build and fly spacecraft for the more complex missions yet to come. Drawing on this experience, the Gemini program was able to produce for its time a highly flexible space vehicle of considerable operational capability. These characteristics enabled a rapid expansion of American flight horizons.

The Gemini program, spanning from 1964 to 1967, marked a pivotal era in NASA's quest to conquer space. Designed to bridge the achievements of the Mercury missions with the ambitions of Apollo, Gemini spacecraft pushed the boundaries of human exploration. Each launch, propelled by a Titan II rocket—a modified Cold War missile adapted for peaceful missions—lifted the spacecraft skyward. Standing 5.8 meters long and with a diameter of 3 meters, the Gemini craft resembled its predecessor, the Mercury capsule, but expanded its capabilities.

During Gemini missions, astronauts demonstrated feats previously unimaginable. On June 3, 1965, Ed White embarked on the first American spacewalk, tethered to the spacecraft by a 7.6-meter umbilical cord. His 23-minute journey outside the craft illustrated humanity's adaptability in microgravity. Later, on November 12, 1966, Buzz Aldrin undertook an extravehicular activity during the Gemini 12 mission, showcasing precise maneuvering near the Agena workstation.

In June 1965, during Gemini 4, astronaut Ed White embarked on America's first spacewalk, demonstrating the feasibility of extravehicular activity (EVA). This milestone not only showcased the ability to work outside the spacecraft but also laid the groundwork for future exploration and construction in space.

Beyond technological achievements, Gemini provided invaluable flight experience for astronauts, honing their piloting skills and operational acumen in the unforgiving environment of space. Each mission tested and refined guidance systems, communications equipment, and navigation techniques, ensuring that every component of the Apollo spacecraft could function flawlessly during the demanding lunar missions ahead.

Gemini missions were not merely about spacewalks and extravehicular activities; they pioneered critical techniques for space exploration. Gemini 5 set a record for the longest manned spaceflight at the time, lasting eight days. Gemini 7 and 6 achieved the first rendezvous of two manned spacecraft in orbit, a vital precursor to Apollo's lunar missions. These milestones in spaceflight—culminating in Gemini 12's automated reentry into Earth's atmosphere—solidified NASA's expertise in orbital maneuvering, docking procedures, and spacecraft systems.

The most significant achievements of Gemini involved precision maneuvering in orbit and a major extension of the duration of manned space flights. These included the first rendezvous in orbit of one spacecraft with another and the docking of two spacecraft together. The docking operation allowed the use of a large propulsion system to carry men to greater heights above Earth than had been previously possible, thereby enabling the astronauts to view and photograph Earth over extensive areas. Precision maneuvering was also employed during the very high speed reentry back to the surface of Earth, enabling accurate landings to be made. The length of our manned space flights was extended to 14 days, which has yet to be exceeded as of this writing, although this was accomplished about three years ago.

Of great general interest were the investigations into the operations of an astronaut outside the confines of his spacecraft, protected from the hard vacuum of space by his pressurized space suit. These extravehicular activities did, in fact, produce some difficulties, but, in the end, highly successful operations were conducted.

These activities have contributed significantly to expanding activities in space that we now have underway or will be forthcoming. In Apollo, the program involved landing men on the lunar surface, and the crews had to be transported roughly 240,000 miles to the Moon and then back to Earth. This trip took a week or more. The Apollo spacecraft had to perform a rendezvous not near Earth but out at lunar distances for this mission to be successful. Once again, the astronauts must leave their spacecraft and, in their pressure suits, step out onto the lunar surface so that scientific exploration can be conducted. The fact that all of these things were initially demonstrated and then investigated further in a number of the Gemini missions dramatically aids the development of the more difficult missions we are about to undertake.

Perhaps the most significant aspect of the Gemini program was how the astronauts contributed to the success of each mission. In the flying of the spacecraft, in the management of the systems, in the overcoming of problems, and the aid to the attainment of significant scientific and technological information, their presence greatly enhanced the program's success. They were backed up by a large, dedicated team of people on the ground who designed, developed, and checked out the vehicles and controlled the flights. The Chronology is presented herein as a factual presentation of events taken primarily from official program documentation. It, therefore, cannot reflect many of the "behind the scenes" activities so crucial to the conduct of a successful program involving exploratory endeavors. The high motivation to make the Gemini program work, the rapid reaction in overcoming difficulties, large and small, and the attention to detail all contributed to the ten successful manned flights, which provided nearly two thousand man-hours of direct space flight experience.

In June 1965, during Gemini 4, astronaut Ed White embarked on America's first spacewalk, demonstrating the feasibility of extravehicular activity (EVA). This milestone showcased the ability to work outside the spacecraft and laid the groundwork for future exploration and construction in space.

Beyond technological achievements, Gemini provided invaluable flight experience for astronauts, honing their piloting skills and operational acumen in the unforgiving environment of space. Each mission tested and refined guidance systems, communications equipment, and navigation techniques, ensuring that every component of the Apollo spacecraft could function flawlessly during the demanding lunar missions ahead.

Chapter 1 - Concept and Design

April 1959 through December 1961

In April 1959, NASA embarked on a critical research phase to advance human spaceflight capabilities beyond the Mercury program. DeMarquis D. Wyatt, Assistant to the Director of Space Flight Development, presented a pivotal request to Congress for $3 million. This funding was earmarked for pioneering research in space rendezvous techniques, laying the groundwork for future missions, including a manned space laboratory.

The objectives outlined were ambitious and multifaceted. Central to the initiative was developing precise methods for positioning two spacecraft about each other—a fundamental requirement for rendezvous in space. This involved pioneering efforts in establishing robust referencing systems capable of fixing relative positions with accuracy unprecedented in aerospace engineering at the time.

Additionally, the project aimed to pioneer lightweight yet highly accurate target acquisition equipment. This technology was pivotal for enabling supply craft to locate and dock with space stations or other spacecraft reliably, thus facilitating complex operations in orbit.

Another critical focus was the advancement of guidance and control systems. The goal was to develop systems capable of precise flight path determination, crucial for navigating spacecraft safely and effectively in the harsh environment of space.

Furthermore, the initiative addressed the need for reliable power sources capable of operating in controlled conditions, essential for sustaining manned missions and onboard systems.

This research phase represented a pivotal period in NASA's history, marking the transition from theoretical concepts to practical applications that would define the capabilities and achievements of the Gemini program in the years to come.

The inaugural meeting of the Goett committee on May 25-26 marked a pivotal moment in NASA's quest for advancing manned spaceflight beyond the Mercury program. Appointed by John W. Crowley, NASA's Director of Aeronautical and Space Research, Harry J. Goett of NASA's Ames Research Center took the helm as chair. As agreed upon from the outset, the committee's mandate was to strategize the long-term objectives of NASA's manned spaceflight initiatives.

Central to their agenda was formulating comprehensive research goals, coordinating research efforts across NASA's various centers, and providing recommendations on spacecraft development and associated research priorities.

A primary focus of the committee's deliberations was planning a manned spaceflight program to succeed Mercury. H. Kurt Strass from NASA's Space Task Group (STG) at Langley Field, Virginia, presented initial concepts for post-Mercury missions. These included proposals such as an enlarged Mercury capsule capable of carrying two astronauts on a three-day orbital mission, a configuration with a two-man Mercury spacecraft coupled with a larger support cylinder for extended two-week missions, and an innovative concept involving a Mercury capsule attached to a cylindrical structure by cables, rotated to simulate artificial gravity.

In its 1960 budget request, NASA sought $2 million to explore methods for constructing a manned orbiting laboratory or adapting the Mercury capsule into a two-man laboratory suitable for prolonged space missions. These initiatives underscored NASA's ambitious vision and commitment to pushing the boundaries of human space exploration during this transformative era.

During a pivotal staff meeting on June 4, Robert R. Gilruth, Director of NASA's Space Task Group (STG), proposed a groundbreaking concept for the next phase of NASA's manned spaceflight program. His suggestion centered on developing maneuverable Mercury capsules capable of executing controlled landings in pre-designated areas.

Gilruth's proposal marked a strategic shift in NASA's approach, aiming to enhance the versatility and operational capabilities of spacecraft beyond the orbital missions of the Mercury program. By enabling precise landings, these maneuverable capsules would open up new avenues for exploring different terrains and conducting targeted scientific research on Earth return missions.

The discussion reflected NASA's ongoing commitment to advancing space exploration technologies, laying the groundwork for future programs that would build upon the achievements and lessons learned from Mercury.

On June 22, H. Kurt Strass, representing the Flight Systems Division (FSD) of NASA's Space Task Group (STG), proposed a significant proposal to advance the agency's manned spaceflight capabilities. Strass recommended the formation of a dedicated committee tasked with developing preliminary designs for a two-man space laboratory.

The proposed committee would comprise representatives from various specialist groups within FSD, collaborating closely with a special projects group. Their collective effort would focus on formulating detailed design specifications tailored for the envisioned two-man Mercury spacecraft.

This initiative underscored NASA's strategic intent to expand the operational scope of its manned missions, moving beyond the Mercury program's single-seat configurations. It reflected a proactive approach to harnessing interdisciplinary expertise within NASA to pioneer new frontiers in space exploration.

The inaugural meeting of the New Projects Panel within NASA's Space Task Group (STG), chaired by H. Kurt Strass, convened on August 12. This panel's primary focus was to address critical challenges related to spacecraft reentry at velocities nearing escape speed, atmospheric and space maneuvers, and recovery via parachute for Earth landing.

Alan B. Kehlet, representing STG's Flight Systems Division, spearheaded a pioneering program to develop a second-generation capsule. This advanced spacecraft was envisioned to accommodate three astronauts and boast enhanced capabilities for both orbital maneuvering and atmospheric flight.

Key features of the proposed spacecraft included a primary reentry system optimized for water landings, with secondary provisions for land-based landings. At the subsequent meeting on August 18, Kehlet presented initial design concepts and recommendations for further development.

The panel's deliberations culminated in a consensus to draft a comprehensive specifications list as the initial step towards formalizing the engineering requirements for this ambitious new spacecraft. This marked a pivotal moment in NASA's quest to evolve beyond the confines of the Mercury program, aiming to pioneer new technologies and capabilities that would redefine the future of manned space exploration.

On September 1, McDonnell Aircraft Corporation of St. Louis, Missouri, released a comprehensive report detailing their studies involving modified Mercury capsules for advancing spaceflight capabilities beyond the initial manned missions of Project Mercury. This extensive 300-page report outlined six proposed follow-on experiments to address various challenges and opportunities in space exploration.

The outlined experiments included:

- Touchdown control
- Orbital maneuverability
- Self-contained guidance systems
- Extended 14-day missions
- Manned reconnaissance missions
- Reentry from lunar orbit

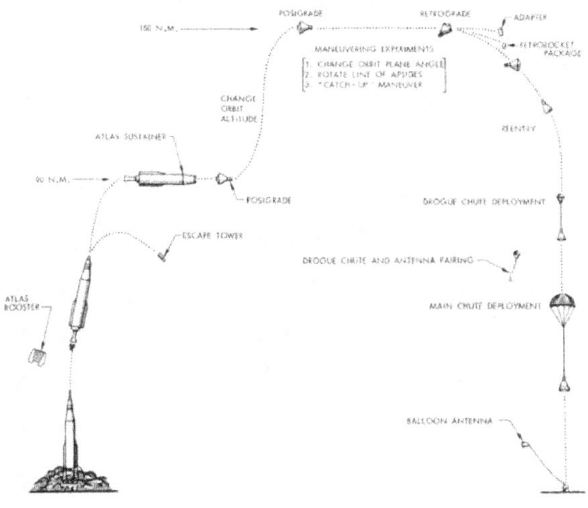

Proposed mission for modified Mercury capsule.

While these concepts were presented more as technically supported suggestions rather than finalized proposals, McDonnell Aircraft Corporation affirmed that all six experiments could feasibly be conducted with practical modifications to the existing Mercury capsule designs.

This report underscored the aerospace industry's ongoing efforts to innovate and expand manned spaceflight capabilities, laying the groundwork for future developments that would ultimately shape the trajectory of human exploration beyond Earth's atmosphere.

On September 28, the New Projects Panel of NASA's Space Task Group (STG) engaged in pivotal discussions regarding proposals put forth by McDonnell Aircraft Corporation for utilizing modified Project Mercury capsules in advanced spaceflight experiments. Chaired by H. Kurt Strass, the panel deliberated on these proposals' feasibility and strategic implications, ultimately acknowledging their relevance within the panel's jurisdiction.

Recognizing both the potential and limitations of McDonnell's suggestions, the panel directed further studies to explore actionable recommendations for STG. This directive set the stage for intense deliberations at the panel's subsequent meeting on October 5, which centered on refining McDonnell's proposals.

While identifying shortcomings in the individual proposals, the panel identified some aspects of significant value that could be consolidated into a unified concept—this consolidated proposal aimed to enhance spacecraft performance and expedite the evaluation of advanced mission concepts.

Anticipating the need for increased orbital capabilities, the panel advised immediate studies on the specifications for a second-stage propulsion system capable of restarts and precise thrust control. Additional studies were recommended for developing a second-stage guidance and control system to achieve orbital altitudes up to 150 miles and ensure controlled reentry within the heat protection limits of the existing or slightly modified capsules.

Furthermore, the panel highlighted the importance of investigating orbital maneuvering for rendezvous experiments and atmospheric reentry control experiments, underscoring NASA's commitment to pushing the boundaries of manned space exploration during this transformative period.

On January 7, 1960, a crucial meeting convened with representatives from NASA's Engineering and Contracts Division, alongside the Flight Systems Division (FSD), to strategize future wind tunnel testing requirements for advanced Mercury projects. Alan B. Kehlet initiated the discussion by outlining the available test facilities, setting the stage for significant deliberations led by Caldwell C. Johnson and H. Kurt Strass.

Johnson primarily focused on enhancements to the existing Mercury configuration, concentrating efforts on refining aspects such as the afterbody design, landing systems featuring rotors for impact control, and improvements to the retro-escape system. Meanwhile, Strass proposed classifying advanced work into two categories: modifications aimed at refining the current Mercury design and entirely new concepts for configuration design. This categorization was widely endorsed by the assembled team.

In response, Johnson agreed to develop models for both program categories, leveraging FSD's Aerodynamics Section to facilitate necessary tests for evaluating modification proposals and advanced configurations. Strass proposed a modification involving an enlarged heatshield diameter capable of deploying half-ringed flaps from the afterbody section near the heatshield, enhancing subsonic lifting capabilities.

The meeting also underscored the critical need for comprehensive aerodynamic data on an advanced Mercury configuration under consideration by Strass' team and insights into a lenticular vehicle concept proposed by the Aerodynamics Section. These discussions marked a pivotal step forward in NASA's quest to refine and innovate upon the foundational technologies of the Mercury program, anticipating advancements that would shape the future of manned space exploration.

On April 5, the Space Task Group (STG) issued preliminary specifications for modifying the Mercury capsule, focusing on integrating a reentry control navigation system. This enhancement aimed to imbue the capsule with a modest lifting capability, achieving a lift-over-drag ratio of approximately 0.26. Central to the proposed modifications was implementing a self-contained navigation system comprising a stable platform, a digital computer, a potential star tracker, and requisite electronic components.

An essential performance requirement stipulated that the modified capsule exhibit a dispersion of less than 10 miles from the predicted impact point upon reentry. The development timeline outlined ambitious milestones: delivery of a prototype for NASA testing by February 1961, the initial qualified system (Modification I) by August 1961, and the final qualified system (Modification II) by January 1962.

STG projected a need for four navigational systems beyond the prototype and qualification units, emphasizing the initiative's scale and urgency. This endeavor underscored NASA's commitment to advancing manned spaceflight capabilities through systematic upgrades to existing technology, paving the way for more precise and controlled missions in Earth's orbit and beyond.

On May 16-17, representatives from NASA's research centers convened at Langley Research Center for a pivotal conference focused on space rendezvous. The gathering served as a platform for presenting papers on ongoing programs related to rendezvous and exploring future research directions in this critical area of space exploration.

The first day of the conference featured presentations detailing current initiatives at Langley, Ames, Lewis, Flight Research Center, Marshall Space Flight Center, and Jet Propulsion Laboratory. These presentations highlighted various approaches and methodologies pursued across NASA's diverse research landscape.

The second day of the conference was dedicated to a roundtable discussion, where consensus quickly emerged among participants. There was a shared conviction that rendezvous techniques would soon become indispensable for future space missions. Participants stressed the urgency of developing these techniques immediately, emphasizing their potential to enable complex missions and expand the scope of human spaceflight.

Throughout the conference, a unanimous call resonated for NASA to prioritize rendezvous experiments to refine these techniques and establish their feasibility. This collective vision underscored NASA's proactive approach to advancing space exploration capabilities, setting the stage for groundbreaking developments that would shape the trajectory of manned missions in the years to come.

In June, the Space Task Group (STG) released significant guidelines to advance manned spaceflight programs. This comprehensive document synthesized insights from five papers presented by STG personnel during meetings held with NASA Headquarters and various field installations throughout April and May.

Central to the discussions was the exploration of a manned circumlunar mission, also known as lunar reconnaissance. Charles J. Donlan, Associate Director (Development), highlighted an intermediate program envisioned to bridge the gap between the conclusion of the Mercury program and the initiation of flight tests for a multimanned vehicle. He articulated the concept of a potential flight-test program involving the reentry unit of this future multimanned vehicle, sometimes likened to a lifting version of the Mercury capsule.

The specifics of such a vehicle remained uncertain, necessitating thorough investigation and data gathering. STG proposed conducting extensive wind tunnel studies at Langley, Arnold Engineering Development Center, and Ames facilities. These studies would focus on exploring the dynamics of a reentry vehicle with a nonzero lift-over-drag ratio, encompassing aspects such as static and dynamic stability, pressure dynamics, and heat transfer characteristics.

The initiative underscored NASA's proactive approach to preparing for advanced manned missions beyond Earth's orbit, reflecting a commitment to rigorous scientific inquiry and technological innovation essential for pushing the boundaries of human exploration in space.

On August 24, McDonnell Aircraft Corporation put forward a bold proposal for a one-man space station, integrating a Mercury capsule with a cylindrical space laboratory. This innovative concept aimed to accommodate one astronaut in a comfortable shirtsleeve environment for up to 14 days in Earth orbit.

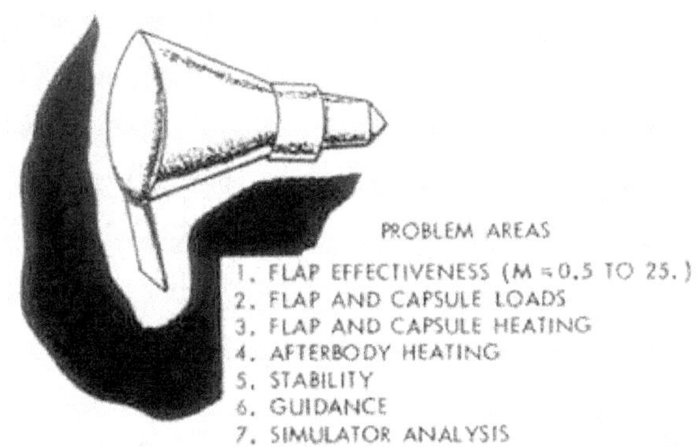

Early version of "lifting" Mercury capsule.

The proposed vehicle, combining the Mercury capsule with the laboratory module, was anticipated to have a gross weight of 7259 pounds at launch. By comparison, the Mercury capsule alone weighed 4011 pounds as of October 25, 1960. The configuration would allow for an 1100-pound payload dedicated to laboratory testing, positioned in a 150-nautical-mile orbit with the assistance of an Atlas-Agena B launch vehicle.

This concept, described as a "minimum cost manned space station," underscored McDonnell Aircraft Corporation's innovative approach to leveraging existing technology and infrastructure to expand human presence and capabilities in space. The proposal represented a significant step towards realizing extended-duration missions and scientific research opportunities in Earth's orbit, reflecting early space exploration endeavors' growing ambition and capability.

NASA's Space Exploration Program Council convened in Washington for a pivotal meeting focusing on the prospects of manned lunar landing. The outcome of this meeting was significant: the council reached a consensus that NASA should initiate planning for an earth-orbital rendezvous program. This program was intended to operate independently while also contributing to the broader goals of the manned lunar landing program.

The decision underscored NASA's strategic approach to developing concurrent missions that could advance technological capabilities and operational readiness for ambitious lunar exploration missions. By prioritizing earth-orbital rendezvous, NASA aimed to refine crucial techniques and operational procedures for rendezvous and docking maneuvers in space.

This directive marked a critical juncture in NASA's trajectory, laying the groundwork for coordinated efforts that would ultimately culminate in historic achievements in manned space exploration, including the eventual lunar landings of the Apollo missions. The meeting's outcomes reflected NASA's commitment to systematic planning and innovation, setting the stage for the next phase of humanity's journey into space.

On January 20, 1961, the management of NASA's Space Task Group convened a crucial Capsule Review Board meeting. Central to the agenda was the discussion of a follow-on program to the Mercury missions, marking a pivotal moment in shaping the future of manned spaceflight initiatives.

During the meeting, several types of missions were deliberated, each representing potential avenues for advancing beyond the achievements of the Mercury program. These included missions focused on long-duration flights, rendezvous operations, experiments in artificial gravity, and flight tests incorporating advanced equipment and technologies.

A critical consensus emerged from the discussions: while there was a collective recognition of the need for a follow-on program, it was deemed essential to define its objectives and specifications in greater detail. This strategic approach aimed to ensure that subsequent missions would build upon the successes and lessons learned from Mercury while pushing the boundaries of human space exploration further.

The outcomes of the Capsule Review Board meeting set the stage for intensified planning and development efforts within the Space Task Group, laying the groundwork for formulating concrete plans to shape NASA's future manned space missions. This commitment to continuous advancement and innovation underscored NASA's evolving role in pioneering new frontiers in space exploration during a transformative era in human history.

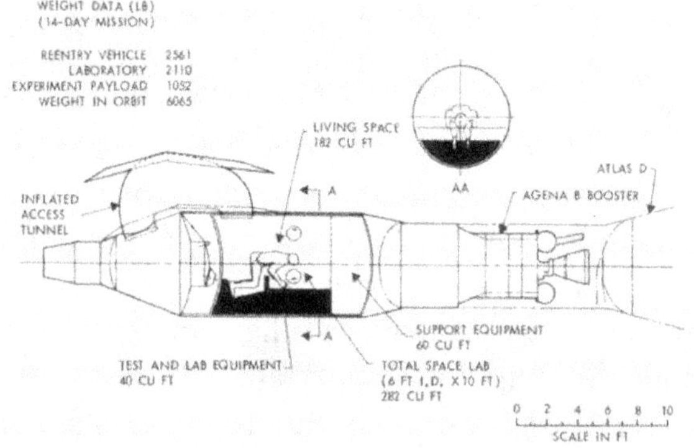

One of two versions of a one-man space station proposed by McDonnell. In this version, access to the laboratory was through an inflated tunnel connecting the Mercury-type capsule (in which the astronaut rode into orbit) with the laboratory proper (the forward section of an Agena booster attached to the capsule).

In early February 1961, NASA and McDonnell Aircraft Corporation initiated discussions regarding an advanced Mercury spacecraft, marking a significant milestone in the evolution of manned spaceflight capabilities.

McDonnell had been exploring concepts for a maneuverable Mercury spacecraft since 1959, and these discussions were formalized on February 1 following a directive from Space Task Group (STG) Director Robert R.

Gilruth. James A. Chamberlin, Chief of STG's Engineering Division and a seasoned collaborator with McDonnell on Mercury projects was tasked with leading studies to enhance the Mercury spacecraft for future manned spaceflight programs. These efforts encompassed exploring various versions of the spacecraft, ranging from incremental improvements to radical redesigns.

By early March, the possibility of conducting extravehicular operations (EVA) prompted Maxime A. Faget of STG to discuss with John F. Yardley of McDonnell the potential for a two-man version of the improved Mercury spacecraft. Yardley subsequently raised this concept with Walter F. Burke, a McDonnell vice president, who promptly ordered the preparation of design drawings for a two-man Mercury variant.

Progress on these initiatives was reported to Abe Silverstein of NASA Headquarters during a meeting at Wallops Island, Virginia, held from March 17-20. James T. Rose of STG, collaborating with Chamberlin, focused on studying the potential objectives for the advanced Mercury, emphasizing mission planning, trajectory analysis, and performance enhancements.

These developments underscored NASA's proactive approach in advancing manned space exploration capabilities, leveraging partnerships with industry leaders like McDonnell to pioneer new frontiers in spaceflight. The collaborative efforts aimed to expand the scope and capabilities of the Mercury spacecraft, laying the groundwork for future missions that would push the boundaries of human space exploration.

Gus Grissom played a significant role in developing the Gemini spacecraft, influencing its design to suit his physique. The Gemini capsule, designed by James A. Chamberlin, was tailored to carry a crew of two astronauts. Chamberlin, formerly the chief aerodynamicist on Avro Canada's CF-105 Arrow program, joined NASA and a group of senior Avro engineers after the Arrow program was canceled. He became head of the U.S. Space Task Group's engineering division, overseeing the Gemini project.

McDonnell Aircraft Corporation was the prime contractor for the Mercury and Gemini capsules. Grissom's involvement in the Gemini spacecraft's design was notable enough that his preferences and physical dimensions influenced early design choices. His fellow Mercury astronauts jokingly referred to the spacecraft as the "Gusmobile," highlighting its tailored design around Grissom's 5'6" frame.

However, as NASA discovered in 1963 that most astronauts did not fit comfortably in the original design, significant interior redesigns were necessary to accommodate broader astronaut dimensions. Grissom's dedication to the Gemini program intensified after realizing that his Mercury flights were concluded, focusing all his energies on contributing to the upcoming Gemini missions. His insights and commitment were instrumental in shaping the early stages of the Gemini spacecraft's development and mission planning.

The Gemini program, a pivotal phase in NASA's manned spaceflight efforts, was overseen by the Manned Spacecraft Center (MSC) based in Houston, Texas. This center operated under the guidance of the Office of Manned Space Flight at NASA Headquarters in Washington, D.C. Dr. George E. Mueller played a crucial role as the Associate Administrator of NASA for Manned Space Flight and acted as the director of the Gemini program during its early stages.

William C. Schneider, the Deputy Director of Manned Space Flight for Mission Operations, assumed the role of mission director for all Gemini flights from Gemini 6A onwards. His responsibilities included coordinating and overseeing the operational aspects of the missions and ensuring the smooth execution of each flight's objectives and tasks.

Under Mueller and Schneider's leadership, the Gemini program achieved several historic milestones, including crucial advancements in space rendezvous, extravehicular activity (EVA), and long-duration missions. Their strategic oversight and operational management were instrumental in paving the way for NASA's subsequent ambitious missions, including the Apollo program's lunar landings.

Guenter Wendt played a pivotal role in NASA's early manned spaceflight programs, serving as a McDonnell engineer responsible for overseeing launch preparations for the Mercury and Gemini programs. His expertise extended into the Apollo program, where he continued supervising crewed mission launches. Wendt's team was crucial in executing the intricate pad close-out procedures immediately before spacecraft launches, ensuring all systems were ready for flight.

Known for his meticulous attention to detail and authoritative demeanor, Wendt became the last person astronauts interacted with before closing the hatch. His role involved ensuring the spacecraft's readiness and safety, a responsibility astronauts deeply appreciated. Despite the pressure of launch preparations, Wendt developed a good-humored rapport with the astronauts, fostering a positive and supportive atmosphere during critical mission moments. His contributions were instrumental in maintaining the operational integrity and success of NASA's early space missions.

The Astronauts

Deke Slayton played a pivotal role as NASA's director of flight crew operations during the Gemini program. He was responsible for assigning crews to each mission. His approach included designating primary and backup crews for each flight, with the backup crew rotating into primary status three missions later.

Slayton's initial priority was to assign mission commands to the four remaining active members of the Mercury Seven astronauts: Alan Shepard, Gus Grissom, Gordon Cooper, and Wally Schirra. John Glenn had retired earlier, while Scott Carpenter was on leave for the Navy's SEALAB project and later grounded due to an arm injury from a motorbike accident. Slayton himself was grounded due to a heart condition.

During the Gemini Project, Shepard's participation was affected by Menière's Disease, which caused inner ear issues and initially prevented him from flying. Despite corrective surgery, Shepard did not fly on Gemini missions but later returned to spaceflight as the commander of Apollo 14. This careful crew selection process under Slayton's guidance ensured that each Gemini mission crew was well-prepared and capable of achieving program objectives.

During the Gemini program, the crew positions and titles for the left-hand (command) and right-hand (pilot) seats were adapted from U.S. Air Force pilot ratings. The commander was called the Command Pilot, while the pilot occupied the right-hand seat. This terminology reflected the military aviation background of many astronauts involved in the program.

In total, sixteen astronauts flew aboard ten crewed Gemini missions. These missions were pivotal in advancing NASA's capabilities in spaceflight, conducting crucial experiments, developing techniques for rendezvous and docking, and laying the groundwork for subsequent Apollo missions that aimed to land humans on the Moon. Each mission required precise coordination between the crew members in their respective roles to achieve mission success and gather essential data for future space exploration endeavors.

In April 1961, NASA initiated study contract NAS 9-119 with McDonnell, marking a pivotal moment in the evolution from Project Mercury to the next phase of American space exploration: Project Gemini. McDonnell swiftly assembled a dedicated project group tasked with enhancing the capabilities of the Mercury spacecraft, which had served as a foundational platform for human spaceflight.

The study focused on substantial improvements to transform the Mercury spacecraft from a prototype into a fully operational vehicle. Key to this evolution was the reevaluation of component placement, shifting from the interior of the pressure vessel—where components were housed in Mercury—to external positions, prioritizing accessibility and operational efficiency.

Under the amended contract, McDonnell was authorized to procure critical long-lead-time items to advance the spacecraft's design. Despite these ambitious plans, budget constraints imposed by NASA's Space Task Group limited initial expenditures to $2.5 million—a reflection of the fiscal discipline necessary for managing such pioneering endeavors.

At its inception, the McDonnell project team comprised 30 to 40 engineers tasked with refining spacecraft technology to meet the rigorous demands of future manned missions. This marked the beginning of a new era in space exploration, where innovation and strategic planning would lay the groundwork for the success of Project Gemini and America's continued journey into space.

On April 24, Major General Don R. Ostrander, NASA's Director of Launch Vehicle Programs, delivered a crucial presentation to the House Committee on Science and Astronautics. His address focused on NASA's ambitious plans to develop orbital rendezvous techniques, a pivotal capability required for advancing space exploration initiatives. This topic became a centerpiece of discussions during the committee's hearings on NASA's proposed 1962 budget, highlighting its strategic importance in shaping future space missions.

Subsequently, on May 23, the Committee reconvened to hear further insights from Harold Brown, Director of Defense Research and Engineering, and Milton W. Rosen, Deputy to General Ostrander. Their presentations delved into the critical need for orbital rendezvous capabilities, outlining the technical methodologies and support infrastructure necessary to achieve this ambitious goal. These discussions underscored the collaborative efforts between NASA and defense sectors in advancing aerospace technology and preparing for the challenges of manned space missions.

On May 1st, anticipating a burgeoning era in manned spaceflight, the Space Task Group (STG) put forth a visionary proposal: establishing a dedicated center for developing manned spacecraft. As the operational hub for the Mercury program, STG had already laid the groundwork necessary for such an ambitious undertaking.

However, the envisioned scale of future manned missions necessitated a significant expansion across multiple fronts. This expansion encompassed the augmentation of personnel and infrastructure and enhanced organizational frameworks and management protocols. The proposal aimed to consolidate and streamline efforts towards a unified goal, ensuring that the evolving demands of space exploration could be met effectively and efficiently.

By advocating for creating a specialized development center, STG positioned itself at the forefront of advancing technology and innovation in aerospace. This initiative marked a pivotal moment in NASA's trajectory, setting the stage for unprecedented achievements in manned spaceflight and expanding the boundaries of human exploration beyond Earth's atmosphere.

On May 5th, the Gemini program emerged as a defining chapter in NASA's ambitious odyssey into space exploration from 1964 to 1967. Designed as a bridge between the foundational Mercury missions and the future aspirations of Apollo, the Gemini spacecraft represented a leap forward in human technological prowess. Each mission was launched atop a Titan II rocket, originally designed for military use during the Cold War but repurposed for peaceful space endeavors. Standing 5.8 meters long with a diameter of 3 meters, the Gemini capsules retained a familiar silhouette inherited from their Mercury predecessors while significantly enhancing operational capabilities.

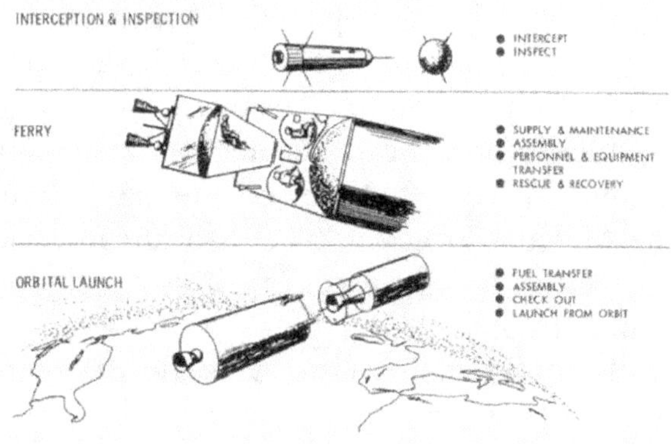

The classes of orbital operations which a NASA Headquarters working group felt would be required in any future space program and which thus made a rendezvous development program necessary.

Throughout the Gemini missions, astronauts achieved feats that were previously the realm of science fiction. On June 3, 1965, Ed White embarked on America's first spacewalk, tethered to the spacecraft by a 7.6-meter umbilical cord, showcasing humanity's adaptability in the microgravity environment of space. Later, during the Gemini 12 mission on November 12, 1966, Buzz Aldrin performed a precision extravehicular activity near the Agena workstation, further refining techniques crucial for future missions.

Beyond these historical milestones, Gemini missions pioneered essential techniques for space exploration—Gemini 5 shattered records with the longest manned spaceflight at the time, lasting eight days. Gemini 7 and 6 achieved the first rendezvous of two manned spacecraft in orbit, a pivotal achievement laying the groundwork for Apollo's lunar missions. These milestones culminated in Gemini 12's successful automated reentry into Earth's atmosphere, solidifying NASA's expertise in orbital maneuvering, docking procedures, and spacecraft systems.

(3 June 1965) — Astronaut Edward H. White II, pilot for the Gemini-Titan 4 (GT-4) spaceflight, floats in the zero-gravity of space during the third revolution of the GT-4 spacecraft. White wears a specially designed spacesuit. His face is shaded by a gold-plated visor to protect him from unfiltered rays of the sun. In his right hand he carries a Hand-Held Self-Maneuvering Unit (HHSMU) that gives him control over his movements in space. White also wears an emergency oxygen chest pack; and he carries a camera mounted on the HHSMU for taking pictures of the sky, Earth and the GT-4 spacecraft. He is secured to the spacecraft by a 25-feet umbilical line and a 23-feet tether line. Both lines are wrapped together in gold tape to form one cord. Astronaut James A. McDivitt, command pilot, remained inside the spacecraft during the extravehicular activity (EVA). NASA

Simultaneously, a NASA Headquarters working group led by Bernard Maggin put forth a comprehensive staff paper advocating for an integrated orbital operations program.

Estimated at approximately $1 billion through 1970, the proposed initiative aimed to develop and apply orbital operations capabilities across three main categories: inspection, ferry, and orbital launch. The group emphasized the necessity of initiating this program independently, coordinating efforts with the Department of Defense, and establishing a dedicated project office to oversee its implementation—a strategic move reflecting NASA's forward-thinking approach in advancing space technology and operations.

Titan II: A Catalyst for Lunar Ambitions

On May 8th, Martin Company personnel conducted a significant briefing for NASA officials in Washington, D.C., focusing on the Titan II weapon system. This pivotal meeting stemmed from earlier discussions initiated by Albert C. Hall of Martin, who had proposed the Titan II as a potential launch vehicle for NASA's lunar landing program. Initially met with skepticism, Hall's proposal prompted NASA's Associate Administrator, Robert C. Seamans, Jr., to arrange a formal presentation.

Impressed by Martin's briefing, Abe Silverstein, director of NASA's Office of Space Flight Programs, took proactive steps. He tasked Director Robert R. Gilruth and the Space Task Group with thoroughly assessing the Titan II's suitability for various aerospace applications. Shortly after that, Silverstein communicated to Seamans the promising prospect of using the Titan II to launch an upgraded Mercury spacecraft.

On May 17th, the Space Task Group (STG) embarked on a significant initiative by issuing a Statement of Work for a Design Study of a Manned Spacecraft Paraglide Landing System. This pioneering study aimed to define and evaluate critical aspects while establishing design parameters for a system enabling spacecraft maneuverability and controlled energy descent using aerodynamic lift.

At the time, McDonnell was already modifying the Mercury spacecraft, and the concurrent paraglide study was intended to integrate this innovative landing system as an integral subsystem. STG Director Robert R. Gilruth played a pivotal role by identifying three companies with prior experience in the paraglide concept: Goodyear Aircraft Corporation in Akron, Ohio; North American Aviation, Inc., Space and Information Systems Division in Downey, California; and Ryan Aeronautical Company in San Diego, California. Each company was awarded a contract worth up to $100,000, with a stipulated completion timeframe of two and a half months from the contract award date.

Gilruth anticipated that one of these companies would subsequently be selected to develop and manufacture the paraglide system further based on the approved design concept. Remarkably, within less than three weeks, contracts were finalized with all three companies. By the end of June, the design study formally transitioned into Phase I of the Paraglider Development Program, marking a critical milestone in NASA's pursuit of advanced landing technologies for manned spacecraft.

On June 9th, James A. Chamberlin, Chief of the Engineering Division at the Space Task Group (STG), conducted a pivotal briefing for Director Robert R. Gilruth, senior STG staff, and NASA Headquarters representatives George M. Low and John H. Disher. The briefing focused on McDonnell's innovative advancements in capsule design, which represented a departure from the traditional Mercury capsule approach.

McDonnell's proposed design emphasized enhanced accessibility to components and systems, streamlined manufacturing and checkout processes, simplified pilot insertion and emergency egress procedures, heightened reliability, and compatibility with a paraglide landing system. Notably, the design differed significantly from the Mercury capsule by relocating most components outside the pressure vessel and improving the performance of retrograde and posigrade rockets.

Initially cautious about adopting what appeared to be a complete redesign of the Mercury spacecraft, the group decided to reconvene on June 12th to further deliberate on the most beneficial features of the new design. Following extensive discussions during the subsequent meeting, the group opted to task McDonnell with studying a modified capsule that could accommodate an extended 18-orbit capability, while minimizing overall modifications.

This decision marked a strategic balance between innovation and continuity, ensuring that future spacecraft designs could incorporate advancements while leveraging existing operational frameworks. It underscored STG's commitment to evolving spacecraft technology in preparation for ambitious manned missions and lunar exploration goals.

On June 12th, representatives from the Space Task Group (STG) and McDonnell gathered in St. Louis for a critical meeting focused on paraglider engineering and operational challenges. Central to their discussions were several pressing issues that needed resolution to advance the development of the paraglider landing system.

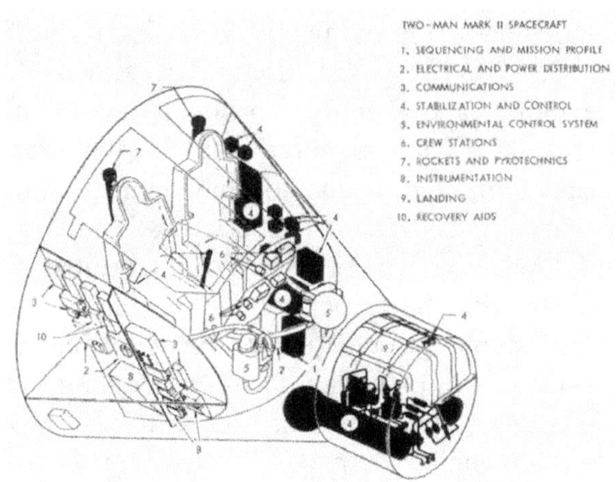

McDonnell-proposed two-man Mercury spacecraft. Shown is the interior arrangement of spacecraft equipment.

Foremost among the concerns was the need to prevent the spacecraft from "nosing in" during the landing phase—a critical safety consideration to ensure a controlled and stable descent. This problem demanded innovative solutions to stabilize the craft and optimize its aerodynamic performance during the final approach and touchdown.

Additionally, increased stowage areas within the spacecraft were recognized as necessary to accommodate the paraglider system effectively. This necessitated careful integration of storage compartments without compromising the spacecraft's structural integrity or operational capabilities.

Another significant topic of discussion was the development of a reliable method for the emergency escape of the pilot post-paraglider deployment. This aspect was crucial for ensuring the safety of astronauts in unforeseen circumstances during mission operations.

On July 7th, Walter F. Burke of McDonnell presented the results of the company's studies on the redesigned Mercury spacecraft to the senior staff of the Space Task Group (STG). McDonnell's evaluations encompassed three configurations tailored to enhance spacecraft capabilities and operational flexibility.

The first configuration proposed by McDonnell involved minimal modifications to the existing capsule design. Primarily focused on improving accessibility and handling, this approach included the addition of an adapter for supplementary items such as extra batteries. This configuration aimed to preserve the Mercury spacecraft's core functionalities while optimizing its utility for extended missions.

The second configuration centered on a reconfigured capsule with an ejection seat featuring most equipment placed outside the pressure vessel on readily accessible pallets. This layout aimed to streamline crew operations and enhance maintenance procedures, bolstering spacecraft reliability and mission readiness.

The third configuration explored by McDonnell was a two-man capsule variant, similar to the reconfigured design but tailored for dual-operator functionality. This adaptation required specific modifications to accommodate the presence of two astronauts and was designed to be deployed using two Mercury-type main parachutes for descent. Including an ejection seat in this configuration was a redundant safety measure, underscoring McDonnell's

commitment to astronaut safety and mission success.

In assessing the trajectory and performance capabilities of the two-man capsule, McDonnell relied on data from the Atlas-Centaur booster, aligning their design considerations with anticipated launch vehicle specifications. These studies represented McDonnell's proactive approach to advancing spacecraft design, integrating cutting-edge technologies to meet the evolving demands of NASA's manned spaceflight program.

On July 27-28, representatives from NASA and McDonnell convened to strategize the direction of McDonnell's efforts regarding the advanced Mercury spacecraft. After extensive discussions and deliberations, a definitive course of action was agreed upon.

McDonnell was tasked with focusing all its resources and efforts on developing two distinct versions of the advanced spacecraft. The first version aimed to minimize changes while ensuring the capability to support a single astronaut in space for up to 18 orbits. This configuration prioritized continuity with existing technology while extending mission duration capabilities, aligning closely with NASA's strategic goals for prolonged manned space missions.

Concurrently, McDonnell was directed to develop a more ambitious two-man spacecraft version. This variant required more substantial modifications and enhancements to accommodate the operational needs of advanced missions. These modifications were intended to significantly expand the spacecraft's capabilities significantly, enabling more complex maneuvers and activities in space.

This strategic decision underscored McDonnell's pivotal role in advancing spacecraft technology and capabilities under NASA's manned spaceflight program. It reflected a balanced approach to incremental improvements and transformative innovations, positioning NASA and

McDonnell at the forefront of manned space exploration during this pivotal era.

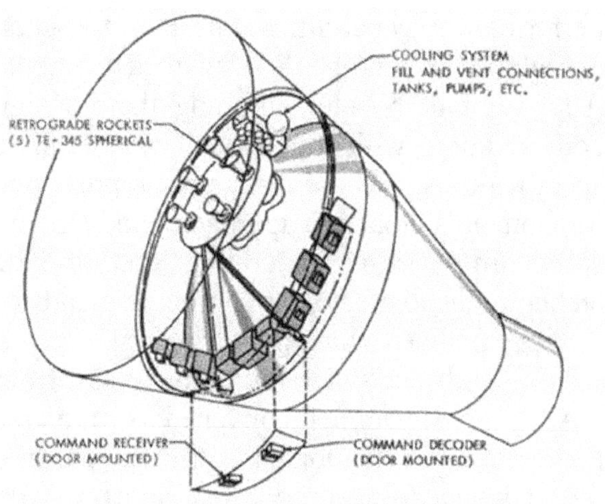

The adapter section of McDonnell's proposed two-man Mercury spacecraft.

Proposal for Lunar Mission Adaptation

In July, engineers James A. Chamberlin and James T. Rose of the Space Task Group made a significant proposal regarding adapting the improved Mercury spacecraft for a lunar mission. Their proposal centered on configuring the spacecraft to accommodate a payload weighing up to 35,000 pounds, notably including a 5,000-pound lunar lander. This ambitious payload configuration was envisioned using a Saturn C-3 rocket, employing the lunar-orbit-rendezvous mode.

This proposal presented a compelling alternative within the ongoing discussions surrounding lunar exploration strategies. It stood in contrast to the Apollo program's prevailing proposals, which favored a direct lunar landing approach. The Apollo mission architecture aimed at landing a payload of approximately 150,000 pounds using a Nova-class vehicle boasting around 12 million pounds of thrust.

Chamberlin and Rose's proposal reflected a strategic divergence in lunar mission planning, emphasizing the feasibility and advantages of utilizing the Mercury spacecraft platform for lunar exploration missions. Their approach sought to leverage existing technology and operational capabilities while optimizing mission efficiency and cost-effectiveness, presenting a viable alternative pathway to achieving lunar exploration objectives during this dynamic period of space exploration planning.

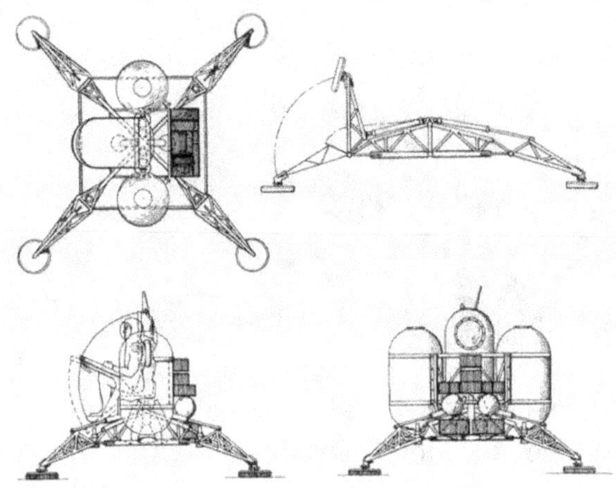

Engineering drawing of the proposed "lunar lander" to be used with an advanced version of the Mercury spacecraft.

In July, James L. Decker from Martin Company proposed a Mercury spacecraft boosted by the Titan rocket. This initiative, known as the Mercury-Titan program, was designed around an 18-month flight schedule leveraging the Air Force's extensive development and testing of the Titan ballistic missile system. The proposal aimed to capitalize on the Air Force's prior work, particularly in adapting the vehicle for manned spaceflight through the Dyna-Soar program.

The Titan rocket offered significant advantages over the Mercury-Atlas configuration. Its first stage boasted a sea-level thrust rating of 430,000 pounds, supplemented by 100,000 pounds from the second stage. This superior thrust capability enabled the Titan to lift much heavier spacecraft payloads than the Mercury Atlas. Additionally, the Titan utilized a hypergolic propulsion system, which relied on storable liquid propellants, simplifying logistical and operational complexities compared to the cryogenic systems used by the Atlas rockets.

Decker emphasized Titan's reputation for reliability, highlighting its redundant flight control systems, including a three-axis reference system, autopilot, servomechanisms, and robust electrical and hydraulic systems. These features were critical for ensuring mission success and astronaut safety during all phases of flight.

The proposal's timeline was contingent upon the availability of Launch Pad 19 at Cape Canaveral, which was earmarked for conversion to accommodate the Titan II configuration. Unlike the other Titan I pads, Pad 19 was initially designed for space applications, making it better suited for the rigorous prelaunch testing and preparations required for manned space missions.

On August 3rd, representatives from Martin Company provided a detailed briefing to Director Robert R. Gilruth and senior staff members of the Space Task Group regarding the Titan II rocket's technical characteristics and anticipated performance capabilities. This briefing was pivotal in assessing the rocket's suitability for manned spaceflight missions.

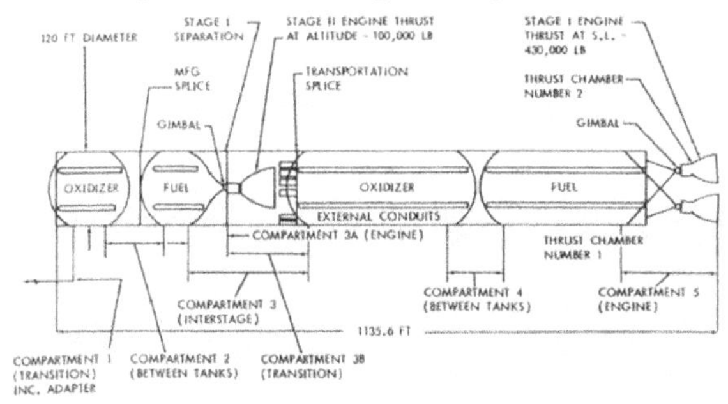

The modified Titan II booster that was to launch the advanced Mercury spacecraft.

During a subsequent senior staff meeting on August 7th, Gilruth highlighted the Titan II rocket's significant potential for advancing manned spaceflight capabilities, particularly its ability to launch larger payloads into orbit than the Atlas rocket. This capability rendered the Titan II highly attractive as a preferred booster for future two-man spacecraft missions.

Martin Company's projections included an estimated cost of $47.889 million for procuring and launching nine Titan II boosters, spanning fiscal years 1962 through 1964. This financial outline encompassed the boosters' cost and ancillary equipment necessary to support the launch operations.

On August 14th, Fred J. Sanders and three colleagues from McDonnell arrived at Langley Research Center to collaborate with James A. Chamberlin and other engineers from the Space Task Group (STG). They were tasked with refining a report detailing the enhanced Mercury spacecraft concept, now referred to as Mercury Mark II. This initiative aimed to build upon the existing Mercury program's success while incorporating significant improvements and adaptations for future missions.

Working closely with Warren J. North from NASA Headquarters Office of Space Flight Programs, the STG team began outlining a preliminary Project Development Plan. This plan was a foundational document intended for submission to NASA Headquarters, outlining the overarching goals, technical specifications, and operational strategies for the Mercury Mark II program.

Although the plan underwent six revisions before its final submission on October 27th, the core principles and initial concepts established in the first draft remained largely intact throughout the iterative development process. This collaborative effort marked a crucial step in advancing the Mercury Mark II program, solidifying its strategic direction and laying the groundwork for future advancements in manned space exploration under NASA's auspices.

On October 27th, James A. Chamberlin, Chief of the Space Task Group (STG) Engineering Division, anticipating imminent approval of the Mark II spacecraft program, urged STG Director Robert R. Gilruth to initiate preparations for McDonnell, the proposed manufacturer, to transition to the new program swiftly. Chamberlin emphasized the need for McDonnell to establish a dedicated organizational structure, assign personnel, and ensure adequate staffing to support the Mark II spacecraft project effectively.

Chamberlin proposed amending the existing letter contract with McDonnell to facilitate a prompt response once the program received official approval. This amendment would direct McDonnell to focus its efforts over the next 30 days on organizing and preparing to assume its role in the Mark II program. This proactive approach aimed to streamline the transition process and position McDonnell to commence work promptly upon formal program approval, ensuring readiness to efficiently meet project milestones and objectives.

On October 27th, the Space Task Group (STG), with assistance from George M. Low, NASA's Assistant Director for Space Flight Operations, and Warren J. North from Low's office, finalized a comprehensive Project Development Plan outlining a robust program of manned spaceflight spanning 1963 to 1965. Central to this plan was developing the two-man Mercury spacecraft, to be launched using a modified Titan II booster. Concurrently, the Atlas-Agena B combination would deploy the Agena B as a rendezvous target in orbit.

The proposed plan leveraged extensive Mercury technology and components, aiming for continuity and efficiency in spacecraft development. A pivotal recommendation included negotiating a sole-source, cost-plus-fixed-fee contract with McDonnell Aircraft Corporation for the Mark II Mercury spacecraft. Launch vehicle procurements were to be arranged through partnerships: General Dynamics/Astronautics for Atlas launch vehicles, Martin-Marietta Space Systems Division for modified Titan II launch vehicles, and Lockheed Missiles and Space Company for Agena target vehicles.

A dedicated project office would oversee the program and plan, direct, and supervise operations. Manpower projections anticipated a team of 177 personnel by the end of fiscal year 1962, reflecting the scale and complexity of the undertaking. The program's estimated cost was approximately $530 million, justified by STG as a crucial step beyond Project Mercury toward long-duration missions and orbital rendezvous techniques.

The objectives of the Mark II program were ambitious and multifaceted, aiming to expand mission capabilities significantly:

- Conduct longer-duration missions beyond Mercury's 18-orbit limit, necessitating multi-man crews.

- Develop operational techniques and equipment for extended spaceflights, gathering crew psychology and physiology data in prolonged space environments.

- Pioneer orbital rendezvous techniques essential for future lunar missions and advanced space operations.

- Implement controlled land landings to streamline spacecraft recovery and establish routine spaceflight procedures.

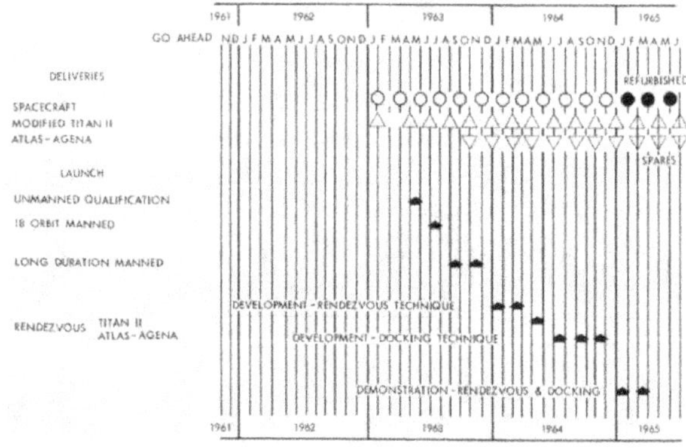

The launch schedule that accompanied the final version of the Mark II Project Development Plan.

The Mark II project was designed for efficiency and speed, utilizing mostly modified existing hardware and modularized equipment to maintain mission flexibility with minimal disruption. Twelve planned flights were outlined, commencing with an unmanned qualification flight in May 1963 and culminating in March 1965. These missions would progressively test and validate spacecraft performance, crew capabilities, and advanced space operation techniques, setting the stage for NASA's ambitious lunar exploration goals in the years to come.

In October 1961, Martin Company received promising signals from the Air Force concerning the Titan II rocket's selection as NASA's successor to the Mercury program. This pivotal moment marked the dawn of Project Gemini, a new era of space exploration.

Collaborating closely with the Air Force and NASA, Martin Company swiftly initiated plans to transform Complex 19 at Cape Canaveral. The goal is to adapt it to the advanced Mercury Mark II configuration, which would evolve into Gemini. This modification was crucial, enabling the Titan II rocket to launch astronauts into space with enhanced capabilities and safety features compared to its predecessor.

On November 1st, 1961, a pivotal day in the annals of space history, leaders from the Space Task Group's Engineering Division convened at NASA Headquarters. They aimed to present the ambitious Mercury Mark II proposal to Robert C. Seamans, Jr., a key figure at NASA. Anticipation hung in the air, as approval for this next phase of space exploration was expected imminently.

Simultaneously, the Space Task Group underwent a transformative rebranding, now assuming the mantle of the Manned Spacecraft Center. This renaming signified more than a mere administrative change; it heralded a new era under the stewardship of Robert R. Gilruth, who stepped into the director role. This transition underscored NASA's commitment to expanding its capabilities beyond the pioneering Mercury program, laying the groundwork for what would soon become Project Gemini.

By November 15th, 1961, McDonnell had delivered a comprehensive set of specifications for the Mercury Mark II spacecraft to the newly established Manned Spacecraft Center. This milestone represented a significant step forward in the evolution of American space technology.

The Mercury Mark II resembled its predecessor in shape and key features, including the iconic escape tower, heat shield, and impact attenuation system. However, McDonnell's design incorporated several notable advancements to enhance mission capabilities and crew safety.

Key innovations included integrating an onboard navigation system, crew ejection seats for emergencies, and bipropellant thrusters to facilitate precise orbital maneuvers. Additionally, the spacecraft was equipped with fuel cells to generate electrical power autonomously, a crucial capability for extended missions lasting up to seven days.

The Paraglider Development Program

On November 20th, 1961, the Manned Spacecraft Center took a significant step forward in aerospace innovation by directing North American Aviation to embark on Phase II-A of the Paraglider Development Program. This strategic initiative was pivotal in refining the paraglider landing system for integration with the upcoming Mercury Mark II project, heralding a new era of precision landing capabilities in space exploration.

Phase II-A focused on optimizing the paraglider system's performance and operational feasibility. Its objectives were clear: to advance beyond theoretical concepts and lay the groundwork for practical implementation through prototype development and manned flight testing.

Integrating the paraglider into the Mercury Mark II spacecraft promised enhanced landing accuracy and reusability, crucial factors in expanding the scope and safety of manned space missions. This forward-thinking approach underscored NASA's commitment to leveraging cutting-edge technology to achieve ambitious goals in space exploration.

As Phase II-A commenced, aerospace engineers and scientists embarked on a journey that would redefine landing methodologies and pave the way for future lunar and planetary missions. The Paraglider Development Program exemplified NASA's pioneering spirit, setting the stage for the groundbreaking advancements that would characterize Project Gemini and beyond.

On November 20th, 1961, NASA's Office of Manned Space Flight took a decisive step towards advancing space exploration by presenting crucial recommendations on orbital rendezvous to Director D. Brainerd Holmes. This pivotal moment marked the recognition of orbital rendezvous as a cornerstone capability for future space missions, prompting an urgent call to action for its development.

Milton W. Rosen, a key figure in NASA's strategic planning, underscored the critical importance of mastering orbital rendezvous techniques. This capability was essential for assembling spacecraft in orbit, facilitating crew transfers, and supporting ambitious lunar and planetary exploration endeavors.

The recommendations highlighted the need for dedicated efforts and resources to accelerate the development of rendezvous capabilities. NASA's proactive stance reflected its commitment to pushing the boundaries of human spaceflight, laying the groundwork for complex missions that would characterize the agency's future endeavors.

At the culmination of November 1961, a collaborative assembly of minds from North American Aviation, Langley Research Center, Flight Research Center, and the Manned Spacecraft Center convened on November 28-29. Their mission was to synchronize efforts for Phase II-A of the Paraglider Development Program. This strategic initiative aimed to refine and integrate the paraglider landing system with the evolving requirements of the Mercury Mark II spacecraft.

The gathering emphasized a comprehensive approach to paraglider research and development, leveraging cutting-edge wind tunnel tests and rigorous flight trials. These efforts were essential to validate the system's performance and reliability under real-world conditions, setting a robust foundation for future manned space missions.

Setting the Stage for Project Gemini

In December 1961, significant strides were achieved in advancing what would soon be known as Project Gemini, building upon the foundations laid by the Mercury Mark II program. On December 5th, pivotal decisions were made based on recommendations from the Large Launch Vehicle Planning Group.

Robert C. Seamans, Jr., and John H. Rubel played instrumental roles in recommending that Secretary of Defense Robert S. McNamara approve the Titan II rocket as the primary launch vehicle for the upcoming rendezvous missions of the Mercury Mark II spacecraft. This endorsement followed initial discussions in August and subsequent meetings in November, underscoring the Titan II's suitability for NASA's ambitious objectives.

Formal recommendations were now in motion, supported by ongoing dialogues with NASA Administrator James E. Webb. This collaborative effort between governmental and space agency leaders marked a crucial milestone in America's space program, paving the way for enhanced capabilities in orbital rendezvous and paving the way for future lunar missions.

The approval of the Titan II rocket not only solidified its role in Project Gemini but also exemplified the collaborative spirit and strategic foresight that defined NASA's approach to pioneering manned space exploration in the early 1960s.

On December 6th, 1961, Robert R. Gilruth, Director of the newly christened Manned Spacecraft Center, took decisive action by submitting a detailed procurement plan for the Mark II spacecraft to NASA Headquarters. This pivotal document outlined a comprehensive strategy encompassing the scope of work, contract administration procedures, negotiation strategies, and precise procurement timelines.

Under the stewardship of D. Brainerd Holmes, Director of Manned Space Flight at Headquarters, urgency underscored the request to immediately release $75.8 million in the fiscal year 1962 funds. These funds were crucial for initiating spacecraft contracts and facilitating essential modifications to launch vehicles, setting the stage for the ambitious objectives of Project Gemini.

December 7th, 1961 marked a pivotal moment in NASA's history as Associate Administrator Robert C. Seamans, Jr. officially approved the Mark II project development plan. This momentous decision signaled the agency's formal endorsement of an ambitious new phase in space exploration.

Colonel Daniel D. McKee, in an accompanying memorandum, underscored the Mark II spacecraft's transformative potential. McKee highlighted its pivotal role in advancing NASA's goals, particularly by developing advanced rendezvous techniques essential for future Apollo missions, including lunar landings.

Director Robert R. Gilruth announced ambitious plans at the Manned Spacecraft Center in Houston in a simultaneous move of strategic foresight and innovation. These plans centered around developing a two-man Mercury capsule, a significant enhancement poised to propel America's space exploration capabilities to new heights.

The enhanced spacecraft, slated to launch atop a modified Titan II booster, aimed to achieve orbital rendezvous with an Agena stage launched by an Atlas rocket. This visionary program encompassed both unmanned and manned flights, strategically designed to progressively test system compatibility and operational readiness for extended missions and complex docking maneuvers in space.

The estimated program cost was $500 million, underscoring NASA's commitment to investing in cutting-edge technology and infrastructure essential for pioneering manned space missions. Targeting the launch of the first manned flights between 1963 and 1964, the two-man Mercury capsule program represented a bold step towards realizing America's lunar exploration ambitions.

On December 7th, 1961, Robert C. Seamans, Jr., and John H. Rubel presented crucial recommendations to Secretary McNamara regarding the division of responsibilities between NASA and the Department of Defense (DOD) in the Mark II program. This strategic alignment aimed to maximize resources and expertise in advancing America's ambitious goals in manned space exploration.

Under the proposed framework, NASA would assume primary responsibility for the Mark II program's management and strategic direction. This would include overseeing crucial aspects such as mission planning, spacecraft design, and overall program coordination.

Concurrently, DOD resources, particularly from the Air Force, would play a pivotal role in supporting procurement and operational facets of the program. This collaboration was envisioned to leverage the military's extensive experience and capabilities in aerospace technology, fostering invaluable insights and advancements in manned spaceflight.

By delineating clear roles and responsibilities, NASA and DOD aimed to synergize their efforts effectively, ensuring the Mark II program's success and laying a robust foundation for future collaborative endeavors in space exploration. This strategic partnership underscored America's commitment to harnessing collective expertise and resources toward achieving historic milestones in the space race.

Guidelines for the Two-Man Spacecraft Development

On December 11th, 1961, NASA solidified its vision for developing the two-man spacecraft with definitive guidelines to push the boundaries of manned space exploration. These guidelines outlined ambitious objectives that underscored the program's strategic importance and technological aspirations.

The spacecraft, designed by McDonnell, was tasked with achieving several key milestones. First and foremost, it aimed to support extended Earth-orbital missions, marking a significant leap forward in mission duration and operational capability. Additionally, the spacecraft was slated to pioneer operational rendezvous and docking maneuvers, critical for assembling spacecraft in orbit and supporting future lunar missions.

Innovative design features included modular systems architecture, which promised flexibility and scalability in spacecraft operations. Onboard maneuvering capabilities were also prioritized, enabling precise control and navigation in space. Furthermore, the spacecraft would integrate advanced controlled land landing systems, enhancing safety and reusability.

The program's ambitious timeline targeted completion by October 1965, underscoring NASA's commitment to accelerating technological advancements in manned spaceflight. These efforts aimed to achieve scientific and exploratory goals and solidify America's leadership in the global space race.

December 22nd, 1961 marked a pivotal moment in America's space exploration efforts as McDonnell accepted a letter contract from NASA to initiate an extensive research and development program for the two-man spacecraft. This significant agreement underscored NASA's commitment to advancing manned spaceflight capabilities and accelerating progress toward ambitious space exploration goals.

The contract outlined McDonnell's responsibilities, including producing 12 spacecraft units, adapters, and essential hardware crucial for the program's success. Stringent delivery schedules were set to ensure alignment with evolving program milestones, emphasizing NASA's rigorous approach to project management and execution.

McDonnell's role in the program extended beyond mere production; it encompassed integrating cutting-edge technologies and innovative design features to enhance spacecraft performance and safety. This included modular systems integration, onboard maneuvering capabilities, and advanced landing systems to support extended missions and operational rendezvous.

December 26th, 1961, marked a significant development in America's aerospace history. The Manned Spacecraft Center took decisive action by directing the Air Force Space Systems Division to commence work on adapting the Titan II rocket for the Mercury Mark II program. This strategic initiative leveraged the Titan II's robust capabilities to support NASA's ambitious goals in manned space exploration.

The adaptation efforts were driven by the need to enhance the Titan II's configuration to accommodate the requirements of the Mercury Mark II spacecraft. This included modifications to ensure compatibility with advanced rendezvous and docking maneuvers, essential for assembling spacecraft in orbit and facilitating future lunar missions.

These efforts culminated in a letter contract on January 19, 1962, formalizing arrangements for Gemini launch vehicles and related ground support equipment. This contractual agreement underscored NASA's commitment to collaboration with the Air Force, harnessing their expertise and resources to propel America's space program forward.

Launching Project Gemini

On December 29th, 1961, NASA marked a definitive milestone in space exploration by issuing the Gemini Operational and Management Plan. This comprehensive document formalized the roles and responsibilities of both NASA and the Department of Defense (DOD) in the ambitious Gemini program, setting the stage for unprecedented advancements in manned spaceflight.

Under the plan, NASA undertook primary responsibilities, including spacecraft development, integration of rendezvous equipment, and overall program direction. These critical tasks aimed to propel technological innovations and operational readiness necessary for extended manned missions and complex orbital maneuvers.

Conversely, DOD assumed pivotal roles in managing launch vehicles, overseeing range operations, and providing crucial recovery support. This collaboration capitalized on the DOD's extensive experience and infrastructure, ensuring the seamless execution of launch and recovery operations essential for mission success.

These December developments laid a solid foundation for Project Gemini, positioning it as a pivotal program that would enhance American capabilities in manned spaceflight and pave the way for the ambitious Apollo lunar missions. By delineating clear roles and fostering collaboration between NASA and DOD, the Gemini Operational and Management Plan underscored America's commitment to achieving historic milestones in space exploration.

As the nation looked towards the stars with renewed determination, December 1961 heralded the beginning of an era defined by technological innovation, strategic collaboration, and the relentless pursuit of knowledge beyond Earth's boundaries.

Chapter 2 - Concept and Design

Throughout 1962, the Gemini program, originally named the Mercury Mark II, made significant progress in its conceptualization and design:

On January 3, 1962, "Gemini" officially became the program designation for the Mercury Mark II. The name was chosen to symbolize the twin stars Castor and Pollux of the Gemini constellation, representing the program's two-man crew, rendezvous mission objectives, and its evolution from the Mercury program. This designation aligned neatly with the Roman numeral "II," signifying the Mark II iteration of the spacecraft.

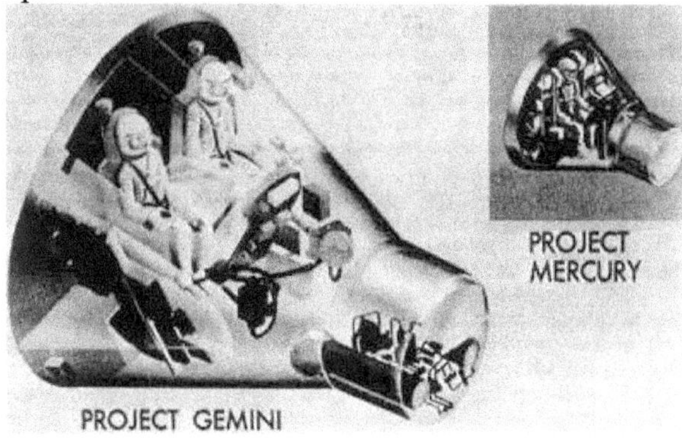

The first illustration of the Gemini spacecraft to be released publicly. It was distributed at the same time NASA announced that the project was to be named "Gemini." (NASA Photo S-62-88, released Jan. 3, 1962.)

On January 3rd, the Manned Spacecraft Center (MSC) prepared a detailed Statement of Work for the Air Force Space Systems Division (SSD), outlining the procurement requirements for 15 Titan II launch vehicles. These vehicles were essential for launching the Gemini spacecraft into orbit. The procurement process involved minimal modifications to the Titan II missile, primarily focused on enhancing crew safety. Initial budgetary constraints required careful planning and management, leading to establishing the Gemini Launch Vehicle Directorate under Colonel Richard C. Dineen by January 11th.

By January 5th, MSC had published the first analysis of the Gemini spacecraft schedule, adjusting launch dates due to delays in program approval. The revised timeline aimed for the first manned flight in late July or early August 1963, with subsequent flights spaced at intervals designed to culminate in a final mission by late April or early May 1965. The integration of the Agena rendezvous mission was scheduled for late February or early March 1964, reflecting the program's phased approach to testing and operational readiness.

On January 15th, Robert R. Gilruth appointed James A. Chamberlin as Manager of the Gemini Project Office (GPO) within MSC, overseeing all technical and contractor activities—this organizational shift streamlined decision-making and project management within the Gemini program.

Concurrently, MSC analyzed power sources for the Gemini spacecraft, identifying fuel cells from General Electric as a promising technology due to their simplicity, weight efficiency, and compatibility with Gemini's operational requirements. This analysis culminated in a $9 million subcontract awarded to General Electric on March 20th to develop fuel cells tailored for the Gemini spacecraft.

On January 26th, Martin Baltimore initiated a study to address potential malfunctions in the modified Titan II/Gemini launch vehicle. This study aimed to enhance flight safety and reliability by integrating backup systems, malfunction detection mechanisms, and selective electrical redundancies. The approach prioritized minimal modifications to preserve the launch vehicle's core reliability while accommodating the Gemini spacecraft as its payload.

By January 31st, MSC communicated to the Marshall Space Flight Center the procurement requirements for 11 Atlas-Agena vehicles essential for rendezvous exercises in the Gemini program. These vehicles underwent modifications to incorporate radar and visual navigation aids, multiple restart-capable main engines, secondary propulsion and stabilization systems, a rendezvous docking unit, and an aerodynamic fairing for launch. The procurement plan aimed to deliver the first vehicle within 20 months, with subsequent deliveries at 60-day intervals, supporting the program's ambitious schedule for orbital rendezvous demonstrations.

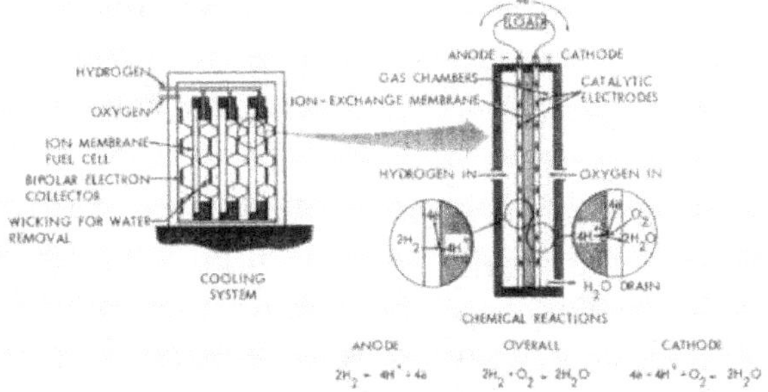

The operating principle of the fuel cell designed by General Electric, adopted for use in the Gemini spacecraft.

Throughout 1962, these foundational steps laid the groundwork for the Gemini program's subsequent achievements, setting the stage for advanced space missions and paving the way for NASA's eventual lunar exploration efforts.

In February 1962, the Gemini program continued to make significant strides in its development.

On February 15th, the Air Force Space Systems Division (SSD) issued a Technical Operating Plan to Aerospace Corporation, officially contracting them to support managing the Gemini Launch Vehicle Program. Based in El Segundo, California, Aerospace Corporation assumed responsibility for systems engineering and technical oversight of the launch vehicle and its subsystem development. This initiative was part of NASA's strategy to leverage external expertise in advancing the Gemini program's critical infrastructure.

February 19th marked a pivotal decision by Howard W. Tindall, Jr., of the Flight Operations Division, to consolidate all Gemini computer programming and operations at NASA's Manned Spacecraft Center (MSC) in Houston. This centralization aimed to streamline trajectory control complexity required for orbital rendezvous missions. It also addressed the need for specialized computer programming, crucial for coordinating spacecraft operations in real time. This decision shifted programming responsibilities away from Goddard Space Flight Center, the primary computing center for Mercury flights, underscoring Houston's emerging role in manned space missions.

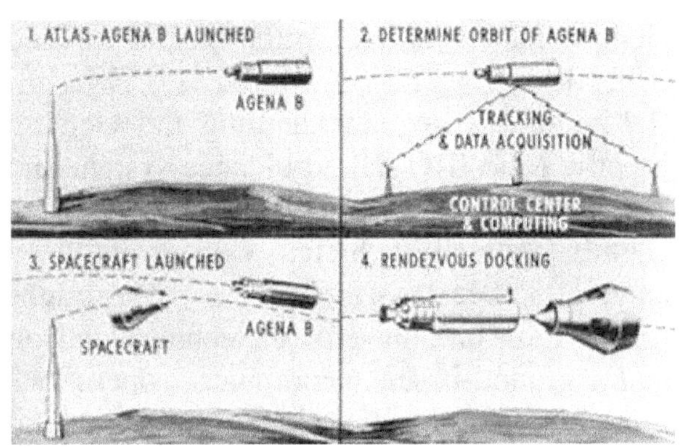

Four stages in a rendezvous mission as conceived early in 1962. (NASA Photo S-62-82, c. Jan. 3, 1962.)

Also on February 19th, AiResearch Manufacturing Company, a division of the Garrett Corporation based in Los Angeles, California, secured a $15 million subcontract from McDonnell to manufacture the Environmental Control System (ECS) for the Gemini spacecraft. Modeled after the ECS used in Project Mercury, AiResearch's system was integral for maintaining suit and cabin atmosphere, controlling temperatures, and managing crew hydration and waste disposal during all flight phases.

On February 19th, MSC held its initial coordination meeting with McDonnell, outlining policy statements and setting the stage for regular business meetings starting March 5th. These meetings, held three times a week, became pivotal in resolving technical and contractual issues between NASA and McDonnell, ensuring alignment on spacecraft system development and integration.

On February 20th, McDonnell issued detailed specifications for the crew-station system within the Gemini spacecraft. This system encompassed displays of spacecraft functions, control interfaces, and crew integration mechanisms, delineating responsibilities for each crew member. Clear specifications were crucial for ensuring operational efficiency and crew safety during missions.

On February 22nd, Martin-Baltimore submitted its proposal for redundant flight control and hydraulic subsystems for the Gemini launch vehicle. Authorized to proceed with design work on March 1st, Martin-Baltimore focused on adapting the General Electric Mod IIIG radio guidance system and Titan I three-axis reference system for the Gemini launch vehicle. Concurrently, a contract was awarded to Burroughs Corporation on July 3rd for technical liaison and ground-based computer operations essential for launch vehicle guidance and control.

February 24th marked a significant subcontract award of $32 million from McDonnell to North American Aviation's Rocktdyne Division in Sacramento, California. This contract was for building the Liquid Propulsion Systems (LPS) required for the Gemini spacecraft. Comprising the Orbit Attitude and Maneuvering System (OAMS) and the Reaction Control System (RCS), these propulsion systems provided critical functions such as orbit insertion, spacecraft attitude control, and reentry maneuvering capability. The RCS, designed with redundancy in critical components, ensured mission reliability during the spacecraft's reentry phase.

On February 28th, representatives from McDonnell, North American Aviation, MSC, and NASA Headquarters initiated discussions to coordinate the interface between the Gemini spacecraft and the paraglider landing system. Key challenges included optimizing the stowage volume within the spacecraft for the paraglider system and seamlessly integrating it into the spacecraft's recovery compartment geometry.

Also, on February 28th, MSC allocated $5.2 million to the Marshall Space Flight Center to procure Atlas-Agena vehicles crucial for Project Gemini's rendezvous exercises. This procurement initiative underscored the collaborative efforts between MSC and Marshall in supporting the program's ambitious schedule, with the first Atlas-Agena launch anticipated by March 15, 1964.

These developments throughout February 1962 underscored the Gemini program's robust progress, laying essential groundwork for subsequent milestones in manned space exploration.

Training Innovations for Gemini Astronauts

On March 5th, Harold I. Johnson, Head of the Spacecraft Operations Branch at NASA's Manned Spacecraft Center, distributed a memorandum outlining the comprehensive training regimen for Project Gemini. Central to this program were advanced training devices designed to prepare astronauts for the complexities of space missions.

Key among these was the flight simulator, a replica of the Gemini spacecraft capable of simulating the full range of mission conditions, including visual, auditory, and vibrational cues. Integrated within the mission simulator complex, this device provided crucial training for flight crews and ground controllers alongside the mission control center and remote site displays.

Specialized equipment was deployed for specific aspects of mission training. The centrifuge at the Naval Air Development Center in Johnsville, Pennsylvania, played a pivotal role in training for launch and re-entry. Equipped with a mock-up of the Gemini spacecraft interior, astronauts experienced gravitational forces akin to those encountered during spaceflight.

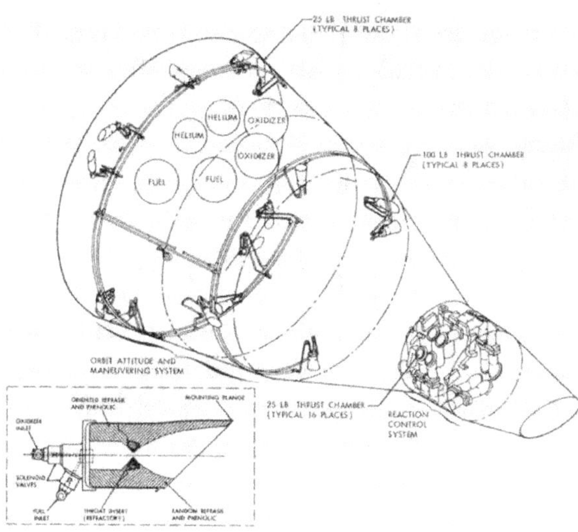

The general arrangement of liquid rocket systems (OAMS and RCS) in the Gemini spacecraft. The insert displays a typical thrust chamber assembly.

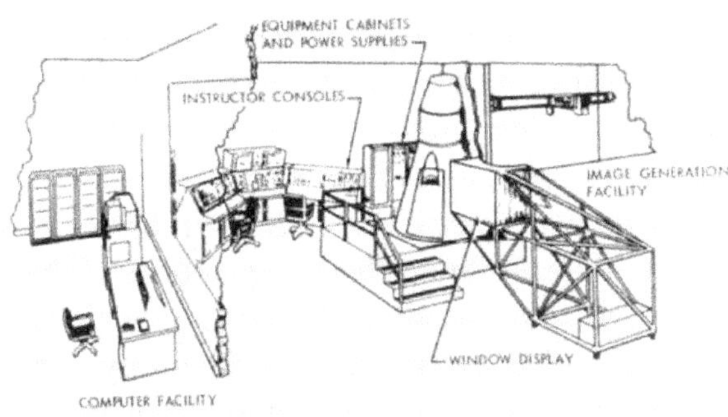

The two major types of simulators to be used in training crews for Gemini missions. The Gemini flight trainer (above) would simulate the entire mission, while the docking trainer (below) would simulate the final stages of rendezvous.

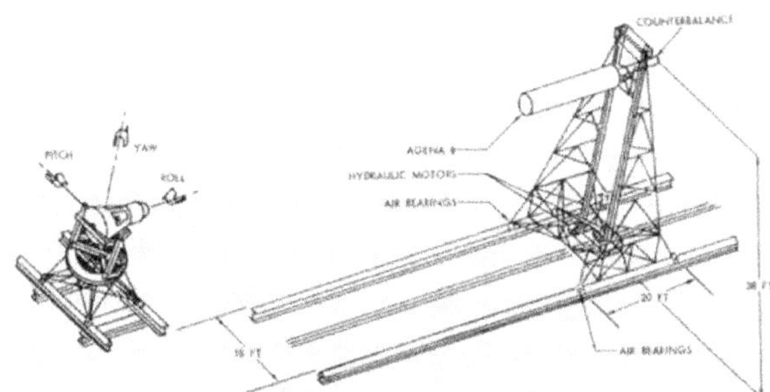

A static spacecraft model was an egress trainer, enabling crews to practice normal and emergency exit procedures following land or water landings. To prepare for land landings, a boilerplate spacecraft equipped with a full-scale paraglider wing underwent flight testing via drops from a helicopter.

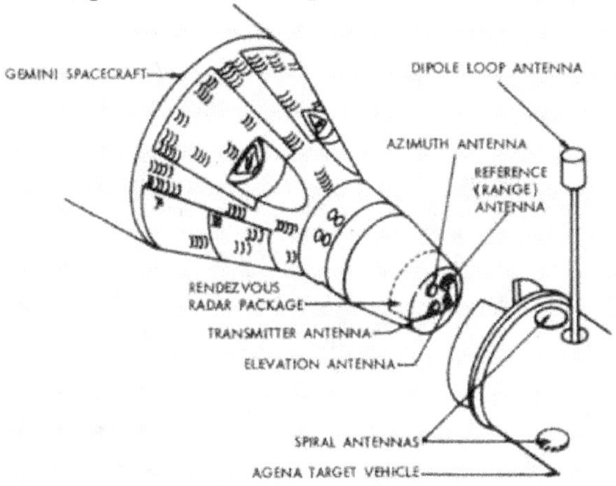

The location of the main elements of the rendezvous radar system on the Gemini spacecraft and the Agena target vehicle.

Docking operations, critical to mission success, were rehearsed on a docking trainer equipped with authentic hardware and crew displays. This trainer, capable of motion in six degrees of freedom, simulated the complexities of spacecraft docking maneuvers.

Additionally, trainers simulated the spacecraft's major systems, providing targeted instruction on specific flight tasks.

On the same day, Westinghouse Electric Corporation in Baltimore, Maryland, secured a $6.8 million subcontract from McDonnell to furnish the Gemini spacecraft's rendezvous radar and transponder system. Positioned within the spacecraft's recovery section, the rendezvous radar tracked the target vehicle during rendezvous maneuvers. Meanwhile, the transponder, located aboard the Agena target vehicle, automatically transmitted signals upon receiving interrogation signals.

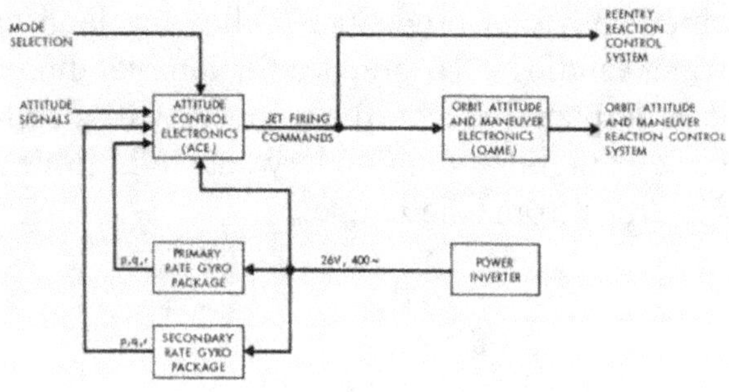

A functional block diagram of the attitude control and maneuvering electronics system of the Gemini spacecraft.

On March 7th, McDonnell, the prime contractor for Project Gemini, awarded a $6.5 million subcontract to Minneapolis-Honeywell Regulator Company in Minneapolis, Minnesota. The contract aimed to integrate the attitude control and maneuvering electronics system into the Gemini spacecraft. This critical system enabled astronauts to control the spacecraft's propulsion systems, linking their manual operations directly to the firing of thrusters within the orbit attitude and maneuvering system and the reaction control system.

Simultaneously, on March 7th, the Gemini Project Office approved McDonnell's preliminary design of the spacecraft's main undercarriage, specifically for landings. This milestone authorized McDonnell to proceed with detailed design work. Scheduled to commence around April 1st, dynamic model testing of the undercarriage would ensure its performance and reliability under various landing conditions.

North American Aviation's Emergency Parachute System Development

On March 8th, 1964, the Manned Spacecraft Center directed North American Aviation to initiate the design and development of an emergency parachute recovery system. This system was intended for both the half-scale and full-scale flight test vehicles required under Phase II-A of the Paraglider Development Program. Subsequently, North American was authorized to subcontract the emergency recovery system to Northrop Corporation's Radioplane Division in Van Nuys, California. The subcontract, valued at $225,000, was awarded to Radioplane on March 16th. This subcontract was part of two significant Phase II-A efforts led by North American, the other being a $227,000 contract awarded to Goodyear to research materials and test fabrics for inflatable structures.

Gemini's Target Vehicle Program

On March 12th, 1964, the Marshall Space Flight Center delivered an Agena procurement schedule dated March 8th to the Gemini Project Office. The Air Force Space Systems Division (SSD) was tasked with contracting Lockheed for 11 target vehicles as part of the Gemini Agena target vehicle program. SSD assigned management of this program to its Ranger Launch Directorate, which oversaw programs utilizing Agena vehicles. Additionally, Marshall reported that a qualified multiple-restart main engine would be delivered within 50 weeks. This development marked a significant improvement, removing it as the limiting factor in the Agena scheduling process. These initiatives underscored NASA's coordinated efforts in advancing technology and procurement strategies critical to the success of the Gemini missions.

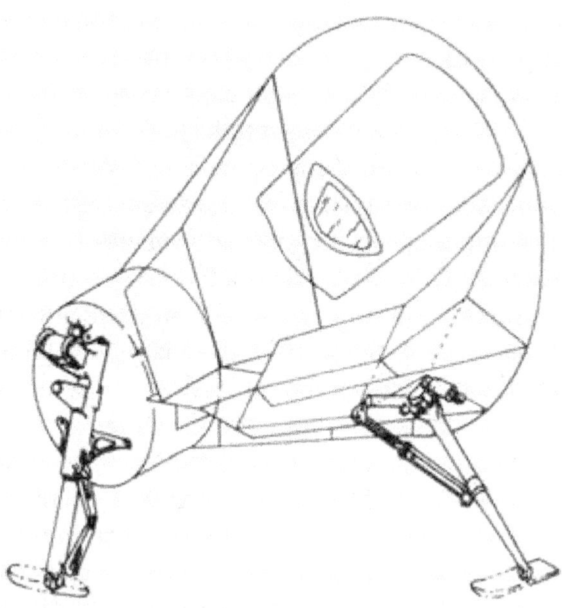

Gemini landing gear: part of the land landing system along with the paraglider.

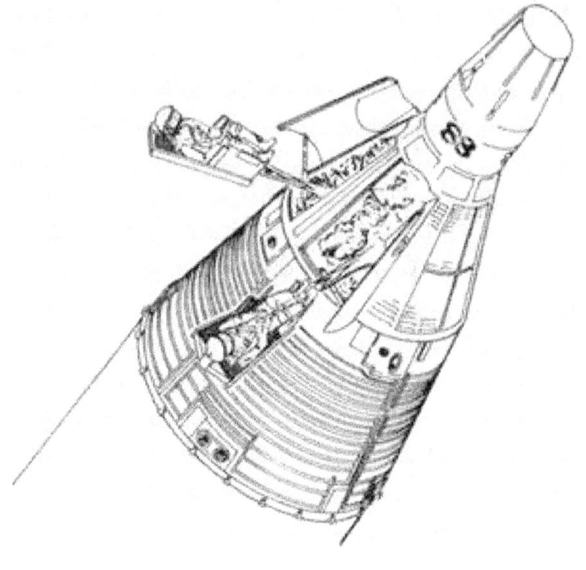

An artist's version of the use of ejection seats to escape from the Gemini spacecraft. The seats were to be used before launch (off-the-pad abort) or during the first phase of powered flight (to about 60,000 feet) if the launch vehicle malfunctioned.

Gemini's Ejection Seat System

On March 14th, 1964, the Gemini Project Office (GPO) made a critical decision regarding designing and implementing the spacecraft's ejection seat system. It was determined that seat ejection would be manually initiated initially, with provisions for future automatic initiation if required. Both seats were designed to eject simultaneously upon activation of either ejection system. This system was crucial for providing astronauts with a means of emergency escape from the Gemini spacecraft under various scenarios: during pre-launch preparations on the launch pad, throughout the initial powered flight phase up to approximately 60,000 feet, or in the event of paraglider failure after reentry into the Earth's atmosphere.

The escape system consisted of several integrated components: a hatch actuation system to open the spacecraft hatches before ejection, a rocket catapult mechanism to propel the ejection seats clear of the spacecraft, a personnel parachute system to safely lower the astronaut after seat separation, and survival equipment to support the astronaut post-landing.

Recognizing the ejection seat system's complexity and critical nature, a collaborative effort was established among McDonnell, the Gemini Project Office, the Life Systems Division, and the Flight Crew Operations Division. This team regularly monitored the ejection seat's development, its associated components, and the comprehensive testing required to ensure reliability and safety.

While ejection seats had been well-established in military aviation, the Gemini project's specific requirements, particularly for off-the-pad abort capabilities, exceeded the capabilities of existing flight-proven systems. In response, McDonnell subcontracted Weber Aircraft, a division of Walter Kidde and Company, Inc., located in Burbank, California, with a $1.8 million contract on April 9th, 1964, to develop and produce the Gemini ejection seats. Additionally, Rocker Power, Inc. in Mesa, Arizona, received a $741,000 subcontract on May 15th, 1964, to manufacture the escape system's rocket catapult.

On March 14th, the Manned Spacecraft Center issued its second Gemini program schedule analysis, incorporating spacecraft and launch vehicle considerations. Unlike previous assessments, this analysis included insights into the recently initiated procurement of the Agena target vehicle, albeit with limited scope due to its recent commencement.

A critical strategy in the Gemini program's engineering development was fabricating various test articles. This approach aimed to avoid delays similar to those encountered in the Mercury program, where the absence of adequate test articles had hindered progress. While the construction of these test articles might initially impact the Gemini spacecraft's construction timeline, the invaluable data they would yield was expected to more than compensate for any temporary delays.

The analysis highlighted no significant issues in the development of launch vehicles. However, the schedule allotted minimal contingency time for unforeseen challenges. As per the updated schedule, the first unmanned qualification flight was slated for April 1964, followed by the second manned flight for late 1964. The first Agena flight was planned for mid-1964, with subsequent missions following regularly until mid-1965. The flight mission objectives remained consistent with those outlined in previous analyses.

On March 15th, the Gemini Project Office reiterated its decision to utilize hardware and subcontractors originally developed for Project Mercury in the Gemini program. This approach aimed to capitalize on existing technologies and partnerships established during the Mercury era, emphasizing continuity and efficiency in spacecraft development and mission execution.

The restatement emphasized that any deviation from this plan would require thorough justification for each specific item of equipment or subcontractor. This stringent requirement ensured that any proposed changes would align with program goals, maintain compatibility with existing systems, and justify any potential additional costs or adjustments in development timelines.

By leveraging proven technologies and established partnerships, NASA sought to streamline the Gemini spacecraft's development process while maintaining the high reliability and performance standards necessary for manned space missions. This strategic reuse of Mercury-era assets underscored NASA's commitment to maximizing efficiency and minimizing risks in the evolving field of space exploration.

Titan II ICBM Launch

On March 16th, a significant milestone was achieved as the United States Air Force successfully launched a Titan II intercontinental ballistic missile (ICBM). This launch marked the inaugural full-scale test of the Titan II, demonstrating its capability by flying 5,000 miles over the Atlantic Ocean.

The successful test flight underscored the Titan II's advanced technological capabilities as an intercontinental ballistic missile, showcasing its range, accuracy, and reliability. This achievement was pivotal for military applications and its role as a robust and powerful launch vehicle in NASA's space exploration endeavors.

Titan II's successful deployment laid a solid foundation for its future role in carrying manned missions under NASA's Gemini program, highlighting its importance as a key component of America's space exploration efforts during the 1960s.

On March 17th, McDonnell awarded AiResearch a significant $5.5 million subcontract to supply the reactant supply system for the Gemini spacecraft's fuel cells. This system was crucial for providing oxygen and hydrogen to power the spacecraft's fuel cells, ensuring continuous operation throughout missions.

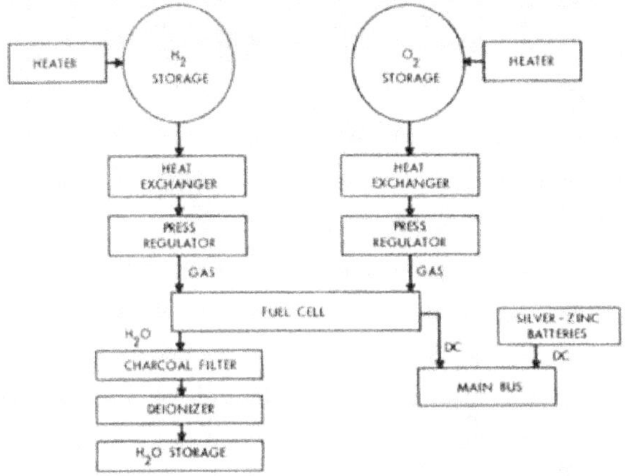

Block diagram of the reactant supply system for the Gemini spacecraft fuel cells.

The reactants, stored in specialized double-walled, vacuum-insulated spherical containers in the spacecraft's adapter section, were maintained as single-phase fluids. This unique storage method stabilized the reactants using supercritical pressures at cryogenic temperatures. Heat exchangers converted these reactants into gaseous forms and delivered them to the fuel cells at operational temperatures.

On March 19th, 1962, McDonnell took a pivotal step forward in developing the Gemini spacecraft by awarding several critical subcontracts. These contracts heralded significant advancements in the spacecraft's technological capabilities, propelling the United States deeper into the space race.

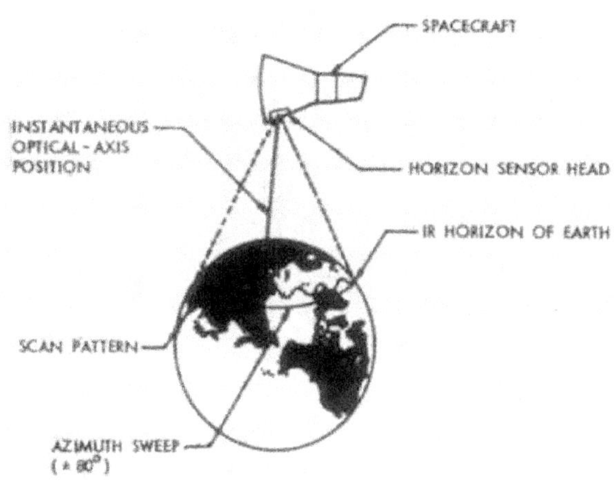

Illustrating the operation of the horizon sensor for the Gemini spacecraft

The subcontract awards represented a crucial phase in the Gemini program, aimed at refining the spacecraft's design and enhancing its operational capabilities. Among the key advancements were guidance and navigation systems, crucial for precision control and orbital maneuvers. These systems laid the groundwork for the spacecraft's ability to rendezvous and dock in space, a pivotal capability for future lunar missions.

Additionally, advancements in life support systems were integral to sustaining astronauts during extended missions. Improved thermal protection systems were also crucial, safeguarding the spacecraft and its crew from the extreme conditions of space travel.

The subcontract awards underscored McDonnell's commitment to pushing the boundaries of space exploration. Each technological enhancement brought the Gemini spacecraft closer to its ultimate goal: paving the way for NASA's ambitious lunar missions and expanding humanity's reach into the cosmos.

On March 19th, 1962, a pivotal moment unfolded in developing the Gemini spacecraft as McDonnell awarded Advanced Technology Laboratories, Inc. of Mountain View, California, a $3.2 million subcontract for the horizon sensor system. This critical technology represented a significant leap forward in the spacecraft's guidance and control capabilities.

The horizon sensor system comprised two essential components—a primary sensor and a standby sensor—designed to monitor the infrared radiation gradient between Earth and space meticulously. This gradient, Earth's infrared horizon, was a precise reference point for the spacecraft's orientation and trajectory.

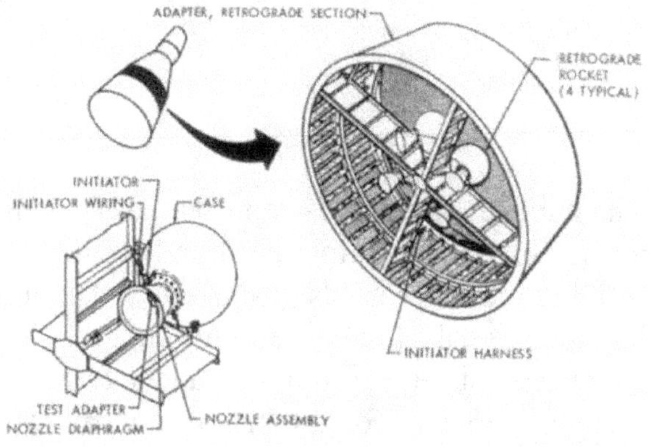

Location and arrangement of the retrograde rocket system in the Gemini spacecraft.

These sensors played a vital role in the spacecraft's navigation by scanning, detecting, and tracking the Earth's infrared horizon. They provided crucial data to align the inertial platform and generated error signals essential for the spacecraft's attitude control and maneuver electronics. This functionality enabled precise adjustment of the spacecraft's pitch and roll axes, ensuring optimal orientation and stability during critical maneuvers and orbital operations.

The subcontract to Advanced Technology Laboratories, Inc. underscored McDonnell's commitment to integrating cutting-edge technology into the Gemini program. Enhancing the spacecraft's ability to autonomously navigate and maintain its orientation in space laid the groundwork for future missions, including rendezvous and docking maneuvers essential for lunar exploration.

Beyond technical achievement, the horizon sensor system represented a strategic milestone in the United States' space exploration endeavors during the Cold War era. It exemplified the nation's dedication to technological superiority and its determination to assert leadership in conquering the final frontier.

Thiokol Chemical Corporation of Elkton, Maryland, secured a pivotal $400,000 subcontract from McDonnell to provide retrograde rockets in a significant advancement for the Gemini spacecraft. These solid-propellant retrorockets were integral components in the spacecraft's adapter section, serving dual roles critical to mission success.

The retrograde rockets played a dual role in the Gemini spacecraft's operational capabilities. Primarily, they functioned as deceleration mechanisms during the crucial reentry maneuver. This capability was essential for safely slowing down the spacecraft and facilitating its safe return through Earth's atmosphere, ensuring the astronauts' safe descent and recovery.

Beyond reentry maneuvers, these retrorockets possessed another critical capability—they could accelerate the spacecraft in the event of a high-altitude, suborbital abort scenario. This functionality allowed for rapid separation from the launch vehicle, enhancing crew safety and mission flexibility during early ascent phases.

The subcontract included a comprehensive modification and qualification program to enhance the retrorockets' performance and safety standards. These efforts were crucial in ensuring that the retrorockets met stringent requirements for manned space missions, reflecting McDonnell's commitment to meticulous engineering and safety in space exploration.

Thiokol Chemical Corporation's subcontract highlighted the United States' strategic advancements in space technology during the Cold War era. By developing and refining critical propulsion systems like retrograde rockets, the nation solidified its position at the forefront of manned space exploration, demonstrating technical prowess and leadership in the global space race.

On March 21st, 1962, pivotal contracts marked significant strides in the Gemini program, notably with the Air Force Space Systems Division awarding a crucial letter contract to Aerojet-General Corporation in Azusa, California. This contract, valued for researching, developing, and procuring 15 propulsion systems, represented a cornerstone in preparing the Gemini launch vehicle.

Aerojet-General Corporation was tasked not only with the design and development of these propulsion systems but also with creating essential aerospace ground equipment necessary for their deployment. The commencement of engine work as early as February 14, 1962, underscored the urgency and meticulous planning involved in ensuring the readiness and reliability of the launch vehicle's propulsion systems. The final engine delivery slated for April 1965 highlighted the comprehensive timeline aimed at meeting stringent mission requirements and operational demands.

Simultaneously, McDonnell awarded a substantial $4.475 million subcontract to the Western Military Division of Motorola, Inc., headquartered in Scottsdale, Arizona. This subcontract focused on designing and constructing the digital command system (DCS) explicitly tailored for the Gemini spacecraft.

The DCS comprised a sophisticated receiver/decoder package and three relay packages, pivotal components integral to the spacecraft's operational architecture. Functionally, the system received digital commands transmitted from ground stations, meticulously decoded them, and efficiently routed them to the corresponding spacecraft systems. These commands encompassed real-time instructions essential for controlling various spacecraft functions and stored program commands crucial for updating the time reference system and digital computer. Such capabilities were pivotal for ensuring precise mission management, operational flexibility, and the overall success of the Gemini missions.

These contracts not only propelled the Gemini program forward but also underscored the United States' commitment to advancing space exploration capabilities during the intense geopolitical climate of the Cold War. By integrating cutting-edge propulsion systems and digital command technologies, the nation positioned itself at the forefront of manned spaceflight, demonstrating technological prowess and leadership in the global race to conquer space.

On March 23rd, 1962, the Air Force Space Systems Division unveiled the "Development Plan for the Gemini Launch Vehicle System," a pivotal document that set forth crucial details for executing the ambitious project. Rooted in lessons learned from the Titan II and Mercury programs, this comprehensive plan projected a budget totaling $164.4 million, reflecting the scale and complexity of the Gemini launch vehicle's development.

Central to the plan was a meticulous approach to resource management underpinned by strategic foresight and financial prudence. Drawing on insights from past programs, the budget included a substantial 50 percent contingency allowance, amounting to $82.2 million. This buffer was strategically allocated to accommodate potential cost increases and unforeseen adjustments during development, ensuring financial stability and operational continuity.

The development plan emphasized the integration of advanced space launch technologies while maintaining flexibility to address evolving challenges and requirements. This approach was crucial for navigating the complexities of developing a sophisticated launch vehicle system capable of supporting manned space missions. By leveraging insights and experiences from previous programs, the plan aimed to streamline processes, enhance efficiency, and mitigate risks associated with technological innovation and operational execution.

Beyond financial allocation, the plan underscored a commitment to mission success and national leadership in space exploration. It outlined strategic milestones, technical objectives, and performance benchmarked to advance the Gemini program's capabilities and achieve operational readiness within specified timelines. This disciplined approach aimed to meet immediate program goals and positioned the United States at the forefront of manned spaceflight, showcasing its technological prowess and commitment to pushing the boundaries of human exploration in space.

The release of the "Development Plan for the Gemini Launch Vehicle System" marked a significant milestone in the United States' quest for space exploration dominance during the Cold War era. By outlining a comprehensive strategy for managing resources, mitigating risks, and advancing technological frontiers, the plan laid a robust foundation for the Gemini program's success. It exemplified NASA's dedication to precision planning and operational excellence, setting the stage for groundbreaking achievements in space exploration and human achievement beyond Earth's atmosphere.

On March 23rd, 1962, the Air Force Space Systems Division unveiled the comprehensive "Development Plan for the Gemini Launch Vehicle System," marking a significant milestone in America's ambitious space exploration efforts. Drawing on invaluable insights gleaned from the Titan II and Mercury programs, this detailed roadmap provided a strategic framework for executing the complex project.

Central to the development plan was a meticulously crafted budget totaling $164.4 million. This financial projection underscored the scale and ambition of the Gemini launch vehicle's development, encompassing essential components such as propulsion systems, avionics, and ground support equipment. The budget included a substantial 50 percent contingency allowance of $82.2 million. This contingency was earmarked to address potential cost escalations and unforeseen adjustments that could arise during the rigorous phases of development, ensuring financial stability and project continuity.

By leveraging experiences from previous programs, the development plan prioritized operational excellence and efficiency. It aimed to optimize resource allocation, streamline development processes, and enhance technological integration to meet stringent mission requirements and timeline commitments. This disciplined approach reflected NASA's commitment to learning from past successes and challenges, guiding the Gemini program toward achieving new milestones in manned spaceflight.

Beyond financial planning, unveiling the development plan signaled America's steadfast commitment to advancing space exploration capabilities during the Cold War era. It outlined strategic objectives, technical milestones, and performance benchmarked to position the United States as a global leader in space innovation. The plan aimed to propel human space exploration beyond Earth's orbit through meticulous planning and forward-thinking strategies, paving the way for historic achievements and scientific discoveries.

On March 29th, 1962, a pivotal subcontract marked a significant advancement in the Gemini program as McDonnell awarded the Aeronautical Division of Minneapolis-Honeywell in St. Petersburg, Florida, an impressive $18 million contract. This contract entrusted them with the critical task of providing the Inertial Measuring Unit (IMU) for the Gemini spacecraft, underscoring the nation's commitment to pioneering advancements in space exploration.

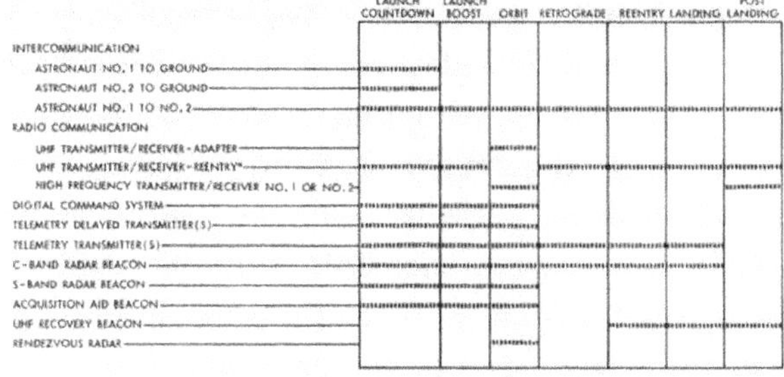

Illustrating the stages of a mission during which various elements of the Gemini spacecraft communications system would be used.

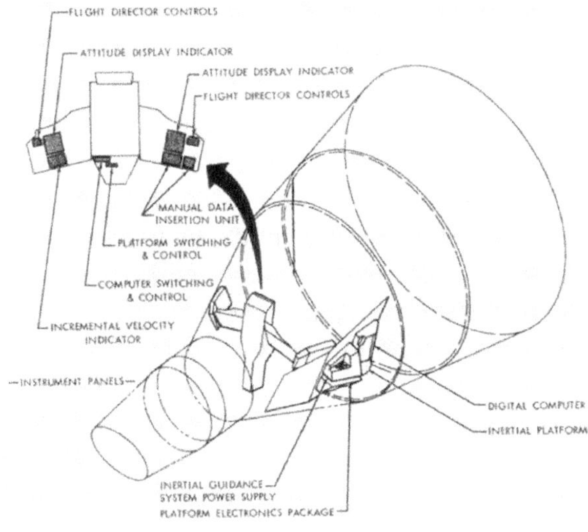

The Gemini spacecraft inertial guidance system.

The Inertial Measuring Unit (IMU) comprised a sophisticated stabilized inertial platform housing an electronic unit and power supply. This technological marvel was central to the spacecraft's operational capabilities by providing a stable reference frame. It enabled precise determination of spacecraft attitude and facilitated the detection of changes in spacecraft velocity—fundamental functions crucial for navigating and controlling the spacecraft during its missions.

The IMU's ability to maintain a stable reference frame in dynamic space conditions was instrumental in ensuring accurate navigation and control. Continuously monitoring spacecraft orientation and velocity changes provided essential data for real-time adjustments and course corrections, enhancing mission precision and safety.

The $18 million subcontract to Minneapolis-Honeywell highlighted America's leadership in space technology during the Cold War era. It represented a strategic investment in advancing capabilities essential for manned space missions, positioning the United States at the forefront of global space exploration efforts.

Beyond its technical specifications, the IMU subcontract exemplified NASA's collaborative approach with industry leaders to push the boundaries of human achievement in space. By harnessing cutting-edge technologies and expertise, the Gemini program aimed to expand scientific knowledge, pioneer new frontiers in space exploration, and inspire future generations to reach for the stars.

On March 30th, 1962, Martin-Baltimore submitted a seminal document titled "Description of the Launch Vehicle for the Gemini Spacecraft" to the Air Force Space Systems Division. This comprehensive submission marked a pivotal moment in the Gemini program, setting forth essential concepts and strategic philosophies for developing the Gemini launch vehicle subsystems.

The document was a foundational blueprint guiding the design and development efforts crucial to supporting the Gemini spacecraft missions. It outlined detailed descriptions and operational philosophies for each proposed launch vehicle subsystem, reflecting meticulous planning and alignment with strategic objectives outlined by the Air Force Space Systems Division.

Integrating cutting-edge technologies and engineering innovations to enhance mission capability and operational efficiency was central to the submission. By delineating clear objectives for each subsystem, Martin-Baltimore aimed to optimize performance, reliability, and safety standards essential for manned space missions.

The submission underscored Martin-Baltimore's collaboration with the Air Force Space Systems Division to advance national interests in space exploration during the Cold War era. It highlighted a shared commitment to achieving technological superiority and leadership in space, emphasizing the Gemini program's role in expanding scientific knowledge and pioneering new frontiers.

Beyond its immediate implications, the submission of the launch vehicle description document laid a critical foundation for future advancements in space exploration. It set forth a visionary path for realizing ambitious goals of lunar exploration and beyond, inspiring a legacy of innovation and discovery that continues to shape humanity's journey into the cosmos.

Crafting the Gemini Spacecraft

On March 31st, the Gemini spacecraft reached a pivotal milestone as its definitive configuration was formally set. This marked the culmination of months of meticulous planning and design efforts by McDonnell, following NASA's greenlight of the project on December 22, 1961. From the outset, McDonnell focused on refining every spacecraft aspect, meticulously detailing its design, performance parameters, and subsystem specifications.

In the preceding months, McDonnell engineers crafted comprehensive specifications that delineated the spacecraft's structure, functionality, and stringent criteria for its major components. Each specification underwent rigorous scrutiny and approval from NASA, ensuring adherence to the program's exacting standards. Moreover, McDonnell developed precise control drawings that standardized the spacecraft's

construction and guided the procurement process with subcontractors.

On April 3rd, a pivotal meeting convened representatives from the Manned Spacecraft Center, Ames Research Center, Martin, and McDonnell to discuss the critical Gemini wind tunnel program led by Ames Research Center—this collaborative effort aimed to comprehensively test and refine the Gemini spacecraft's design across multiple fronts.

The objectives of these rigorous wind tunnel tests were manifold. Firstly, using a scaled six percent model of the spacecraft and launch vehicle, engineers assessed the structural loads during launch, particularly focusing on the influence of hatches on launch stability. This analysis was crucial for ensuring the spacecraft's robustness under the intense lift-off conditions.

Secondly, the tests delved into booster tumbling characteristics. Engineers explored how large angles of attack, Reynolds number variations and retrorocket jet effects affected the booster's stability and attachment loads. These insights were essential for optimizing the spacecraft's ascent dynamics and ensuring the reliability of its launch vehicle.

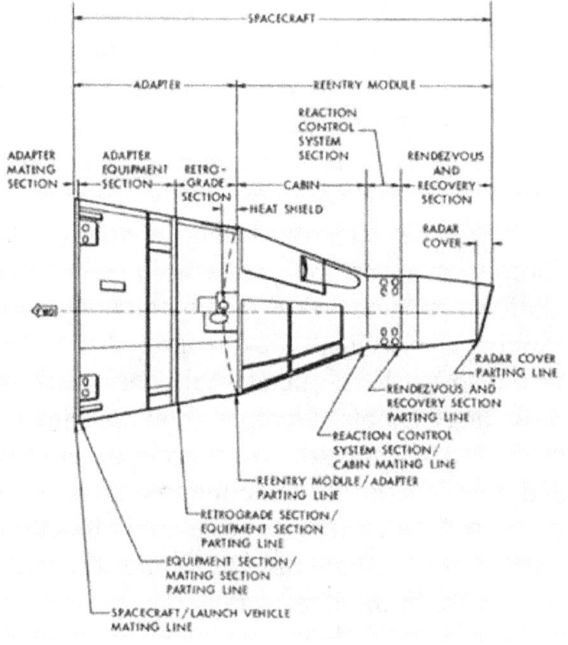

Gemini spacecraft nomenclature.

Thirdly, the team scrutinized spacecraft exit characteristics, simulating various scenarios to understand how the spacecraft behaved during exit maneuvers. This aspect was pivotal for refining emergency procedures and ensuring astronaut safety during critical moments.

Lastly, the wind tunnel tests thoroughly examined the aerodynamic and thermal properties of the reentry module during atmospheric reentry. This analysis provided crucial data to enhance the spacecraft's reentry capabilities, ensuring it could withstand the extreme conditions of returning to Earth.

On April 4th, a significant development unfolded at the Manned Spacecraft Center as it awarded a pivotal contract worth $209,701 to the Aerospace and Defense Products Division of B.F. Goodrich Company in Akron, Ohio. This contract, structured as a cost-plus-fixed-fee arrangement, aimed to spearhead the design, development, and fabrication of prototype pressure suits crucial for the impending Gemini program. Concurrent contracts were also awarded to the Arrowhead Products Division of Federal-Mogul Corporation in Los Alamitos, California, and Protection, Inc., in Gardena, California, highlighting the collaborative effort across multiple specialized firms.

The B.F. Goodrich Company initiated preliminary work on this contract on January 10, 1962, laying the groundwork for two distinct pressure suit development initiatives. The first focused on producing four successive prototypes of an advanced full-pressure suit, while the second aimed to deliver two prototypes of a partial-wear, quick-assembly, full-pressure suit. Initially, these efforts were not explicitly tied to any specific manned space flight program.

However, recognizing the critical need for tailored spacesuit solutions in the context of Project Gemini, the contract underwent a pivotal amendment on September 19, 1962.

This amendment explicitly aligned the ongoing development efforts with the Gemini program's requirements, emphasizing the need for advanced pressure suits to support astronauts through the unique challenges of manned space missions.

On April 7th, a pivotal subcontract valued at $1 million was awarded by McDonnell to the ACF Electronics Division, part of ACF Industries, Inc. based in Riverdale, California. This subcontract was designated to supply crucial C- and S-band radar beacons essential for the Gemini spacecraft. These radar beacons were central to the spacecraft's sophisticated tracking system, ensuring precise monitoring throughout its missions.

The C- and S-band radar beacons, although differing in frequency characteristics, served similar functions within the spacecraft's tracking architecture. Both variants were designed to respond to interrogation signals from ground stations, facilitating accurate tracking of the spacecraft's position and trajectory. This capability was vital for mission control to maintain constant awareness of the spacecraft's whereabouts and to execute maneuvers as needed.

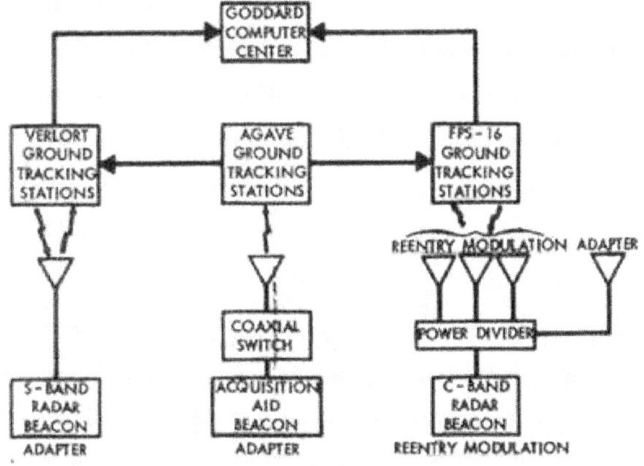

Gemini spacecraft tracking aids (beacon system)

On April 9th, Earl Whitlock, representing McDonnell, delivered a comprehensive "Gemini Manufacturing Plan" to the Gemini Project Office (GPO), marking a significant step forward in the spacecraft's production timeline. Presented on April 6th, this plan meticulously outlined the manufacturing schedule, initially prioritizing the production of spacecraft No. 1 followed by static article No. 1.

Recognizing the critical importance of achieving optimal quality in the first production item, the GPO made a strategic request. They directed McDonnell to adjust the schedule, moving up the commencement of static article No. 1 to around May 15, 1962, ahead of spacecraft No. 1. This adjustment allowed ample time for thorough ground testing and validation before initiating full-scale spacecraft production.

Under McDonnell's contractual obligations, four static articles were mandated for development. These units were engineered explicitly for rigorous ground testing, meticulously crafted to replicate the construction and materials used in flight-ready spacecraft. This approach ensured comprehensive validation of spacecraft components and systems under simulated operational conditions, mitigating mission readiness risks.

Preparing for Extended Missions:

On April 12th, a crucial affirmation from the Manned Spacecraft Center solidified the readiness of the Agena spacecraft's systems to support a five-day orbital mission, meeting the stringent operational demands set forth by the Gemini Project. This milestone marked a significant advancement in the project's preparation, ensuring that the Agena spacecraft's capabilities were fully aligned with the mission's duration requirements.

This confirmation underscored the meticulous testing and validation processes undertaken to ensure the reliability and endurance of the Agena spacecraft's systems under extended orbital conditions. Such capability was essential for facilitating complex mission objectives, and maneuvers planned for the Gemini program, highlighting NASA's commitment to achieving sustained operational success in manned space exploration.

On April 13th, Martin-Baltimore and the Air Force Space Systems Division (SSD) delivered critical updates to the Gemini Project Office regarding challenges in implementing the malfunction detection system (MDS). This system was integral for ensuring the reliability of critical subsystems during Gemini missions. In response to these challenges, the Manned Spacecraft Center swiftly mobilized a task force comprising experts from Martin, McDonnell, and Aerospace to tackle the issue head-on.

Their primary objective was to define robust abort criteria that would safeguard mission success in the event of subsystem malfunctions detected by the MDS. By April 23rd, Martin had presented a comprehensive study to SSD, proposing a specific configuration for the MDS. This system was designed to monitor key launch vehicle subsystems in real-time, providing crucial performance data directly to the astronauts onboard. Notably, the decision to initiate an abort would ultimately lie with the astronauts themselves, highlighting the critical role of human judgment in mission safety.

Concurrently, Chance Vought Corporation of Dallas, Texas, completed a simulation study affirming the feasibility and necessity of enabling manual abort initiation as part of the mission protocols. This capability would empower astronauts to respond promptly to unforeseen challenges, enhancing mission flexibility and safety protocols for the Gemini program.

The Recruitment and Training of NASA's Gemini Astronauts

On April 18, NASA made a significant announcement, declaring that applications were open for additional astronauts until June 1, 1962. The agency aimed to augment its current seven-member Mercury astronaut team with five to ten recruits who would play crucial roles in Project Mercury support operations and the piloting of the two-man Gemini spacecraft.

Prospective candidates had to meet stringent criteria to qualify: they needed to be experienced jet test pilots actively engaged in flying high-performance aircraft. Furthermore, they must have achieved experimental flight test status through military service, work in the aircraft industry or NASA, or have graduated from a military test pilot school. Academic qualifications included holding a degree in physical or biological sciences or engineering. Applicants had to be under 35, no taller than six feet, and be recommended by their parent organization.

The selection process involved rigorous stages. In July, candidates meeting these criteria would undergo interviews and written examinations assessing their engineering and scientific knowledge. Those who passed these phases would undergo thorough medical evaluations conducted by specialists.

Successful applicants would embark on an intensive training regimen. This program encompassed collaboration with design and development engineers, simulator training, centrifuge sessions to simulate gravitational forces, additional scientific education, and flights in high-performance aircraft.

On April 19, McDonnell Aerospace awarded a substantial $26.6 million subcontract to International Business Machines (IBM) Corporation's Space Guidance Center in Owego, New York. The contract was for IBM to provide the crucial computer system for the Gemini spacecraft, which was the core of its guidance and control systems.

The computer system included several key components essential for spacecraft operations. These included the digital computer itself, pivotal for executing precise guidance and control functions. Supplementary equipment comprised the incremental velocity indicator, which visually conveyed changes in spacecraft velocity; the manual data insertion unit, facilitating the input and display of data from the computer; and the auxiliary computer power unit, crucial for maintaining stable voltages necessary for the computer's operation.

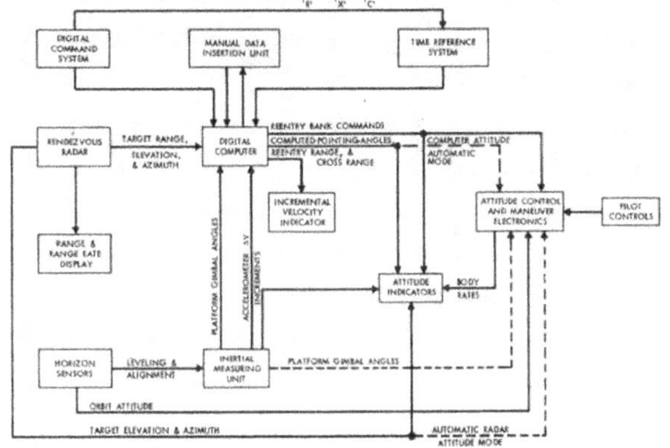

Block diagram of the Gemini spacecraft guidance and control system.

In addition to supplying these critical components, IBM's responsibilities extended to integrating the computer with a complex array of spacecraft systems and components. This integration encompassed electrical connections with the inertial platform, rendezvous radar, time reference system, digital command system, data acquisition system, attitude control and maneuver electronics, launch vehicle autopilot, console controls and displays, and aerospace ground equipment.

On April 25th, McDonnell took a significant step in securing the Gemini spacecraft's safety by subcontracting Studebaker Corporation's CTL Division in Cincinnati, Ohio, for $457,875. This contract aimed to manufacture two backup heatshields for the Gemini spacecraft, ensuring redundancy in critical thermal protection systems. These heatshields were to be constructed using materials and manufacturing techniques akin to those successfully employed in Project Mercury, underscoring NASA's commitment to proven technology for mission-critical components.

The decision to procure backup heatshields from CTL Division stemmed from ongoing evaluations of new materials for the primary heatshield. McDonnell had initiated assessments of four promising materials through advanced screenings for their ablation properties. To bolster these efforts, McDonnell had engaged Vidya, Inc. of Palo Alto, California, on March 16th and later enlisted Chicago Midway Laboratories of Chicago, Illinois, in mid-April to conduct detailed evaluations of these materials' suitability.

On April 26th, Lockheed made significant strides in the Gemini-Agena program by presenting detailed plans for propulsion development during an Atlas-Agena coordination meeting. Their comprehensive program outlined a series of initiatives to enhance the reliability and functionality of the propulsion systems crucial to the Gemini missions.

Central to Lockheed's strategy was a thorough study to optimize the primary propulsion system, ensuring it met the stringent performance requirements for the Gemini spacecraft. Additionally, Lockheed embarked on a multiple-restart development initiative for the primary propulsion system, a critical capability needed to support various mission profiles and orbital maneuvers.

Lockheed also dedicated resources to developing a secondary propulsion system alongside the primary system efforts. This parallel program aimed to provide additional flexibility and redundancy in maneuvering capabilities during mission operations, enhancing overall mission safety and success.

On April 26-27, a critical meeting convened involving representatives from North American Aviation, NASA Headquarters, and several key research centers, including Langley, Flight Research, Ames, and the Manned Spacecraft Center. The focus of this gathering was to meticulously review the design and testing strategy for the Half-Scale Test Vehicle (HSTV) as part of phase II-A in the Paraglider Development Program.

At the heart of the discussions was validating the emergency parachute recovery system, a crucial safety component for the HSTV. Once certified, the vehicle would proceed to rigorous drop tests to evaluate the stability and control of the paraglider configuration with the pre-deployed wing. These tests were designed to gather essential empirical data on the vehicle's performance and systems during deployment, vital in refining and advancing the paraglider technology.

Following an extensive review process, the NASA Half Scale Test Vehicle Design Review Board recommended 21 specific changes to North American Aviation regarding the test vehicle's design and procedural enhancements. These recommendations were pivotal in ensuring that the HSTV met stringent safety and performance standards necessary for the success of future manned spaceflight missions under the Paraglider Development Program.

On May 1st, McDonnell proposed enhancing the capabilities of the Gemini spacecraft by introducing a rendezvous evaluation pod. This innovative concept aimed to assess the spacecraft's rendezvous radar and maneuvering systems during early orbital flights. The proposal received enthusiastic support from the Manned Spacecraft Center (MSC), prompting them to encourage McDonnell to conduct further detailed studies.

By the end of June, McDonnell achieved a significant milestone with official approval from MSC to proceed with the design and development of the rendezvous pod. This pod was meticulously designed to integrate essential components for its operational effectiveness. These included a radar transponder for precise tracking, a C-band beacon to facilitate communication with ground stations, a flashing light for visibility in space, and robust batteries to power its systems autonomously.

The Gemini-Agena Target Vehicle (GATV) Program

On May 1, the Air Force Space Systems Division (SSD) issued a pivotal contract to Lockheed Missiles and Space Company, marking the beginning of the Gemini-Agena Target Vehicle (GATV) program. This initiative aimed to transform eight existing Agena vehicles into sophisticated targets essential for the Gemini spacecraft missions. Each GATV had to fulfill a precise set of operational requirements:

Firstly, it had to achieve and maintain a circular orbit meticulously aligned within specified parameters. This orbit needed stability and precisely controlled to facilitate rendezvous and docking maneuvers with the Gemini spacecraft.

Secondly, the GATV had to be responsive to commands from ground control stations and the spacecraft itself. This capability ensured flexibility and adaptability during missions, allowing real-time adjustments to accommodate varying conditions.

Thirdly, the GATV was essential for executing complex orbital maneuvers autonomously through immediate commands or pre-programmed instructions stored onboard. This feature was crucial when optimal launch conditions were not achieved for the Agena or the spacecraft.

Moreover, the GATV was required to sustain its operational capability in active orbit for a minimum of five days, ensuring extended availability for mission objectives.

Lockheed extensively modifies the Agena vehicles to meet these stringent requirements. This included redesigning the primary propulsion system to enhance performance and reliability. A secondary propulsion system was also integrated, featuring both small 16-pound and larger 200-pound thrusters. These systems provided critical functions such as ullage orientation and minor adjustments to the GATV's orbital trajectory.

The digital command and communications subsystem underwent significant development to enable seamless interaction between the GATV, ground control, and the Gemini spacecraft. This subsystem incorporated advanced technologies, including a programmer, controller, pulse-code-modulated telemetry system, and an onboard tape recorder. These innovations ensured robust communication and data management capabilities essential for mission success.

Furthermore, Lockheed implemented specific modifications to support the GATV's unique guidance and control requirements. These adaptations were crucial for maintaining precise orbital alignment and facilitating successful rendezvous operations with the Gemini spacecraft.

Finally, an auxiliary forward equipment rack was added, featuring an interface to support the docking adapter essential for the spacecraft's docking maneuvers with the GATV.

Following directives from the Air Force Systems Command Headquarters, SSD authorized Lockheed to proceed with the Gemini-Agena program on March 19, underscoring the program's critical role in advancing manned spaceflight capabilities during the Gemini missions.

On May 1, following a comprehensive briefing by Lockheed on pulse-code-modulation (PCM) instrumentation systems, representatives from Goddard Space Flight Center and Manned Spacecraft Center (MSC) initiated a collaborative effort. They aimed to explore the feasibility of upgrading the Gemini telemetry system to a fully integrated PCM system. PCM technology offered significant advantages over traditional telemetry systems, including increased data channels, faster transmission rates, enhanced accuracy, and reduced equipment weight per data channel.

Goddard had already evaluated several PCM ground station proposals, affirming the system's capability to support future NASA programs effectively. The participants in the briefing unanimously agreed that implementing a full PCM telemetry system, both airborne and ground-based, was feasible within the timeline required to support the Gemini program.

In response to this agreement, the Gemini Project Office sanctioned the establishment of the MSC-Gemini PCM Instrumentation Working Group. This group was tasked with overseeing the seamless integration and compatibility of PCM systems across both airborne and ground-based platforms for the Gemini missions.

On May 4, the Manned Spacecraft Center (MSC) released its latest assessment of the Gemini program schedule, marking significant progress and challenges ahead. Ground test plans for spacecraft were finalized, and construction of necessary test hardware was already underway. Two boilerplate spacecraft were integrated into the program to facilitate rigorous ground testing.

A key milestone was the planning for Flight No. 2, which aimed to debut the paraglider landing system. However, the paraglider program required meticulous oversight to avoid any potential delays. Contingency plans were set in motion to substitute a parachute landing system if needed, ensuring flexibility in mission execution.

The spacecraft's production schedules faced threats primarily due to delayed deliveries of critical components from vendors. For spacecraft No. 1, components crucial to the instrument and recording systems and the inertial platform posed the most significant risk. Similarly, spacecraft No. 2 encountered potential delays in delivering communication and electrical system components. Despite these challenges, no issues were foreseen with the booster system.

The program analysis affirmed that the launch schedule remained unchanged despite the logistical hurdles. This report underscored the concerted efforts to mitigate risks and maintain momentum towards achieving key milestones in the Gemini program, a crucial step towards advancing manned space exploration.

On May 10-11, pivotal developments within the Gemini Project highlighted meticulous preparations for upcoming manned space missions with exceptional detail and foresight.

The Gemini Project Office entrusted McDonnell with a critical task: determining the operational intricacies of opening and closing spacecraft hatches in the challenging space environment. Concurrently, the Manned Spacecraft Center's Life Systems Division embarked on an equally crucial initiative to define the specialized features needed in pressure suits. These enhancements were crucial to enable crew members to perform extravehicular activities (EVAs) lasting up to 15 minutes, anticipating future needs for spacewalks and expanding the capabilities of manned spaceflight.

In parallel, the Life Systems Division proposed seven parameters to monitor crew health throughout all Gemini missions. These included blood pressure—supported by electrocardiogram and phonocardiogram backups—electroencephalogram, respiration rate, galvanic skin response, and body temperature. The bio instrumentation required for these measurements weighed approximately three and a half pounds per astronaut, consumed about two watt-hours, and utilized six telemetry channels.

Following careful review, the Gemini Project Office refined the set of measurements to include electrocardiogram, respiration rate and depth, oral temperature, blood pressure, phonocardiogram, and nuclear radiation dosage. These vital signs would be crucial in monitoring astronaut health during missions, ensuring their well-being in the demanding space environment.

However, the development of biomedical measurement devices remained a critical task. These devices needed meticulous design, rigorous reliability testing, qualification for space operations, and subsequent procurement. This highlighted the stringent standards and comprehensive preparations required for the pioneering Gemini missions, emphasizing NASA's commitment to astronaut safety and mission success.

On May 10-11, plans for the postlanding survival kit for Gemini crew members were solidified, building upon the successful model used in Project Mercury. Each survival kit, weighing approximately 24 pounds, was meticulously prepared and allocated for every crew member. These kits were essential to ensure crew members had access to vital supplies and equipment for survival and recovery operations immediately after landing.

Drawing from the lessons learned in Project Mercury, the survival kits were designed to contain crucial items such as food, water, first aid supplies, communication devices, signaling equipment, and tools for basic survival tasks. The allocation of these kits underscored NASA's commitment to enhancing crew safety and readiness for any potential contingencies upon returning from space missions.

On May 11, the Manned Spacecraft Center (MSC) strategically decided to establish a liaison office at Martin-Baltimore. This initiative aimed to strengthen communication and collaboration between MSC and Martin, a key aerospace contractor involved in the Gemini program. Scott H. Simpkinson was initially appointed to the role on May 15, but he was later succeeded by Harle Vogel, who continued in the position for the duration of the program.

Establishing this liaison office proved instrumental in facilitating the exchange of crucial information and promoting seamless coordination between MSC and Martin. This collaboration was essential for supporting the timely progress and success of Gemini missions. The liaison office played a pivotal role in overcoming challenges and optimizing efficiency throughout the Gemini program by fostering close communication channels and ensuring alignment on project goals and requirements.

In May 12, James E. Webb, NASA's newly appointed Administrator, undertook a comprehensive review of the Gemini program, evaluating its progress and financial health. At this juncture, the projected costs for Project Gemini had swelled significantly beyond initial estimates. The total estimated expenditure had surged to $744.3 million, a marked escalation from the original projection of $250 million.

The revised cost breakdown revealed substantial increases across key components:

Spacecraft Costs: Originally estimated at $240.5 million, these expenses had risen to $391.6 million.

Titan II Rocket Costs: Projected costs had increased from $113.0 million to $161.8 million.

Atlas-Agena Launch Vehicles: Costs associated had climbed from $88.0 million to $106.3 million.

Supporting Development Efforts: Including initiatives like the paraglider program, costs had grown from $29.0 million to $36.8 million.

Operations Costs: While most categories saw increases, operations costs experienced a slight decrease from $59.0 million to $47.8 million.

James E. Webb's review underscored the financial complexities facing the Gemini program. As NASA pursued ambitious goals in manned spaceflight, his assessment provided critical insights into the program's budgetary landscape. It highlighted the imperative for meticulous management and oversight to ensure the successful execution of Gemini missions amidst evolving costs and priorities.

On May 14-15, a pivotal event unfolded as representatives from McDonnell, Northrop Ventura (formerly Radioplane), Weber Aircraft, and the Manned Spacecraft Center gathered at McDonnell's St. Louis facilities for the first ejection seat design review. This milestone marked a crucial phase in the Gemini program, focusing on developing and assessing ejection seat systems destined for the Gemini spacecraft.

The design review convened experts and stakeholders from leading aerospace engineering firms and NASA, highlighting a collaborative effort to ensure the safety and functionality of the Gemini spacecraft's ejection seat technology. Discussions during the review encompassed detailed examinations of design specifications, performance criteria, safety protocols, and integration challenges within the spacecraft's overall architecture.

Key deliberations likely included:

• Design Specifications: Defining precise parameters for ejection seat functionality and performance under various mission scenarios.

• Performance Requirements: Assessing the seat's capability to operate effectively in emergencies and adverse conditions.

• Safety Protocols: Implementing rigorous safety standards to protect astronauts during critical phases of flight.

• Integration Considerations: Ensuring seamless integration of the ejection seat system with the spacecraft's structural and operational framework.

This inaugural review underscored the technical rigor and collaborative spirit driving the Gemini program and marked a significant step toward enhancing astronaut safety and mission success. As discussions progressed, the groundwork was laid for the meticulous development and refinement of ejection seat technology crucial to the future of manned space exploration.

On May 16-17, critical decisions and initiatives shaped the ongoing evolution of the Gemini spacecraft. One pivotal move was the establishment of the Launch Vehicle-Spacecraft Interface Working Group, signaling a significant stride in fostering collaboration between the Gemini Project Office (GPO) and aerospace contractors.

Comprising members from Martin and McDonnell, with a McDonnell representative appointed as chairman, the working group facilitated the exchange of detailed mechanical, electrical, and structural data. This collaborative effort aimed to ensure seamless integration and compatibility of components between the spacecraft and booster systems. Importantly, the group operated within the framework of technical coordination, focusing solely on technical exchanges rather than policy decisions.

The group's primary function was underscored by its regular reviews and coordination meetings led by the GPO. These sessions provided opportunities to assess progress, address challenges, and maintain alignment across all project phases. By fostering a robust exchange of information and insights, the working group played a crucial role in enhancing the efficiency and effectiveness of integration efforts within the Gemini program.

Concurrently, a pivotal decision emerged during a mechanical systems coordination meeting where McDonnell and GPO representatives resolved to bolster the retrograde rocket motors of the Gemini spacecraft. This decision marked a strategic move towards developing more powerful motors to significantly increase thrust levels—approximately threefold—while maintaining the existing configuration.

The upgraded motors were designed to enhance the spacecraft's ability to perform retrorocket abort maneuvers at altitudes as low as 72,000 to 75,000 feet. This enhancement played a critical role in augmenting crew safety during crucial phases of flight, ensuring the spacecraft's capability to respond effectively to unforeseen contingencies.

Following this decision, McDonnell took proactive steps by terminating its original subcontract with Thiokol and initiating a new contract on July 20 to spearhead the development of the upgraded motors. The projected cost for this developmental effort was estimated at $1.255 million, underscoring the firm commitment to advancing propulsion technology within the Gemini program.

These developments underscored NASA's proactive stance in fortifying spacecraft capabilities and safety protocols as it advanced toward key milestones in manned space exploration. By prioritizing technological advancements and operational readiness, NASA and its industry partners reinforced their dedication to ensuring the success and safety of future Gemini missions.

Parachute System Development in Gemini

On May 18, McDonnell achieved a significant milestone in the Gemini program by subcontracting the parachute landing system to Northrop Ventura at an estimated cost of $1,829,272. This system was designated for use during the inaugural Gemini flight, highlighting its critical role in ensuring the safe and controlled descent of the spacecraft and its crew.

Initially, the Gemini Project Office had opted for a single-chute system featuring an 84.2-foot diameter ring-sail parachute. However, during a subsequent mechanical systems coordination meeting in Houston on May 16-17, it was determined that enhancements were possible by integrating an 18-foot ring-sail drogue parachute. McDonnell proposed deploying the drogue parachute at 10,000 feet, just two seconds after releasing the rendezvous and recovery system. Fifteen seconds later, the main recovery parachute would transition from single-point to two-point suspension, aligning with the initiation of the reaction control system propellant dump process, completed within 105 seconds. The recovery parachute would then be jettisoned shortly after touchdown.

This deployment and recovery operations sequence was thoroughly reviewed and approved during a subsequent coordination meeting on May 23-24 at the Manned Spacecraft Center. This affirmation underscored NASA's meticulous planning and commitment to refining operational procedures, aiming to bolster mission safety and success within the dynamic framework of the Gemini program.

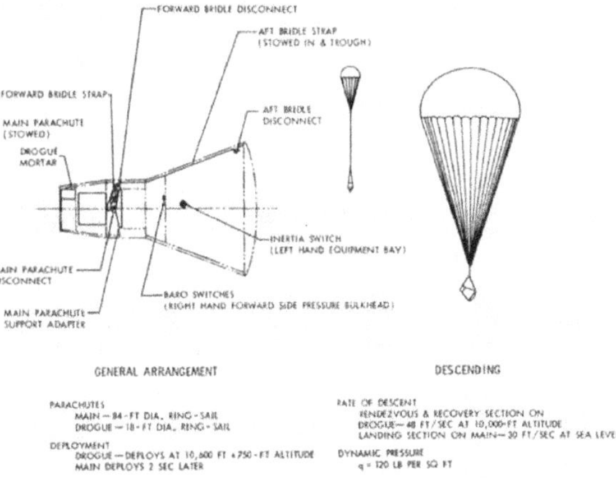

The parachute recovery system to be used instead of paraglider on the first Gemini spacecraft: stowed and deployed modes.

These decisions highlighted NASA's proactive approach in adapting and optimizing technologies to meet the stringent demands of manned spaceflight. By integrating advanced parachute systems and refining deployment strategies, NASA and its contractors reinforced their dedication to enhancing mission readiness and ensuring astronauts' safe return during pivotal Gemini missions.

On May 21, McDonnell made a significant stride in the Gemini program by awarding an $8 million subcontract to Electro-Mechanical Research, Inc. of Sarasota, Florida, to develop the data transmission system for the Gemini spacecraft. This system was pivotal in transmitting telemetry data in digital form using pulse-code modulation (PCM).

The PCM encoding method utilized by the data transmission system involved varying the length of pulses to represent and transmit information accurately. This technology ensured the precise and efficient transmission of measurements from the spacecraft and the target vehicle to ground receiving stations via radio links.

The data transmission system comprised several key components:

PCM Subsystem: Responsible for encoding and decoding telemetry data using pulse-code modulation techniques.

Onboard Tape Recorder: Stored telemetry data for later transmission or playback, enhancing data reliability and accessibility.

VHF Transmitters: Two transmitters that could transmit data in real-time or with a delayed transmission strategy, optimizing communication flexibility during missions.

This subcontract underscored McDonnell's commitment to integrating advanced communication technologies within the Gemini program. By ensuring robust data transmission capabilities, NASA enhanced its ability to monitor and analyze mission-critical data in real time, supporting operational decision-making and ensuring mission success.

The development and integration of the data transmission system exemplified NASA's proactive approach in advancing space exploration capabilities. As Gemini missions progressed, this technology would prove instrumental in facilitating effective communication between astronauts and ground control, marking a pivotal advancement in the era of manned spaceflight.

Chapter 3 - Preparing for Gemini Missions

Transforming Pad 19

Concurrently, Amendment No. 6 to the Gemini launch vehicle procurement contract earmarked $2.609 million to convert Pad 19 at Cape Canaveral, tailored explicitly for Gemini missions. Originally built for the Titan I development program by the Air Force, Pad 19 underwent extensive modifications starting in February, immediately following the final Titan I flight in January.

In April, the Gemini Project Office made crucial decisions regarding Pad 19's adaptation:

Erector System: Opted for an erector system over a gantry, streamlining launch preparation processes and enhancing operational efficiency.

White Room Addition: The upper third of Pad 19 was dedicated to a white room, a meticulously controlled environment that ensures astronaut safety and cleanliness during ingress and egress procedures.

These modifications underscored NASA's commitment to preparing Cape Canaveral facilities for the unique requirements of manned Gemini missions. By investing in advanced infrastructure and tailored facilities, NASA aimed to optimize launch operations, mitigate risks, and ensure the smooth execution of critical mission phases.

The conversion of Pad 19 represented a significant leap forward in NASA's capability to support human spaceflight. Equipped with state-of-the-art amenities and operational enhancements, Pad 19 would be pivotal in facilitating safe and efficient launches, setting the stage for historic achievements in manned space exploration during the Gemini program.

Following the final design review on July 9-10, the Army Corps of Engineers awarded Consolidated Steel of Cocoa Beach, Florida, the construction contract for Pad 19 modifications. Construction swiftly began in September and concluded, with Pad 19 operational on October 17, 1963. These efforts highlighted NASA's meticulous planning and execution in preparing launch facilities explicitly tailored for the requirements of the Gemini missions.

The culmination of the final design review marked a critical step towards ensuring that Pad 19 met all necessary specifications for Gemini launches. Consolidated Steel's swift and efficient construction efforts underscored NASA's commitment to achieving operational readiness within ambitious timelines.

The completion of Pad 19 represented a significant advancement in NASA's capabilities for manned spaceflight. Equipped with state-of-the-art modifications and tailored facilities, Pad 19 was pivotal in supporting safe and efficient launches during the Gemini program. This achievement underscored NASA's readiness to embark on pioneering missions and marked a crucial milestone in America's journey toward exploring the frontiers of space.

On May 23, McDonnell and the Manned Spacecraft Center representatives concluded an intensive series of 24 meetings from April 19 to negotiate critical technical details essential for Project Gemini. These negotiations were pivotal in shaping the Gemini spacecraft's comprehensive support and documentation framework throughout its development and operational phases.

Specifications for Gemini Systems and Subsystems: Detailed discussions defined the technical specifications and requirements for various systems and subsystems integrated within the Gemini spacecraft. This encompassed propulsion systems, life support, and communication systems, ensuring alignment with mission objectives and operational feasibility.

Rigorous criteria were established to address environmental factors and structural design aspects critical for spacecraft durability and operational effectiveness in harsh space conditions. These criteria aimed to enhance spacecraft resilience and longevity during mission operations.

Clear performance metrics and specifications articulated the Gemini spacecraft's expected capabilities and operational limits. This included parameters crucial for astronaut safety and mission success, guiding the spacecraft's operational boundaries.

Comprehensive plans for rigorous testing programs were developed to validate the functionality, reliability, and safety of all spacecraft systems and components. These tests simulated space conditions and operational scenarios to ensure readiness for mission deployment.

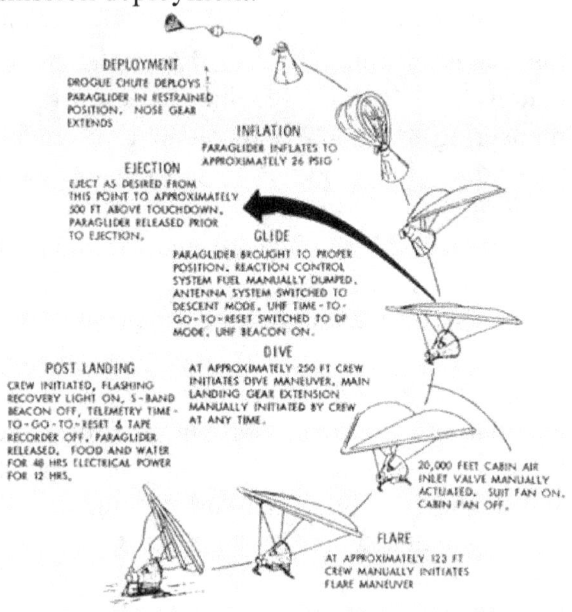

The proposed sequence of events in deploying the paraglider to land the Gemini spacecraft.

Detailed strategies and protocols were established to uphold the reliability and quality of every aspect of the Gemini spacecraft. Stringent quality assurance measures and validation processes were integrated to meet NASA's exacting standards, ensuring robust performance and mission readiness.

On May 23, Ames Research Center embarked on a pivotal wind tunnel test featuring a half-scale inflatable paraglider wing, marking a significant milestone in the Paraglider Development Program. This test, conducted within their full-scale test facility, aimed to gather crucial aerodynamic and load data for the integrated wing and spacecraft system while identifying potential aerodynamic challenges and evaluating design feasibility.

Comprehensive Test Scope:

The test encompassed a series of flight scenarios crucial for paraglider performance evaluation:

• Assessing the wing's stability and functionality upon deployment.

• Examining aerodynamic efficiency and performance during glide.

• Evaluating control and stability during descent and landing preparation.

During the later stages of testing, an unforeseen issue arose when the paraglider's sail unexpectedly tore. Despite this setback, the primary objective of gathering fundamental aerodynamic and load data was already achieved. Notably, the failure occurred under conditions more severe than anticipated in actual flight scenarios, minimizing the impact on the overall test outcomes. As a result, minor corrective measures were deemed sufficient, and the decision was made not to repeat the test, concluding efforts on July 25.

Following the wind tunnel test, a coordination meeting on July 26 affirmed the completion of Ames' test program for the paraglider landing system. This highlighted NASA's proactive approach to conducting rigorous evaluations to enhance the reliability and performance of innovative technologies like the inflatable paraglider, which was crucial for future manned space missions.

On May 23-24, the Manned Spacecraft Center (MSC) approved McDonnell's proposed sequence for the paraglider recovery system designed for Gemini missions. This milestone marked a pivotal advancement in spacecraft recovery technology, outlining a systematic approach to ensure safe and controlled descent following reentry into Earth's atmosphere.

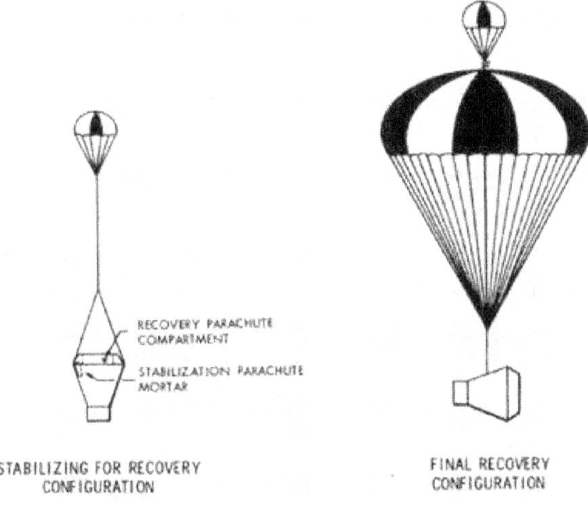

The emergency parachute recovery system for the half-scale paraglider flight test vehicle for Phase II-A of the development program.

The approved sequence detailed the step-by-step recovery process for Gemini missions:

• 60,000 feet: Deployment of the drogue parachute, initiating separation of the recovery compartment from the spacecraft and stripping the paraglider from the recovery system.

• 50,000 feet: Release of the rendezvous and recovery section from the spacecraft.

• 10,000 feet: Dumping of any remaining propellant in the reaction control system (RCS) after paraglider deployment.

• Post-Touchdown: Jettisoning of the paraglider wing as part of the recovery process.

Initially intended for deployment on all Gemini missions except the first, the paraglider system represented a groundbreaking innovation in landing technology. Enabling safe and controlled descent significantly enhanced manned spaceflights' safety and operational capabilities during the Gemini program. This innovative approach underscored NASA's commitment to advancing technologies that would pave the way for future missions beyond Earth's orbit.

On May 24, North American Aviation initiated a crucial phase in the Paraglider Development Program, focusing on certifying the emergency parachute system for the half-scale flight test vehicle in Phase II-A. This pivotal program aimed to validate the reliability and functionality of emergency recovery mechanisms essential for astronaut safety during unforeseen contingencies.

The program commenced with two successful drop tests on May 24 and June 20, affirming the system's capability to deploy and operate effectively under controlled conditions. However, a setback occurred during the third drop test on July 10 when the main recovery parachute failed to deploy as anticipated, necessitating a meticulous analysis of the malfunction.

After this setback, detailed modifications were collaboratively developed and agreed upon during an August 16 meeting between North American Aviation and Northrop Ventura, the responsible subcontractor for the parachute system. These modifications underwent rigorous testing and were proven effective during the fourth test conducted on September 4.

Following the successful validation of the emergency parachute system during the fourth test, the Manned Spacecraft Center (MSC) endorsed North American Aviation's assessment, affirming the system's qualification for the half-scale test program. This achievement highlighted NASA's rigorous standards in ensuring the safety and reliability of critical systems crucial for manned space missions, particularly emphasizing emergency protocols integral to the pioneering Paraglider Development Program.

Development and Testing of the Gemini Spacecraft Ejection Seat

On May 29, a collaborative effort involving representatives from McDonnell, Weber Aircraft, the Gemini Procurement Office, Life Systems Division, Gemini Project Office, and the US Naval Ordnance Test Station at China Lake, California, finalized plans for the development testing of the spacecraft ejection seat. These plans were specifically tailored to address the unique requirements of the Gemini spacecraft, mainly focusing on its capability for off-the-pad aborts.

The testing strategy emphasized initial evaluations from a stationary tower early in the program. These simulated off-the-pad ejection tests aimed to assess the effects of varying the center of gravity on the trajectory of the ejected seat and to refine the timing of the recovery sequence. The tower tests commenced on July 2, beginning a comprehensive testing phase.

Subsequent tests were planned to involve rocket sled ejection trials to explore simultaneous ejection scenarios with open hatches under maximum dynamic pressure conditions. Notably, sled tests were initiated on schedule, beginning November 9, before the tower tests were fully concluded.

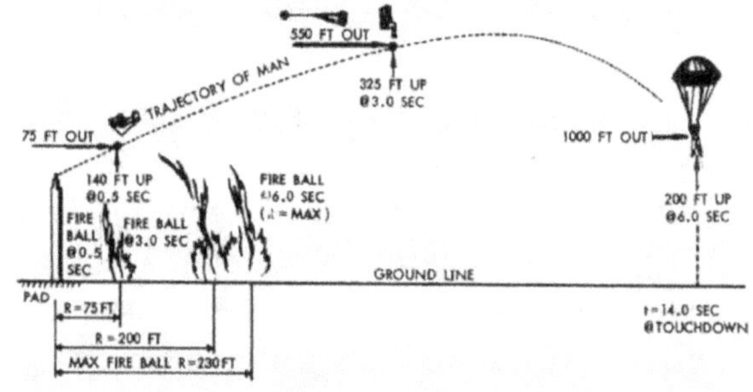

The "off-the-pad" escape mode for an aborted Gemini mission.

This proactive approach in testing highlighted NASA's commitment to ensuring the safety and effectiveness of the spacecraft ejection system, critical for astronaut safety during emergency situations, such as off-the-pad aborts, throughout the Gemini program.

During the first spacecraft operations coordination meeting on June 1, a comprehensive list was unveiled outlining the essential aerospace ground equipment necessary for handling and conducting pre-flight checkouts of the Gemini spacecraft. This meticulously curated equipment ensured that each spacecraft underwent thorough inspection and verification before launch.

The presentation likely detailed specialized tools, testing instruments, and support systems specifically designed to meet the unique requirements of the Gemini spacecraft. Emphasis was placed on the exacting preparation needed to ensure mission readiness and operational reliability.

On June 4, the Air Force School of Aviation Medicine at Brooks Air Force Base, Texas, launched a simulated long-duration Gemini mission. This groundbreaking simulation involved two individuals living for 14 days in an environment meticulously designed to mirror the proposed conditions of the Gemini spacecraft: a 100% oxygen atmosphere maintained at a pressure of 5 pounds per square inch (psi).

The primary objective of this simulation was to meticulously study the physiological and operational effects of prolonged exposure to these conditions, which closely resembled those anticipated during actual Gemini missions. Researchers aimed to gather crucial data on the feasibility and safety of sustaining astronauts in a high-oxygen, low-pressure environment over an extended period.

The insights gleaned from this experiment were pivotal for shaping the design and operational protocols of the Gemini spacecraft. By understanding the impacts of such environments on human health and performance, NASA could ensure astronauts' health, safety, and well-being throughout their missions in space. This simulation underscored NASA's commitment to rigorous preparation and innovation in advancing human spaceflight capabilities during the Gemini program.

On June 6, two pivotal developments marked significant progress in the preparations for Project Gemini. McDonnell was authorized to procure an additional boilerplate spacecraft for parachute landing system tests. Initially, McDonnell had planned to adapt a boilerplate spacecraft from North American Aviation to test the emergency parachute system designed for paraglider drop tests. However, after careful evaluation, McDonnell estimated that modifying the North American boilerplate would incur higher costs, ranging from $17,000 to $19,000.

In contrast, acquiring a new boilerplate spacecraft was projected to be more cost-effective, with estimates between $10,000 and $12,000. This decision underscored McDonnell's strategic approach to optimizing resources while ensuring thorough testing and validating critical systems essential for the success of the Gemini program.

Concurrently, on June 6, significant strides were made in preparing for Project Gemini as Whirlpool Corporation Research Laboratories in St. Joseph, Michigan, secured a contract from NASA's Manned Spacecraft Center (MSC) to develop a comprehensive food and waste management system. This pioneering system encompassed essential components such as water dispensers, food storage facilities, and waste management systems.

The US Army Quartermaster Corps Food and Container Institute in Chicago, Illinois, supplied food and zero-gravity feeding devices as part of this initiative. Under the guidance of MSC's Life Systems Division, the direction and coordination of this crucial development program were meticulously overseen.

On June 20, 1962, NASA's Manned Spacecraft Center (MSC) authorized North American Aviation to proceed with Phase II, Part B(1) of the Paraglider Development Program, formalized under letter contract NAS 9-539. This pivotal program phase entrusted North American Aviation with several crucial objectives:

Firstly, to design, construct, and rigorously test an advanced two-man paraglider trainer, essential for preparing astronauts for the unique landing capabilities of the paraglider system.

Secondly, a comprehensive flight simulation program should be initiated to train pilots in the intricate maneuvers required for successful paraglider landings.

Lastly, finalizing the design of a man-rated Gemini paraglider wing was crucial for enabling the spacecraft's safe re-entry and precision landings.

This authorization represented a significant advancement in developing the paraglider system in Project Gemini. Beyond engineering goals, the program aimed to establish robust training protocols, ensuring that astronauts could effectively pilot spacecraft equipped with this innovative landing system.

Following this authorization, the final contract for Phase II, Part B(1) was officially awarded to North American on October 31, 1962, solidifying the commitment to advance the paraglider development program towards operational readiness for manned missions in space.

On June 21-22, 1962, a significant milestone in the Paraglider Development Program occurred with the Paraglider Full-Scale Test Vehicle Design Engineering Inspection at North American's Space and Information Systems Division in Downey, California. This inspection was conducted by a team from NASA's Manned Spacecraft Center (MSC), which meticulously reviewed the design of the full-scale paraglider wing, the capsule, and all associated equipment. Additionally, they scrutinized the test program and schedules set for Phase II-A of the program.

During the inspection, the MSC team identified 33 suggested changes, primarily on hardware aspects. These recommendations were crucial for refining the design and ensuring that the full-scale test vehicle met the rigorous standards necessary for safe and effective use in upcoming phases of the Paraglider Development Program. The inspection underscored NASA's commitment to meticulous engineering and safety protocols as they advanced toward developing a reliable and innovative landing system for the Gemini spacecraft.

On June 25, 1962, the Gemini Project Office made a pivotal decision following an extensive study of reentry tracking data from earlier Mercury missions, including Atlas 4, 5, 6, and 7. This comprehensive analysis highlighted the necessity for a significant modification: replacing the planned initial S-band radar tracking beacon with a more effective C-band radar tracking beacon in the spacecraft's reentry section.

The adjustment was critical to enhance the reliability and accuracy of tracking the spacecraft's reentry trajectory through the ionization zone of Earth's atmosphere. Adopting the C-band radar beacon was expected to offer superior performance, address challenges identified in previous missions, and improve overall tracking capabilities.

By June 27th, Walter C. Williams, MSC Associate Director, officially communicated NASA's decision to adopt a pulse-code-modulation (PCM) telemetry system for real-time data transmission during the Gemini and Agena missions. This decision marked a significant milestone in telemetry technology, setting the stage for advanced data handling capabilities crucial for manned space missions.

Under the newly approved system, ten strategic sites were designated to host PCM equipment. Each site was equipped with dual acquisition systems, dual digital command capabilities, and pulse coders optimized to distinguish between the manned Gemini spacecraft and the Agena target vehicle when both were in orbit simultaneously. This setup ensured efficient and accurate data transmission, essential for mission control to monitor and command spacecraft operations effectively.

The adoption of PCM telemetry represented a strategic leap forward for the Gemini program, enhancing data reliability, transmission rates, and overall mission safety. This technological advancement underscored NASA's commitment to leveraging cutting-edge innovations to support the success of complex manned space missions.

Integration of Gemini Program Objectives into Mercury-Atlas 8 Mission

On June 27-28, 1962, discussions between the Gemini Project Office and McDonnell culminated in integrating Gemini program objectives into the upcoming Mercury-Atlas 8 mission, piloted by Wally Schirra and named Sigma 7. Recognizing the opportunity presented by this mission, they identified testing of heatshield materials and afterbody-shingle characteristics as paramount investigations beneficial to the Gemini program.

Central to this decision was flying samples of the Gemini heatshield on the Sigma 7 mission. This strategic move allowed NASA to gather invaluable data on the performance and durability of these materials under the extreme conditions of space reentry. By leveraging the Mercury mission platform, NASA accelerated developing and validating technologies crucial for the forthcoming Gemini missions.

Integration of Paraglider System into Gemini Spacecraft Design

On June 28, representatives from McDonnell and North American Aviation convened for their inaugural meeting to exchange detailed technical information regarding installing the paraglider system in the spacecraft. This pivotal gathering marked a significant milestone in coordinating the integration of the innovative paraglider technology into the spacecraft's design.

The meeting focused on ensuring compatibility and operational readiness of the paraglider system for future Gemini missions. Detailed technical exchanges between McDonnell and North American were crucial in aligning the efforts of both contractors and streamlining the integration process of new technologies into the spacecraft configurations.

On June 30, operations commenced at Martin-Baltimore's newly established airborne systems functional test stand in Baltimore. Spanning 3000 square feet, this state-of-the-art facility was dedicated to assembling and meticulously testing all airborne systems integral to the Gemini launch vehicle.

Airborne systems functional test stand at Martin's Baltimore plant.

The facility's scope encompassed various systems, including flight control, hydraulic, electrical, instrumentation, and malfunction detection components. Testing procedures involved assembling these systems on specialized tables and benches, utilizing actual engines with simulated propellant tanks and guidance systems.

Key objectives of the facility included conducting rigorous individual and combined systems tests, validating system design modifications, and effectively troubleshooting issues identified in previous test programs. These efforts were crucial in ensuring that every component and system of the Gemini launch vehicle underwent thorough testing, guaranteeing their readiness for operational deployment in forthcoming missions.

On July 2nd, critical simulated off-the-pad ejection tests commenced at the Naval Ordnance Test Station, marking a pivotal phase in the rigorous development of Project Gemini. Over a month, five meticulously executed ejection tests provided invaluable insights and revealed significant challenges, prompting essential design enhancements.

Two key modifications emerged from these tests:

Drogue-Gun Mechanism for Personnel Parachute Deployment: A drogue-gun mechanism was adopted to improve deployment reliability during emergency ejections. This mechanism ensured a more consistent and controlled deployment of the personnel parachute.

Robust Three-Point Restraint-Harness-Release System: A robust three-point restraint-harness-release system, modeled after systems used in military aircraft, was implemented. This enhancement aimed to enhance astronaut safety and stability during ejection scenarios, ensuring secure release and effective operation under dynamic conditions.

By early August, representatives from the Manned Spacecraft Center and the ejection system contractors convened to assess progress in Project Gemini's ejection system development. They determined that further off-the-pad ejection tests should be postponed until all critical design elements of the ejection seat hardware were integrated and thorough validation testing of the personnel parachute was completed.

Throughout August, meticulous checks and ground firings were conducted to validate these design alterations. These efforts culminated in a pivotal inflight drop test on August 30th, which successfully confirmed the efficacy of the ejection seat and its safety mechanisms using a dummy payload.

Following these successful validations, off-the-pad ejection testing resumed in September, marking a significant milestone in developing safety protocols and technological advancements crucial for Project Gemini's success. These tests were instrumental in enhancing astronaut safety and ensuring the reliability of emergency escape systems during critical phases of space missions.

On July 3rd, the Gemini Project Office convened a pivotal meeting attended by a diverse spectrum of key stakeholders, including representatives from the Manned Spacecraft Center's Flight Operations Division, McDonnell, International Business Machines (IBM), Aerospace Corporation, Air Force Space Systems Division, Lockheed Martin, and Space Technology Laboratories, Inc. in Redondo Beach, California, alongside members from the Marshall Space Flight Center. The primary agenda of this meeting was to delineate critical tasks essential for advancing toward final mission planning.

A coordinating body was established during this gathering, comprising two delegates from each participating agency. This collaborative effort aimed to streamline communication and integration across the multifaceted aspects of the Gemini project. This coordinated approach was crucial for ensuring the program's success by fostering alignment on objectives and milestones.

On July 6th, Martin formulated a comprehensive strategy to initiate flight tests for the malfunction detection system (MDS) designed for the Gemini launch vehicle. This strategic plan was closely intertwined with the development flights of the Titan II weapon system. In response to directives from the Gemini Project Office (GPO), Martin successfully integrated the Gemini MDS into the Titan II engines, a proposal thoroughly reviewed and endorsed by Martin's Systems Division and Aerospace Corporation.

By early August, the GPO granted full approval to proceed with what became known as the "piggyback plan." This innovative approach involved conducting rigorous tests by deploying the Gemini MDS on six Titan II flights. The primary objective was to validate the system's reliability under actual flight conditions, ensuring its readiness for deployment on the Gemini missions.

Mandate for Water Landings in Project Gemini

On July 11th, a pivotal decision solidified the requirement for the Gemini spacecraft to successfully execute water landings, affirming both the parachute landing system and the paraglider landing system as integral components. This mandate underscored the spacecraft's dual imperative: ensuring crew safety and maintaining seaworthiness during and after water landings, encompassing a crucial 36-hour post-landing period.

This stringent requirement reflected the meticulous planning and rigorous standards essential for the Gemini program. It emphasized robust safety protocols and operational readiness in all mission contingencies, ensuring that astronauts could safely return from space missions even under challenging conditions such as ocean landings.

On July 12th, pivotal meetings significantly shaped the Gemini program's operational readiness and launch planning trajectory. Representatives from the Gemini Project Office (GPO), Flight Operations Division, Air Force Space System Division,

Marshall Space Flight Center, and Lockheed convened in Houston for an Atlas-Agena coordination session.

During this crucial meeting, the GPO delineated essential maneuvers that both the Gemini spacecraft and ground command stations would need to command from the Agena. They distributed a preliminary statement outlining the basic mission objectives and requirements for the Atlas-Agena configuration. This collaborative effort aimed to ensure seamless integration and operational efficiency across all facets of the upcoming Gemini missions.

Simultaneously, at the Air Force Missile Test Center in Cape Canaveral, Florida, another pivotal gathering marked the official establishment of the technical team responsible for detailed launch planning under a new banner. Formerly known as the Gemini Operations Support Committee, this group, now named the Gemini-Titan Launch Operations Committee, consisted of representatives from all major stakeholders involved in supporting Gemini-Titan launch operations.

The restructuring aimed to enhance coordination, ensure alignment with mission objectives, and streamline operational consistency as preparations intensified following the conclusion of the Mercury flights. This unified approach underscored NASA's commitment to optimizing launch operations for the Gemini missions, leveraging lessons learned and integrating best practices from previous endeavors.

Integration Initiative for Gemini Spacecraft and Agena Target Vehicle

On July 13th, the Gemini Project Office initiated a critical effort to ensure seamless mechanical and electrical compatibility between the Gemini spacecraft and the Gemini-Agena target vehicle. This initiative established an interface working group comprising representatives from Lockheed, McDonnell, the Air Force Space Systems Division, Marshall Space Flight Center, and the Manned Spacecraft Center.

The primary objective of the interface working group was to facilitate a smooth exchange of data concerning design specifications and physical details between the contractors responsible for the spacecraft and the Agena target vehicle. This collaborative effort aimed to ensure that both components of the Gemini mission were integrated seamlessly, enhancing operational efficiency and readiness for mission success.

On July 19th, the Gemini Project Office and North American Aviation reached a significant agreement outlining design parameters for critical components of the Gemini program. Key among these agreements was the establishment of guidelines for the advanced paraglider trainer, the paraglider system for static test article No. 2, and the operational paraglider system destined for integration into the Gemini spacecraft.

Central to these guidelines was the insistence that all critical operations within the paraglider systems incorporate redundancy. This provision was crucial for enhancing safety and reliability during essential flight phases. It ensured that the Gemini spacecraft would be equipped with robust backup systems capable of effectively managing unforeseen challenges.

Establishing Mission Control

On July 20th, 1962, the future of American space exploration took a decisive turn with NASA Administrator James E. Webb's landmark announcement. At the helm of the National Aeronautics and Space Administration, Webb unveiled plans for a dedicated mission control center at the Manned Spacecraft Center (MSC) in Houston, Texas. This strategic move marked a pivotal shift from the Cape Canaveral control center, which had admirably overseen the pioneering missions of Project Mercury but was now deemed insufficient for the ambitious goals set forth by the forthcoming Gemini and Apollo programs.

The establishment of this new control center represented more than just a relocation—it symbolized NASA's commitment to advancing manned space flight capabilities to unprecedented heights. Situated in Houston, the MSC would serve as the nerve center for all future missions, harnessing cutting-edge technology and the collective expertise of NASA's brightest minds. Here, flight controllers would monitor and manage the complexities of human space missions with unparalleled precision and efficiency.

Webb's announcement underscored NASA's recognition of the evolving demands of Project Gemini, the pivotal precursor to Apollo's lunar ambitions. Unlike Mercury's straightforward orbital flights, Gemini missions would introduce new challenges such as extended duration flights, rendezvous and docking maneuvers, and extravehicular activities—a leap forward in technological complexity and operational sophistication.

Thus, the decision to centralize mission control at the MSC was not merely administrative but strategic, aligning NASA's operational infrastructure with the ambitious goals of the Gemini and subsequent Apollo programs. This pivotal moment laid the groundwork for the unprecedented achievements and historic milestones defining

America's triumphant journey to the moon and beyond.

In April 1962, as NASA geared up for the next phase of manned space exploration, Philco Corporation's Western Development Laboratories in Palo Alto, California, received a pivotal contract. Their task: to spearhead the study and design of the flight information and control functions for NASA's new mission control center. This state-of-the-art facility would be pivotal in guiding the daring missions of Project Gemini and laying the groundwork for Apollo's lunar ambitions.

Under the watchful eye of the US Army Corps of Engineers, renowned for their oversight of major infrastructure projects, construction of the center began at the Manned Spacecraft Center (MSC) in Houston, Texas. This location, chosen for its strategic advantages and NASA's expanding footprint, would become the nerve center for America's manned space flights.

Scheduled for completion by 1964 at an estimated cost of $30 million, the mission control center was designed with precision and foresight. Its development was timed to align seamlessly with the rigorous demands of the upcoming Gemini rendezvous missions. These missions posed unprecedented challenges, including orbital maneuvers, extended duration flights, and the intricate choreography of spacecraft rendezvous—a testament to NASA's expanding capabilities and vision.

On July 25th, 1962, McDonnell Aerospace delivered a critical update on the Gemini propulsion system. The adjustment focused on reducing the rated thrust of the two forward-firing thrusters from 100 pounds to 85 pounds each. This recalibration was a strategic response to potential disturbance torques that could arise during spacecraft maneuvers, especially when one engine might be non-operational.

The decision underscored McDonnell's commitment to enhancing the stability and reliability of the Gemini spacecraft—a pivotal next step in NASA's ambitious manned space exploration program. By fine-tuning the propulsion system, engineers aimed to mitigate operational risks and ensure precise control during the complex orbital maneuvers planned for Gemini missions.

This adjustment reflected a meticulous approach to spacecraft design, where each component's performance was scrutinized to meet the rigorous demands of manned space flight. As NASA prepared for the challenges ahead, McDonnell's adaptation of the propulsion system marked a proactive measure in safeguarding mission success and astronaut safety.

The modified thruster configuration would play a crucial role in the forthcoming Gemini missions, setting a precedent for the meticulous engineering and strategic foresight that defined America's pursuit of space exploration milestones.

From July 25th to 26th, 1962, a pivotal event unfolded at Aerojet-General's Liquid Rocket Plant in Sacramento, California. This facility, crucial for developing engines powering the Titan II launch vehicle, hosted a comprehensive reliability review of its engine system. This review held particular significance as it aimed to surpass the reliability standards set for intercontinental ballistic missile (ICBM) engines, marking a critical step in preparing for the upcoming Gemini missions.

Under the auspices of the Air Force Space Systems Division, Aerojet embarked on augmenting the existing reliability program for ICBM engines. The stringent directive highlighted the imperative to elevate engine reliability to unprecedented levels, a mandate essential for ensuring the success and safety of manned space missions under the Gemini program.

This meticulous review underscored Aerojet's commitment to rigorous testing and quality assurance, essential pillars in pursuing space exploration milestones. By enhancing engine reliability, Aerojet fortified the Titan II's capabilities and contributed decisively to NASA's overarching goal of achieving manned space flight with unparalleled safety and precision.

The outcomes of this reliability review would reverberate throughout the Gemini program, shaping operational readiness and setting new benchmarked in engineering excellence. As NASA prepared for the challenges of orbital rendezvous and extended-duration missions, the reliability of the Titan II's engine system stood as a testament to the meticulous planning and technical prowess driving America's pioneering efforts in space exploration.

August 2nd, 1962, marked a significant juncture in the evolution of NASA's Gemini program as Lockheed presented crucial findings and design recommendations for the Agena D propulsion systems. This milestone event occurred at a gathering of representatives from Marshall Space Flight Center, Manned Spacecraft Center, and the Air Force Space Systems Division in Houston, Texas. The meeting followed NASA and the Air Force's tentative decision in July to upgrade from the Agena B to the more advanced Agena D for the Gemini program.

Lockheed's comprehensive presentation marked the culmination of the analysis phase of the Gemini-Agena initiative. The study encompassed detailed evaluations of both primary and secondary propulsion systems, highlighting key advancements and challenges in the propulsion technology critical to Gemini missions.

Central to Lockheed's presentation was an in-depth assessment of the multiple-restart main engine's start system—a pivotal component for successful rendezvous and docking maneuvers planned for Gemini missions. The presentation also spotlighted recent testing milestones achieved by Bell

Aerosystems Company in Buffalo, New York, the engine's subcontractor. These tests validated critical performance parameters and underscored progress toward achieving operational readiness.

Following thorough deliberations, a pressurize-start tank system was selected in September as the preferred solution, reflecting ongoing efforts to optimize the Agena D's capabilities for the Gemini program's demanding mission requirements. This decision underscored the collaborative effort and meticulous planning essential to advancing propulsion technology crucial for achieving the program's ambitious objectives in space exploration.

In the summer of 1962, North American Aviation embarked on a rigorous testing program to qualify the emergency parachute recovery system for the full-scale test vehicle under Phase II-A of the Paraglider Development Program. The initiative aimed to ensure the reliability of the system crucial for safely recovering Gemini crew members in emergency scenarios.

The initial test on August 2nd demonstrated promising results, validating the system's functionality under simulated recovery conditions. However, subsequent tests encountered setbacks that highlighted the complexities of parachute deployment and recovery. During the second test on August 22nd, the loss of one main parachute post-deployment posed challenges, though fortunately, no significant damage occurred to the test vehicle.

Undeterred by these initial setbacks, testing continued with a third session on September 7th, where the loss of two parachutes resulted in minor damage to the test vehicle. Despite ongoing challenges, the dedication to refining the emergency recovery system remained paramount.

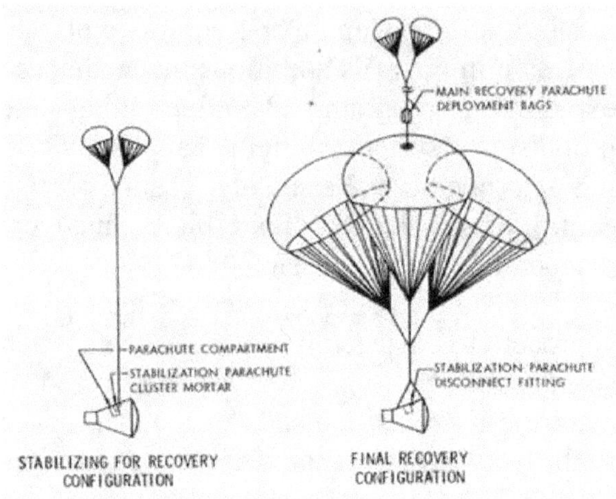

The emergency parachute recovery system for the full-scale paraglider flight test vehicle.

The culmination of the test series on November 15th brought a critical incident: all recovery parachutes detached from the spacecraft immediately after deployment, destroying the test vehicle upon impact. In response, the Manned Spacecraft Center decided to terminate this phase of the test program.

Recognizing the importance of ongoing development, McDonnell was directed to provide North American Aviation with a boilerplate spacecraft for future tests. This directive aimed to facilitate continued refinement and enhancement of the emergency recovery systems vital for crew safety in the Gemini program.

Converting Complex 14 for Gemini Operations

On August 3rd, 1962, a pivotal meeting convened in Los Angeles where the United States Air Force outlined comprehensive plans to the Gemini Project Office. The focus was on converting Complex 14 at the Atlantic Missile Range, Cape Canaveral, Florida—a site steeped in the history of Mercury launches—to support upcoming Project Gemini operations.

Historically, Complex 14 had been instrumental in launching Mercury missions, but its adaptation for Gemini marked a strategic evolution in America's manned space flight capabilities. The site's redesignation as the primary launch facility for the Agena target vehicle underlined its pivotal role in facilitating rendezvous and docking maneuvers—a critical objective of the Gemini program.

The Air Force's plans detailed extensive modifications to enhance the complex's infrastructure to meet the rigorous demands of Gemini missions. These included upgrades to launch pad systems, fueling capabilities, and mission control facilities, ensuring seamless integration with NASA's evolving operational needs.

The decision to repurpose Complex 14 underscored a collaborative effort between NASA and the Air Force, leveraging existing infrastructure while advancing capabilities for future space missions. This strategic alignment positioned Cape Canaveral as a cornerstone of America's ambitious quest for lunar exploration and beyond, symbolizing a new chapter in the nation's pioneering spirit in space.

The Air Force assumed responsibility for funding, designing, modifying, and equipping Complex 14 to accommodate the Atlas-Agena configuration essential for Gemini missions. The timeline for this critical initiative was outlined as follows: preliminary design criteria were to be established by September 1st, with final design criteria scheduled for completion by October 1, 1962.

In parallel, the Mercury Project Office confirmed that Complex 14 would be ready to support Gemini operations by September 1, 1963. This coordinated effort underscored the meticulous planning and infrastructure enhancements necessary to meet the evolving operational demands of the Gemini program, ensuring that facilities were optimized for the successful execution of upcoming space missions.

Training Gemini Flight Controllers

On August 9th, the Flight Control Operations Branch of the Manned Spacecraft Center's Flight Operations Division unveiled a comprehensive training program to prepare Gemini flight controllers for their pivotal roles in upcoming missions. This multifaceted program was structured to ensure thorough readiness across various operational aspects:

A one-month course hosted at McDonnell, cycling through three classes of 10-15 personnel. The training curriculum encompassed three weeks of intensive systems training, one week focusing on hardware, and familiarization with McDonnell's drawing standards.

Flight controllers underwent specialized training in systems and network operations, including updates on system capabilities and practical exercises to enhance operational proficiency.

Team training included site-specific training for personnel supporting the missions, command site teams, and remote site teams. Team training aimed to optimize coordination and response capabilities across all operational levels.

Emphasis was placed on mastering control, communications, and decision-making within the network flight control organization. This involved detailed simulations and checkouts of operational procedures, countdown protocols, systems tests, and network equipment usage.

Drawing from their experience with the Mercury program, existing flight controllers were selected to transition into roles for Project Gemini. Additional personnel were also recruited to augment the team, ensuring adequate manpower to meet the heightened demands of the advanced Gemini missions.

This structured training approach underscored the meticulous preparation and interdisciplinary skills required to manage the complexities of manned spaceflight operations, reflecting NASA's commitment to ensuring mission success through rigorous training and readiness protocols.

On August 14th, North American Aviation commenced flight testing of the half-scale vehicle (HSTV) as part of Phase II-A of the Paraglider Development Program, albeit starting two months behind schedule. The instrumented HSTV, with the paraglider already predeployed, was towed aloft by helicopter for these initial tests.

The primary objectives of these predeployed flights were to assess several key performance metrics:

• Flight performance under different conditions.

• Longitudinal and lateral control characteristics.

• Effectiveness of control systems.

• Capability of executing the flare maneuver with the paraglider.

Despite encountering various minor malfunctions across all five test flights conducted on August 14th, 17th, 23rd, September 17th, and October 23rd, 1962, the results conclusively demonstrated the stability of the wing/vehicle combination during free flight. Moreover, they affirmed the effectiveness of control mechanisms in guiding the HSTV, highlighting the adequacy of these systems for future phases of the Paraglider Development Program.

These successful tests marked a significant milestone in advancing the development of paraglider technology for potential use in manned spacecraft recovery systems, reinforcing the program's progress despite initial setbacks in scheduling.

On August 15th and 16th, the Manned Spacecraft Center (MSC) formally reviewed McDonnell's engineering mock-up of the Gemini spacecraft in St. Louis. McDonnell had initiated construction of the mock-up in January, shortly after being awarded the spacecraft contract. Initially slated for mid-July, the mock-up review was delayed due to informal examinations by MSC representatives, including James A. Chamberlin and several astronauts, who provided feedback and suggested changes.

Two McDonnell technicians examine the engineering mockup of the Gemini spacecraft

During the formal review process, MSC thoroughly evaluated McDonnell's mock-up and issued 167 requests for alterations based on their observations and assessments. These requests aimed to refine and improve various aspects of the spacecraft design to meet the stringent requirements and safety standards set for the Gemini program.

McDonnell implemented the requested changes following the review, and MSC conducted a follow-up inspection of the revised mock-up in November. This meticulous review and iterative improvement process underscored the collaborative effort between NASA and its contractors to ensure the Gemini spacecraft's readiness for the challenges of manned space missions, emphasizing the program's commitment to precision and safety in spacecraft design and development.

On August 16th, a significant agreement was reached between the Air Force and NASA concerning using a standard Atlas space booster for the Gemini program, with both agencies sharing the development costs equally. This decision was pivotal in streamlining operations and enhancing efficiency across the Gemini missions.

Key ground rules were established for the standard Atlas space booster, which was already under development by the Air Force:

• No new development program was to be initiated.

- Equipment on the launch pad would be rearranged for standardization purposes.

- Splices would be eliminated.

- Electrical installations would be consolidated.

- Differences between various programs would be minimized.

- The ballistic missile aspects of the Atlas space booster would undergo specific modifications:

- A fully-qualified engine would be up-rated from 150,000 to 165,000 pounds of thrust.

- Vernier rockets would be eliminated to reduce propellant usage.

- Standard tank pressures and pneumatic pressures would be adopted.

- Retrorockets would be removed.

- A standard range safety package would be incorporated.

The first standardized Atlas vehicle was anticipated to be ready by September 1963. This agreement represented a strategic alignment between NASA and the Air Force to optimize resources and capabilities, ensuring a reliable and efficient launch vehicle for the Gemini program's ambitious objectives in manned space exploration.

On August 16th, the Gemini Project Office conducted a thorough review of the Agena status displays and approved a set of eight crucial displays for mission operations. These displays included seven green lights, each indicating the satisfactory status of various Agena functions. Additionally, an eighth red light was designated to signal any main engine malfunction, providing essential real-time feedback to mission controllers.

During the same session, the Gemini Project Office also finalized and approved the comprehensive list of commands necessary to control specific Agena functions during rendezvous and docking maneuvers with the Gemini spacecraft. These commands were primarily transmitted via radio, ensuring seamless interaction between the spacecraft and its target vehicle.

Further refinement of the command protocols occurred during a subsequent review on September 13th and 14th, identifying 34 minimum commands required for initiating and executing rendezvous maneuvers during Gemini missions. This meticulous planning and validation of command procedures underscored NASA's commitment to precision and operational readiness, crucial for the success of complex space missions involving orbital rendezvous and docking maneuvers.

Coordinating Gemini Missions:

On August 27th, the Gemini Project Office launched a comprehensive program to coordinate and integrate the development efforts for Gemini rendezvous and long-duration missions. This initiative was structured around a mission-planning and guidance-analysis coordination group supported by three specialized working panels.

The mission-planning and guidance-analysis coordination group was the central hub for overseeing mission development's strategic planning and analytical aspects. It facilitated collaboration among various stakeholders involved in defining mission objectives, optimizing mission trajectories, and ensuring the compatibility of spacecraft systems for extended mission durations.

The three working panels, operating under the coordination group's umbrella, were tasked with addressing specific aspects critical to mission success:

Rendezvous Planning Panel: Responsible for devising strategies and procedures for orbital rendezvous maneuvers, ensuring precise alignment and approach between the Gemini spacecraft and its target vehicles.

Long-Duration Mission Panel: Focused on developing protocols and systems to support prolonged stays in space, including life support, crew health monitoring, and spacecraft maintenance strategies.

Integration and Systems Analysis Panel: Tasked with integrating and analyzing data from various subsystems to ensure seamless operation and performance during both rendezvous and extended mission scenarios.

By establishing these coordinated efforts, the Gemini Project Office aimed to enhance mission efficiency, mitigate risks, and advance the capabilities required for achieving successful and sustained human spaceflight beyond Earth's orbit.

On August 28th, during a spacecraft production evaluation meeting, the Gemini Project Office and McDonnell revised the projected launch date for the first Gemini flight, pushing it from August to September 1963. This adjustment was necessary due to delays in delivering critical components from vendors, impacting the spacecraft assembly timeline.

Despite this delay in the unmanned test flight, the schedule for the first manned mission (the second Gemini mission) remained unchanged, targeting a launch in November of the same year. This decision reflected the project's commitment to maintaining the overall mission sequence and readiness for crewed missions while addressing and mitigating challenges in the production and integration phases of the Gemini spacecraft.

On August 31st, the Gemini Project Office detailed its plans for conducting spacecraft checkout operations at Cape Canaveral. The approach for Gemini preflight checkout closely followed the established procedures from the Mercury program. It involved a comprehensive series of end-to-end functional tests designed to assess the spacecraft and its systems thoroughly.

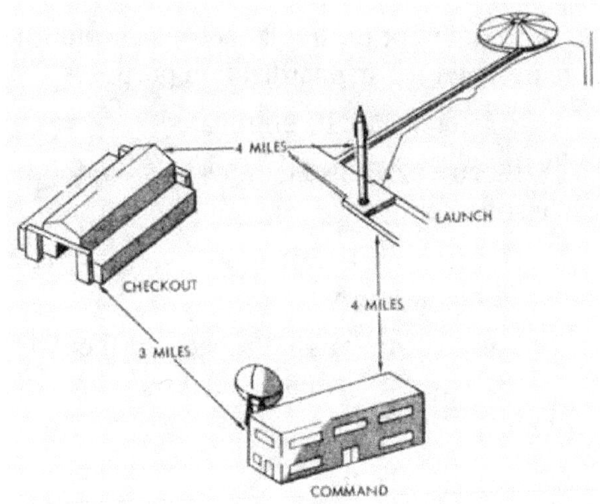

Proposed layout of Gemini facilities at Cape Canaveral.

The checkout process began with independent modular systems tests to validate the functionality of each subsystem. Subsequently, the spacecraft modules were integrated, and a series of integrated tests were conducted to ensure seamless operation and compatibility across all systems. These tests culminated in a simulated flight scenario, serving as the final verification before the spacecraft was prepared for transfer to the launch complex.

The Hangar S complex at Cape Canaveral was slated for expansion to accommodate these rigorous testing requirements. Major test stations were planned to be housed in Hangar AR, an existing facility adjacent to Hangar S, which would support intensive testing and checkout procedures.

The construction and enhancement of these facilities were scheduled to be completed by March 1, 1963, ensuring readiness to commence checkout operations for Gemini spacecraft No. 1 upon its arrival at Cape Canaveral by the end of April 1963. This meticulous planning and infrastructure preparation underscored NASA's commitment to meticulous preparation and thorough testing, essential for the success and safety of the upcoming Gemini missions.

Milestones and Challenges in Gemini Spacecraft Propulsion

Rocketdyne achieved significant milestones in August when developing the Gemini spacecraft's liquid propulsion systems. They completed the design and fabrication of prototype hardware for these systems, marking a crucial step forward in the spacecraft's propulsion capabilities.

Additionally, Rocketdyne initiated testing of the reaction control system (RCS), which plays a vital role in maneuvering and stabilizing the spacecraft in space. However, engineers observed nozzle erosion issues during test firings of the 25-pound thrust chambers. This erosion led to performance degradation, impacting the system's efficiency and operational lifespan. It showed signs of degradation after approximately one-third of the specified burn time.

These findings prompted further analysis and adjustments to ensure the reliability and longevity of the reaction control system, critical for the precise control and maneuverability required for the Gemini spacecraft's orbital operations and rendezvous maneuvers.

On September 1st, George W. Jeffs assumed the role of Program Manager for the Paraglider Development Program at North American, replacing N. F. Witte, who transitioned to Assistant Program Manager. This change marked a significant organizational shift, elevating the paraglider effort from project to program status within North American's Space and Information Systems Division.

The paraglider program was swiftly promoted to operating division status, occurring three months later when George W. Jeffs was appointed Vice President of the Space and Information Systems Division. This progression underscored the growing importance and strategic focus placed on developing the paraglider system, highlighting its role in advancing North American's capabilities within the aerospace industry.

On September 4th, the Gemini Project Office issued directives to McDonnell concerning spacecraft No. 3. Specifically, McDonnell was tasked with outfitting this spacecraft with rendezvous radar capability and providing a rendezvous evaluation pod. This requirement was mandated for missions 2 and 3 of the Gemini program.

To fulfill these specifications, McDonnell was instructed to produce four rendezvous evaluation pods in total:

- One prototype pod

- Two flight-ready pods

- One spare pod for potential mission contingencies

These pods were crucial components designed to enhance the Gemini spacecraft's capabilities, particularly in the context of orbital rendezvous maneuvers, a key operational objective of the Gemini program. This directive underscored NASA's commitment to advancing the spacecraft's technological readiness and operational versatility as it prepared for increasingly complex missions in space.

On September 5th, the Manned Spacecraft Center (MSC) outlined its operational plan for Gemini rendezvous missions. The procedure involved launching the Agena target vehicle first. Under normal conditions, the Gemini spacecraft would be launched the following day. This sequential approach was designed to facilitate rendezvous operations in space, ensuring optimal conditions for the spacecraft to maneuver and dock with the pre-positioned Agena target vehicle.

On September 6th, following the Gemini mock-up review held on August 15-16, a dedicated study group convened to assess the progress of the ejection seat development program. This crucial meeting was convened by McDonnell, who delivered promising news regarding the successful redesign and rigorous testing of the ejection seat system.

This achievement marked a significant milestone, crucially clearing the path for resuming off-the-pad developmental testing.

McDonnell's report highlighted that the primary focus would be determining the dynamic center of gravity for the combined seat and occupant configuration. This task was particularly vital as it aimed to ensure optimal performance under the anticipated acceleration profiles that Gemini astronauts would encounter during various phases of their missions.

The discussion underscored the meticulous engineering efforts underway, emphasizing the meticulous attention to detail necessary for the safety and operational success of the Gemini spacecraft. This development phase underscored the critical role of rigorous testing and design refinement in preparing for the challenges of manned spaceflight.

On September 9th, a pivotal moment unfolded at Launch Pad 14 as workers successfully fired an 11-second static firing of Atlas rocket engines. This critical test aimed to validate a new fuel feed system, essential for ensuring the reliability and performance of the upcoming mission. The meticulous preparation continued the next day as the dedicated team hoisted the Mercury capsule atop the rocket, marking another significant step toward launch readiness.

President Kennedy's Historic Visit to Cape Canaveral

President John F. Kennedy's visit to Cape Canaveral on September 11th marked a significant milestone in the nation's space exploration narrative. As part of his whirlwind tour of pivotal space facilities, including his iconic address at Rice University in Houston, Kennedy personally inspected the assembled rocket at Cape Canaveral. Guided by astronaut Wally Schirra, the President witnessed firsthand the culmination of rigorous engineering and meticulous preparation for America's ambitious space endeavors. This momentous occasion underscored Kennedy's unwavering commitment to advancing the nation's capabilities in space exploration and inspiring the world with America's pioneering spirit.

Meanwhile, on the opposite side of the globe, U.S. Navy recovery forces were mobilizing under the command of the aircraft carrier U.S.S. Kearsarge (CVS-33). Positioned northeast of Midway Island, they awaited the anticipated splashdown of astronaut Schirra's Mercury capsule after its full six-orbit mission. Additional support ships were strategically deployed in the Atlantic Ocean to cover alternate recovery zones, highlighting the meticulous planning and global coordination essential for every facet of America's early manned space missions.

On September 12th, Test No. 6, the rigorous testing regimen of the redesigned Gemini escape system, resumed. This marked a pivotal phase in ensuring the system's reliability in simulated off-the-pad scenarios. The subsequent Test No. 7, conducted on September 20th, continued these efforts, although it uncovered some technical challenges that demanded immediate attention.

Specifically, issues were identified with the seat-structure thrust pad, prompting a thorough reevaluation and subsequent redesign of this critical component. As a result, simulated off-the-pad testing was temporarily suspended until the final configuration of the rocket catapult could be established. A significant breakthrough occurred on January 4, 1963, with a successful rocket motor test that validated the structural integrity of the thrust-pad area. This milestone paved the way for the resumption of simulated pad ejection tests in the following month, marking a decisive step forward in developing a robust escape system for the Gemini spacecraft.

On September 14th, a pivotal coordination meeting focused on mission planning and guidance laid out the ambitious objectives for the first Gemini mission. This inaugural flight was designated as a critical test to assess the maximum heating rates the spacecraft could endure. The mission's overarching goal was to subject as many spacecraft systems as feasible to rigorous testing, paving the way for subsequent manned missions.

Following this meeting on September 18th, a crucial discussion between the Manned Spacecraft Center and McDonnell further refined the parameters for the mission. It was established that the spacecraft would follow a ballistic trajectory spanning approximately 2200 miles. The primary objective centered on gathering essential thermodynamics and structural data crucial for enhancing the spacecraft's design and performance in future missions. The mission aimed to partially qualify various spacecraft systems, marking an incremental step towards manned spaceflight capabilities.

Introducing America's First Manned Spaceflight Pioneers

On September 17th, a significant event unfolded at the University of Houston's Cullen Auditorium, orchestrated by Director Robert R. Gilruth of the Manned Spacecraft Center (MSC). Here, the nine pioneering individuals selected for the MSC flight crew training program for Gemini and Apollo missions were formally introduced to the public.

Among these select few, the lineup reflected a diverse blend of military and civilian expertise. Representing the Air Force were Major Frank Borman, and Captains James A. McDivitt, Edward H. White II, and Thomas P. Stafford, each bringing distinguished service backgrounds and a wealth of aviation experience to the program.

Volunteers from the Navy included Lieutenant Commanders James A. Lovell, Jr., John W. Young, and Lieutenant Charles Conrad, Jr., all known for their exceptional naval careers and readiness for space exploration.

Rounding out the group were two civilians: Neil A. Armstrong and Elliot M. See, Jr. Their inclusion highlighted NASA's inclusive approach, recognizing the invaluable contributions of civilian pilots in advancing the frontiers of human spaceflight.

This introduction marked a pivotal moment in the history of American space exploration. These nine men embarked on a journey defining the future of manned space missions, contributing significantly to the Gemini and Apollo programs.

On September 19th, ACF Electronics reached a critical milestone by delivering an engineering prototype radar beacon to McDonnell. This component represented a pivotal advancement in aerospace technology, specifically designed to enhance the reentry capabilities of spacecraft within the context of the Gemini program.

Before delivery, rigorous testing of an engineering prototype C-band beacon had been conducted at ACF Electronics. These tests simulated reentry conditions, validating the beacon's performance under extreme operational environments. Importantly, the tests confirmed that the beacon maintained optimal functionality without any degradation, demonstrating its reliability and suitability for integration into future spacecraft missions.

This achievement underscored the collaborative efforts between industry leaders like ACF Electronics and aerospace pioneers at McDonnell, highlighting their dedication to advancing the technological frontiers essential for safe and successful manned space missions.

On September 19th, the Life Systems Division reported significant progress concerning critical studies to enable extravehicular operations during Gemini missions. These studies encompassed several key areas essential for the safety and functionality of astronauts outside the spacecraft.

Firstly, advancements in thermal protection were highlighted, particularly the evaluation of a superinsulation coverall designed to be worn over the pressure suit. This innovative garment aimed to shield astronauts from extreme temperature fluctuations encountered during spacewalks, ensuring their comfort and safety in the harsh space environment.

Secondly, rigorous assessments were ongoing regarding ventilation system requirements and the development of corresponding hardware. These efforts were crucial for maintaining astronauts' environmental control and life support systems during extended periods outside the spacecraft.

Lastly, studies focused on refining methods for maneuvering in close proximity to the spacecraft, a fundamental capability for astronauts conducting extravehicular activities (EVAs). These maneuvers required precise coordination and innovative solutions to ensure astronauts could navigate effectively and safely while performing complex tasks in microgravity conditions.

Together, these ongoing studies underscored NASA's proactive approach to advancing the technologies and operational protocols vital for successful extravehicular missions during the Gemini program.

On September 25th, a preliminary design criteria review conference convened in Los Angeles, setting pivotal ground rules for all contractors involved in developing Complex 14. This facility, located at a crucial site, was designated for launching both the Gemini-Agena target vehicle and Mariner spacecraft, underscoring its strategic importance in advancing American space exploration objectives.

Key target dates were established during the conference to ensure operational readiness: July 1, 1963, was designated for stand availability; November 1, 1963, marked the estimated beneficial occupancy date; and February 1, 1964, was set as the deadline for the vehicle to be on the stand. These milestones were essential for aligning efforts across various contractors and agencies, ensuring timely preparations for upcoming missions.

While Complex 14 was slated for dual use, primarily supporting the Gemini program, modifications were outlined to accommodate the unique requirements of launching both the Gemini-Agena target vehicle and Mariner spacecraft. Notably, the Air Force Space Systems Division had previously reviewed a criteria summary report on November 15, 1962, suggesting minor engineering adjustments to optimize the facility for its expanded operational scope.

On October 1st, the Air Force Space Systems Division undertook a significant revision of the Development Plan for the Gemini launch vehicle, marking a pivotal moment in the project's evolution. This revision included a substantial budget adjustment, raising the allocated funds to $181.3 million. Rising costs associated with several critical aspects of the program necessitated this increase.

Earlier in September, escalating expenses related to the construction of the vertical test facility at Martin's Baltimore plant, the conversion of pad 19 at Cape Canaveral, and the development of aerospace ground equipment had already prompted a budget adjustment, bringing the total to $172.6 million. These investments were essential for enhancing the infrastructure and capabilities required to support the Gemini missions effectively.

The revised Development Plan also indicated a revised schedule, with the first launch date now postponed to December 1963. This adjustment reflected the meticulous planning and adjustments needed to ensure the Gemini launch vehicle's safety, reliability, and operational readiness and associated facilities.

These developments underscored the dynamic nature of aerospace projects, where careful budget management and schedule adjustments were crucial to meeting ambitious technological and operational milestones during the Gemini program.

On October 3rd, a pivotal moment marked the progression of the Gemini Program. The Manned Spacecraft Center (MSC) released the Gemini Program Instrumentation Requirements Document (PIRD). This document was the cornerstone for integrating the global Manned Space Flight Network, a critical backbone supporting the ambitious Gemini missions.

The development of the PIRD was a collaborative effort involving multiple NASA installations and Department of Defense components. Together, they constructed, maintained, and operated an extensive network of tracking stations and communication facilities. These facilities were indispensable for monitoring and supporting Gemini missions, ensuring seamless coordination between ground-based operations and spacecraft.

The PIRD provided meticulous specifications and requirements for instrumentation across the network. These guidelines were essential for real-time monitoring, communications, and data acquisition during all phases of mission operations. By establishing clear standards, MSC and its partners laid the groundwork for a robust infrastructure capable of facilitating complex maneuvers and experiments integral to the Gemini program's objectives.

On the same day, significant progress was achieved in spacecraft engineering. At a mechanical systems coordination meeting, McDonnell presented findings from its evaluation of substituting straight tube brazed connections for threaded joints in the spacecraft's propulsion systems. This proposal, initiated with tests starting on June 26, 1962, aimed to enhance reliability and efficiency.

Following McDonnell's presentation, the decision was made to adopt the brazed connections due to their demonstrated advantages. These connections minimized potential leak paths and contributed to a reduction in overall propulsion system weight. Such enhancements were crucial for ensuring the integrity and optimal performance of the spacecraft's propulsion systems during critical mission phases.

Meanwhile, preparations were underway at Cape Canaveral's Hangar S for astronaut Schirra's upcoming Mercury-Atlas 8 mission. Early on October 3rd, his flight surgeon conducted a brief physical examination before Schirra enjoyed a traditional pre-launch steak-and-eggs breakfast. Donning his silver spacesuit, Schirra made his way to the launch pad, where, during the journey, he briefly dozed off. Upon arrival, fellow astronaut Cooper and pad leader Guenther Wendt assisted him as he entered the spacecraft capsule and sealed the hatch.

The countdown proceeded smoothly, momentarily interrupted by a technical issue at the Canary Islands tracking station, which the mission control team swiftly resolved. These meticulous preparations underscored the collaborative efforts and attention to detail critical for successfully executing manned space missions during the Gemini program.

Walter Schirra's mission on October 3rd was meticulously planned and executed, beginning with extensive preflight testing and simulations. Leading up to the mission, Schirra underwent rigorous training both in the Mercury spacecraft simulator and under the guidance of Flight Director Christopher C. Kraft at the Mercury Control Center (MCC) in Cape Canaveral. These simulations were crucial in preparing controllers to manage any potential challenges during the mission.

During the flight, astronaut Donald K. "Deke" Slayton was the capsule communicator (CAPCOM), maintaining direct contact with Schirra from MCC. Backup astronaut Cooper supported Slayton, ensuring robust communication throughout the mission. Additional astronauts were strategically positioned at tracking stations worldwide to support Schirra's journey into space.

The launch itself was witnessed by a notable audience, including eight of the nine newly selected NASA astronauts, who gathered to witness Schirra's historic liftoff. As the countdown concluded and the Atlas rocket carrying Schirra ignited, he reported to CAPCOM Slayton, "I have liftoff. Clock had started, and she feels real nice."

However, early into the ascent, ground controllers observed an unexpected roll in the Atlas rocket, prompting initial concerns of a potential abort scenario. Despite these concerns, the roll stabilized, allowing the mission to proceed as planned. The Atlas booster engines shut down over two minutes after liftoff, and the single sustainer engine continued firing, successfully placing Schirra and the Sigma 7 spacecraft into a precise elliptical orbit—100 by 175 miles—at a remarkable speed of 17,557 miles per hour.

Throughout his mission aboard Sigma 7, Schirra encountered and managed several challenges, including an overheating spacesuit, which he successfully addressed. During his second orbit, he focused on testing the spacecraft's attitude control system, ensuring its functionality for later mission phases. Transitioning the spacecraft into free drift mode during the third orbit helped conserve fuel, allowing Schirra to capture

additional photographs of Earth whenever the spacecraft's orientation permitted.

Schirra's mission exemplified NASA's meticulous planning and operational prowess, demonstrating the capabilities required for subsequent Gemini missions and paving the way for further advancements in manned space exploration.

Following the smooth progression of subsequent orbits, Walter Schirra meticulously prepared for the crucial retro-fire maneuver aboard Sigma 7. Manually orienting the spacecraft for optimal positioning before returning it to automatic control, he ensured everything was set for the scheduled firing of the retrorockets. The precise ignition marked the start of Sigma 7's descent from orbit. As the spacecraft re-entered Earth's atmosphere, a plasma cloud momentarily disrupted communications with ground control for nine minutes—an intense moment that Schirra later described as "thrilling," likening the spacecraft's stability to that of an airplane.

Approximately 150 miles from the intended landing site, the radar aboard the U.S.S. Kearsarge accurately tracked Sigma 7's descent trajectory. Sailors aboard the ship observed as the capsule descended under controlled conditions, deploying its drogue parachute at 40,000 feet to stabilize and slow its speed, followed by the main parachute at 10,000 feet.

With remarkable precision, Sigma 7 splashed down into the Pacific Ocean, landing just four and a half miles from the waiting U.S.S. Kearsarge. Schirra's mission, spanning six orbits, had lasted an impressive nine hours, 13 minutes, and 10 seconds—a duration that set a new record for the longest American crewed spaceflight at that time.

Immediately following splashdown, U.S. Navy frogmen swiftly deployed from helicopters to attach a flotation collar around the spacecraft. Schirra remained inside his capsule during the initial recovery phase, prompting sailors from the Kearsarge to approach via whaleboat and secure a tow line to the spacecraft. Carefully, the carrier maneuvered alongside and hoisted Sigma 7 aboard, ensuring its safe retrieval.

Within just 42 minutes of splashdown, Sigma 7 was securely positioned on the carrier deck. Schirra promptly initiated the spacecraft's side hatch, emerging from the compact confines of the Mercury capsule with the assistance of Navy and NASA personnel. Stepping onto the carrier deck, he briefly inspected the condition of his spacecraft, exchanged handshakes with the ship's captain, and acknowledged the assembled sailors before proceeding to the ship's sick bay for his postflight medical examination.

Walter Schirra's triumphant Project Mercury mission aboard Sigma 7 validated the spacecraft's extended operational capabilities and showcased the precision and professionalism of U.S. space exploration efforts. His achievements set the stage for future longer-duration missions and expanded scientific endeavors in space exploration.

Shortly after arriving on the carrier, Schirra took congratulatory phone calls from President Kennedy, Vice President Lyndon B. Johnson, and his wife Jo and their two children, Marty and Suzanne. Flight surgeons thoroughly examined Schirra and declared him to be fit. They only found two minor issues, temporary orthostatic hypotension, or tendency to faint upon standing – likely caused by a combination of dehydration and being in the capsule for many hours – and a bruise on his knuckle caused by the plunger mechanism to blow the hatch once the capsule arrived on the carrier. The orthostatic hypotension had cleared by the following morning. Schirra spent the next three days aboard the Kearsarge as it steamed toward Pearl Harbor in Honolulu, with a brief stop at Midway Island to offload the Sigma 7 spacecraft for its journey back to Cape Canaveral. During that time, he held debriefs with NASA managers and five fellow astronauts who had arrived aboard the carrier by aircraft.

On October 5th, McDonnell and Lockheed presented critical findings on radiation hazards during a Trajectories and Orbits Coordination meeting for Gemini missions. McDonnell's initial assessment revealed that normal operations with some shielding posed no radiation hazard for Gemini astronauts. However, missions without shielding would only be feasible up to 14 days, limited to an altitude of 115 nautical miles due to radiation concerns. Lockheed emphasized the threat of solar flares at higher altitudes and recommended restricting operations below 300 miles until further data could be gathered on new radiation belts formed by the Atomic Energy Commission's Project Dominic in July 1962.

The following day, on October 6th, Walter Schirra received a warm welcome at Hickam Air Force Base in Honolulu, greeted by a crowd of 2,000 dignitaries and well-wishers. Among them, eight-year-old Kalani Flood presented him with a traditional lei of red carnations, symbolizing honor and hospitality.

From Hawaii, Schirra embarked on a non-stop journey to Houston, arriving at Houston's International Airport (now William P. Hobby Airport) at 1 a.m. on October 7th. Despite the late hour, he was met on the tarmac by Texas Governor M. Price Daniel, Houston Mayor Lewis W. Cutrer, and NASA officials eager to welcome him back. Schirra was reunited with his wife Jo and their two children before heading to their new home near the Manned Spacecraft Center (MSC), now known as NASA's Johnson Space Center, southeast of Houston. This marked their first night in the house, a much-needed rest after Schirra's intense training schedule.

Later that day, the Schirras visited the Farnsworth and Chambers Building, MSC's interim headquarters in southeast Houston. During their visit, child actor Jay North, famous for his role in the television series "Dennis the Menace," was in Houston for a promotional tour and joined the occasion, donning a spacesuit for photographs with the Schirras.

The culmination of the day's events saw an impressive turnout as an estimated 300,000 spectators lined the motorcade route from MSC Headquarters through the streets of Houston to Rice University. There, NASA had scheduled a postflight press conference, underscoring the immense public interest and celebration of Walter Schirra's successful mission aboard Sigma 7.

On October 7th, NASA Administrator James E. Webb and MSC Director Robert R. Gilruth stood alongside Walter Schirra on stage, leading a press conference attended by 300 reporters. Webb opened the session by introducing Gilruth and Schirra, emphasizing the pivotal role of missions like Schirra's in shaping America's future in space—a testament to the collective effort of thousands of individuals. Schirra introduced his family and fellow astronauts, then recounted his mission, describing it as a "textbook flight." He highlighted the smooth communications with Mission Control and ground stations worldwide, expressing confidence in extending future missions, possibly lasting a full day.

Two days later, on October 9th, Schirra returned to Cape Canaveral to review mission data with engineers and scientists, underscoring his commitment to thorough debriefings and post-mission analysis.

On October 12th, Associate Director Walter C. Williams of MSC formed the Project Gemini Management Panel, inviting top-level managers from major government and contractor organizations involved in the program. This panel, chaired by George M. Low of NASA's Office of Manned Space Flight, convened for the first time on November 13, 1962. Including representatives from NASA, the Air Force, and key aerospace companies like McDonnell, Martin, Aerojet-General, and Lockheed, this structure mirrored the successful management approach of Project Mercury, aiming to streamline communication and expedite problem-solving.

Continuing the celebrations, on October 14th, the Schirras traveled to New Jersey for homecoming events in Walter Schirra's birthplace of Hackensack and hometown of Oradell. A motorcade through the streets drew an estimated 40,000 well-wishers, culminating at Oradell football stadium, where 8,000 people gathered. There, NASA Administrator Webb awarded Schirra the NASA Distinguished Service Medal for his successful Mercury-Atlas 8 mission, reading a congratulatory message from President Kennedy. Later that evening, the Schirras flew to Washington, D.C., where President Kennedy hosted them at the White House in an informal ceremony held amidst the unfolding Cuban Missile Crisis. Following the White House visit, Schirra addressed NASA employees, signed autographs, and received his Navy astronaut wings from Secretary of the Navy Frederick H. Korth, capping off a momentous day.

After returning home to Houston, Schirra departed the following morning for Cape Canaveral to complete debriefings and finalize his post-mission report, ensuring thorough documentation of his historic flight aboard Sigma 7.

On October 15th, NASA took a significant step forward in its space exploration endeavors by awarding a pivotal contract to the International Business Machines Corporation (IBM). Valued at $36,200,018, this contract tasked IBM with developing the ground-based computer system essential for Projects Gemini and Apollo. This computer complex would form an integral part of the Integrated Mission Control Center at the Manned Spacecraft Center in Houston.

The decision to engage IBM underscored NASA's commitment to cutting-edge technology and robust infrastructure necessary to support increasingly complex manned space missions. This computer system would be crucial in orchestrating and monitoring missions from liftoff to splashdown, ensuring precision, reliability, and the capability to handle the vast amounts of data required for space exploration at a new frontier.

Chapter 4 - NASA's Budget Cuts and the Gemini Program

On October 19th, Wesley L. Hjornevik, Assistant Director for Administration at NASA's Manned Spacecraft Center (MSC), delivered sobering news to senior staff members regarding budget cuts imposed by NASA Headquarters for fiscal year 1963. The MSC budget was reduced from $687 million to $660 million, with the entire $27 million cut to be borne by the Gemini program.

Hjornevik expressed deep concern over the implications of such a significant budget cut, fearing that it could only be managed by drastic measures within the Gemini program. Specifically, he outlined plans to potentially eliminate the paraglider, Agena spacecraft, and all rendezvous equipment from the program due to funding limitations. Already, these constraints had forced contractors Martin and McDonnell to scale back their activities.

The impact of these budget cuts necessitated reevaluating and reprogramming the Gemini schedule. The first unmanned Gemini flight was rescheduled to December 1963, followed by the second manned mission three months later. Subsequent flights were planned at two-month intervals, with the introduction of the Agena spacecraft delayed until the fifth mission in August or September 1964. This four-month delay reflected the challenges imposed by budget constraints, requiring a significant reduction in the scale of planned test programs.

By December 20th, detailed reprogramming efforts had been finalized under the guidance of GPO Manager James A. Chamberlin, confirming December 1963 as a realistic target for the first Gemini flight despite the budgetary setbacks. The total Gemini funding allocated for fiscal year 1963 amounted to $232.8 million, underscoring the financial pressures and strategic adjustments necessary to navigate the complexities of early manned space exploration efforts.

On October 25th, the Manned Spacecraft Center (MSC) notified Lockheed about necessary budget adjustments for the Gemini program, specifically focusing on the Gemini-Agena program. These adjustments required extensive reprogramming efforts, which were discussed further in subsequent meetings on November 2nd and November 20th.

During these meetings, the primary goal was to implement the Gemini-Agena program with minimal cost impacts. The overall test program for the Agena spacecraft and its propulsion systems was significantly scaled back. However, the fundamental scope and requirements of the Agena program remained unchanged.

One of the main outcomes of the reprogramming was a four-month delay in the scheduled launch date of the first Agena spacecraft, pushed to September 1964. Still, it was notably shorter than initially anticipated at the start of the reprogramming efforts, which had forecasted a longer delay.

Additionally, Lockheed was instructed to continue its program activities at a reduced level for the remainder of 1962, approximately six weeks, before resuming normal operations on January 1, 1963. These adjustments underscored the challenges and strategic maneuvers required to navigate the ambitious Gemini program's financial and logistical complexities during its developmental stages.

On October 31st, significant orbital parameters were defined for the Gemini program. The basic spacecraft orbit model's apogee (highest point) was set at 167 nautical miles, indicating the furthest distance from Earth in its elliptical path. Conversely, this elliptical orbit's perigee (lowest point) was established at 87 nautical miles, marking its closest approach to Earth.

In addition to the spacecraft's orbit, specific altitude parameters were also determined for the circular orbit of the target vehicle. This orbital altitude was fixed at 161 nautical miles, providing a stable reference point for rendezvous and docking maneuvers crucial to the Gemini program's objectives.

In October, Minneapolis-Honeywell achieved a significant milestone by delivering two engineering prototype attitude control and maneuver electronics systems to the prime contractor involved in the Gemini program. The prime contractor, McDonnell, promptly integrated one of these systems into the electronic systems test unit (ESTU). This unit was vital in the testing and validation process, designed to simulate and monitor all electronic components in their intended flight configurations.

The integration process included comprehensive compatibility checks with prototype horizon scanners, essential devices for determining spacecraft orientation relative to the Earth's horizon. This phase marked the beginning of rigorous testing, which officially commenced on November 19th. The ESTU, acting as a simplified spacecraft mock-up, was critical in ensuring that all electronic systems met the stringent performance standards required for actual space missions.

On November 5th, the Goddard Space Flight Center made a significant announcement regarding contracts awarded to enhance NASA's Manned Space Flight Tracking Network. These contracts, totaling approximately $12 million, aimed to adapt the network to better support long-duration missions and rendezvous operations within the Gemini program.

The contracts were awarded to several key suppliers:

• Canoga Electronics Corporation, Van Nuys, California, received $1.045 million for the tracking antenna acquisition aid system.

• Radiation, Inc., Melbourne, Florida, was awarded $1.95 million for digital command encoders.

• Collins Radio Company, Dallas, Texas, received $1.725 million for the radio frequency command system.

• Electro-Mechanical Research, Inc., Sarasota, Florida, secured the largest portion of the contract, amounting to $7,376,379, for the pulse code modulation system.

These contracts represented a strategic investment in upgrading essential components of the tracking network, ensuring it could effectively support the complex operational requirements of Gemini missions. The enhancements included advanced technologies for tracking, command encoding, radio frequency management, and digital data transmission, crucial for maintaining communication and control over spacecraft during critical phases of their missions.

Evolution of Spacesuit Development for the Gemini Program

On November 6th, B. F. Goodrich delivered a prototype partial-wear, quick-assembly, full-pressure suit to NASA's Manned Spacecraft Center (MSC) for evaluation by the Life Systems Division. This suit prototype was designed to meet the specific demands of long-duration missions planned for the Gemini program. The partial-wear feature of the suit allowed for detachable components such as sleeves, legs, and helmets, facilitating easier donning and doffing by astronauts during extended missions.

This delivery initially marked the second of two partial-wear suit prototypes as per the original contract. However, due to evolving requirements and after evaluating the prototype (originally designated G-2G, then G-2G-1), MSC requested B. F. Goodrich produce an additional 14 suits based on this design. These additional suits, designated from G-2G-2 to G-2G-15, varied only in size while adhering to the October 10, 1962 specifications.

Following evaluations and subsequent design changes, MSC continued refining the suits. The final model produced under this contract was G-2G-8, which B. F. Goodrich delivered to MSC on January 21, 1963. Despite these developments, MSC later opted for a different suit design offered by the David Clark Company, Inc., based in Worcester, Massachusetts. This new suit design incorporated B. F. Goodrich helmets, gloves, and additional hardware, marking a strategic shift in the choice of spacesuit for the Gemini program.

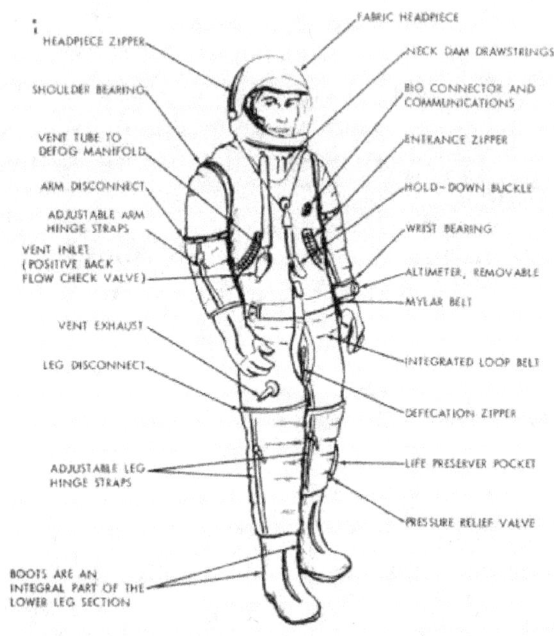

The B. F. Goodrich partial-wear full-pressure suit being developed for the Gemini program.

Sled Ejection Test No. 1

On November 9th, a pivotal moment unfolded at the Naval Ordnance Test Station, marking the initiation of the Gemini project's ambitious testing phase. The event, officially termed Sled Ejection Test No. 1, embarked on a journey defining the spacecraft's aerodynamic prowess and structural resilience in the face of adversity.

The test vehicle, a rocket-propelled sled mounted with boilerplate spacecraft No. 3, mirrored the aerodynamic profile of its flight-ready counterpart. Designed to gather crucial data rather than eject seats, this inaugural test aimed to measure aerodynamic drag and validate structural integrity, laying crucial groundwork for future escape system trials.

Despite meticulous planning, the course of events took an unexpected turn. A motor on the sled broke loose, striking the heatshield of the boilerplate spacecraft. The resultant damage ignited a fire that consumed instrumentation and equipment, casting a shadow of uncertainty over the mission's immediate future.

Undeterred by the setback, engineers undertook a comprehensive assessment and repair effort. The boilerplate spacecraft, though severely compromised, was earmarked for restoration. Simultaneously, plans were set in motion to overhaul or potentially rebuild the sled itself.

In the face of adversity, the Gemini Project Office exhibited unwavering resolve. Despite the unforeseen repairs and modifications the incident required, there remained a steadfast commitment to maintaining momentum in the sled test program. The incident underscored the inherent risks of pushing the boundaries of aerospace engineering while reinforcing the project's dedication to overcoming challenges with ingenuity and perseverance.

On November 16th, pivotal discussions unfolded within the Gemini Project Office, led by Andre J. Meyer, Jr., signaling a strategic shift in spacecraft adapter design. At the forefront was a pioneering study conducted by Space Technology Laboratories under NASA Headquarters' auspices—a study centered on the feasibility of integrating a "T-back" pod into the Gemini spacecraft adapter.

Unlike its Agena counterpart, renowned for its multifaceted capabilities, the proposed "T-back" pod was envisioned primarily as a stable rendezvous target. While lacking in translational mobility, this design distinction offered a streamlined alternative. Notably, despite approaching near parity in cost with the Agena, the "T-back" pod promised to circumvent the logistical complexities associated with separate launch configurations.

The study's emergence underscored NASA's ongoing quest for precision and efficiency in space rendezvous operations. By exploring innovative alternatives to established norms, the Gemini Project reaffirmed its commitment to advancing space exploration through strategic foresight and technological innovation.

As discussions unfolded and feasibility assessments continued, the prospect of integrating the "T-back" pod into future missions stood poised to redefine operational paradigms, paving the way for new frontiers in orbital rendezvous techniques.

November 21st marked a pivotal juncture in the Gemini Project's quest for efficiency and reliability, as McDonnell and the Manned Spacecraft Center representatives convened for a critical mechanical systems coordination meeting.

During this session, a decisive resolution emerged: McDonnell, in collaboration with the Manned Spacecraft Center, opted to terminate its subcontract with the CTL Division of Studebaker. This bold move was propelled by two primary factors: mounting confidence in McDonnell's innovative new heatshield design and the persistent challenges faced by CTL in fabricating Heatshield No. 1.

The decision to sever ties with CTL was underscored by technical considerations and financial prudence. Discontinuing the subcontract projected an estimated savings of $131,000, a testament to the project's commitment to optimizing resources without compromising on safety or quality.

This strategic pivot exemplified the Gemini Project's proactive approach to managing partnerships and resources. It reflected a steadfast dedication to leveraging internal expertise and resolving operational hurdles swiftly and decisively. As the project forged ahead, propelled by these strategic realignments, it reaffirmed its trajectory toward achieving ambitious milestones in manned space exploration.

As November drew to a close, the Gemini Project Office confronted pivotal challenges in advancing its liquid propellant rocket systems—a cornerstone of the spacecraft's maneuverability and precision in space.

The primary area of concern was the development of a specialized 25-pound thruster. This critical component met stringent operational specifications, particularly sustaining optimal performance over an extended burn time of five minutes. While shorter-duration chambers for the reaction control system (RCS) had succeeded in testing phases, the more demanding long-duration chambers essential for the orbit attitude and maneuver system (OAMS) had encountered significant setbacks.

Compounding these technical hurdles, Rocketdyne, the designated contractor, faced delays in developmental testing. Specifically, they lagged three weeks behind schedule in validating RCS and OAMS components, with an additional five-week delay in comprehensive systems testing.

Amid these challenges, the Gemini Project reaffirmed its commitment to meticulous testing and refinement, prioritizing reliability and precision in every development phase. The concerted efforts to overcome technical obstacles underscored the project's resilience and determination to achieve breakthroughs in manned space exploration.

Integration and Expansion at Cape Canaveral

On November 30th, the Gemini Project Office unveiled pivotal revisions to its facilities strategy, marking a strategic evolution in the preflight checkout procedures for the Gemini spacecraft at Cape Canaveral.

Central to the revised plans was a decisive shift away from the exclusive use of Hangar S complex. Originally designated as the primary site for Gemini spacecraft preflight checkouts, the project's evolving schedule and strategic considerations necessitated a broader integration with planned Apollo facilities on Merritt Island.

The decision to integrate Gemini facilities with those slated for Apollo missions stemmed from several factors, including schedule synchronization and the resolution of compatibility issues between the spacecraft's fuel-oxidizer and cryogenic systems. These advancements paved the way for a consolidated approach, ensuring seamless operations and optimal resource utilization.

As originally planned, the first two Gemini spacecraft would undergo preflight checkouts at Hangar AF. However, with the completion of Merritt Island facilities projected for the first quarter of 1964, the entire preflight checkout operation would transition to this expansive new site. This transition underscored the project's commitment to operational efficiency and readiness as it advanced towards ambitious manned space missions.

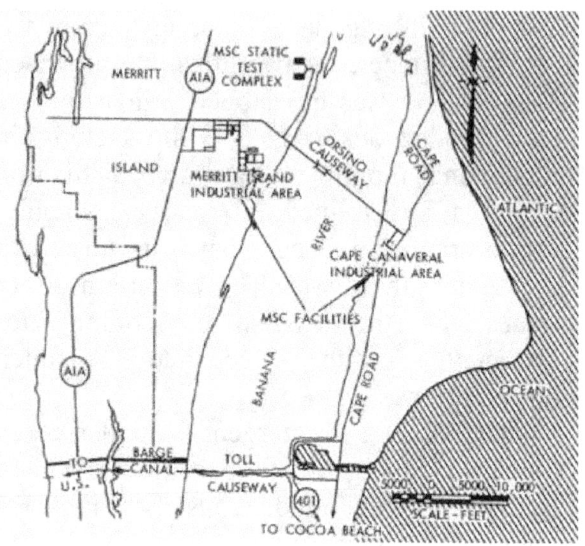

Location of Manned Spacecraft Center facilities at Cape Canaveral and Merritt Island.

The strategic realignment at Cape Canaveral exemplified the Gemini Project's adaptability and foresight, setting the stage for integrated, streamlined operations that would pave the way for future milestones in manned space exploration.

Throughout November, pivotal negotiations unfolded between the Air Force Space Systems Division and Martin-Baltimore, culminating in a landmark agreement for Phase I of the Gemini launch vehicle program.

The negotiations, spanning the first three weeks of the month, resulted in formulating a comprehensive cost-plus-fixed-fee contract. This pivotal agreement, valued at $52.5 million with a fixed fee component of $3.465 million, laid the groundwork for advancing the Gemini program's ambitious launch vehicle initiatives.

The contract encompassed crucial developmental phases, including the initial development and procurement of the program's inaugural launch vehicle. Furthermore, it set in motion critical preparations for the subsequent manufacturing and procurement of an additional 14 vehicles essential to the Gemini program's mission objectives.

This milestone agreement underscored a collaborative commitment to innovation and excellence in aerospace engineering. It positioned the Gemini launch vehicle program on a trajectory of advancement, poised to redefine capabilities and pave the way for historic achievements in manned space exploration.

December 10th marked a pivotal milestone in Phase II-A of the Paraglider Development Program, as North American initiated deployment flight testing of the half-scale test vehicle (HSTV).

Strapped beneath a helicopter, the HSTV embarked on a series of tests aimed at refining the intricate mechanics of deployment—from the ejection and inflation of the paraglider wing to the seamless transition into glide mode. These tests were crucial in identifying and addressing critical areas of improvement in the deployment sequence.

The initial flight offered promising insights, substantiating the feasibility of the basic deployment sequence. However, the journey was not without challenges. Emergency recovery procedures became essential, underscoring the program's commitment to safety and resilience in adversity.

Subsequent tests in January and March of 1963 presented formidable obstacles. During the second test, the paraglider sail suffered a catastrophic failure, marking a setback in developmental progress. In the third test, scheduled for March 11th, the rendezvous and recovery canister failed to separate as anticipated. Efforts to salvage the situation with emergency recovery systems were thwarted when the main parachute failed to deploy, resulting in the loss of both vehicles upon impact.

Gemini paraglider half-scale test vehicle slung beneath an Army helicopter at the beginning of the second deployment flight test. (NAA-SID Photo 277/4, Jan. 4, 1963.)

These trials underscored the inherent risks and complexities of pioneering aerospace technology. Yet, they also fueled an unwavering commitment to innovation and continuous improvement within the Paraglider Development Program. Each setback catalyzed refinement and reevaluation, steering the program toward achieving its ambitious objectives in manned space exploration.

On December 10-11, a consortium of aerospace experts convened a pivotal Design Engineering Inspection for the full-scale test vehicle (FSTV) as part of Phase II-A of the Paraglider Development Program. Representatives from the Manned Spacecraft Center, NASA Headquarters, Flight Research Center, Langley Research Center, and Ames Research Center gathered to assess this pioneering aerospace innovation's critical components and capabilities.

Conceived initially during Phase I of the program, the FSTVs—two stipulated in the contract—aimed to achieve dual objectives. Firstly, they were instrumental in refining systems and techniques for deploying paraglider wings. Secondly, they were crucial platforms for evaluating flight performance and control characteristics during glide phases.

The inspection encompassed a comprehensive review of flight test objectives, vehicle hardware, and the intricate electrical and electronic systems array. Following rigorous evaluation, the inspecting team meticulously compiled 24 requests for alterations and enhancements, underscoring their commitment to optimizing the FSTV's operational efficiency and safety.

This collaborative endeavor highlighted the Paraglider Development Program's dedication to precision engineering and iterative refinement. Each recommendation for improvement represented a pivotal step forward in advancing aerospace technology, setting the stage for future breakthroughs in manned space exploration.

December 14th marked a significant milestone in Project Gemini as the 10-percent fluctuating-pressure spacecraft model concluded its rigorous exit configuration test program. Spanning the Mach number range from 0.6 to 2.5, this testing phase focused on evaluating the spacecraft's performance under maximum dynamic pressure conditions.

Shortly thereafter, on January 15, 1963, another critical testing milestone was achieved with the completion of the Gemini spacecraft dynamics stability model. This model provided essential dynamic stability coefficients crucial for ensuring spacecraft reentry stability at higher Mach numbers ranging from 3.0 to 10.

These accomplishments culminated all originally scheduled wind tunnel testing for Project Gemini. However, the program's commitment to innovation and excellence remained unwavering, evidenced by the initiation of three additional test programs:

Testing of the spacecraft's 20-percent ejection seat model: This initiative aimed to enhance safety and efficiency in emergency ejection scenarios.

The 10-percent model of the Gemini spacecraft used in wind tunnel testing at McDonnell.

Testing of the astronaut ballute model: Essential data gathered from this model would inform the design and implementation of the astronaut stabilization system, crucial for safe reentry and landing.

Testing of the rigid frame paraglider model: Focused on optimizing sail configuration, this testing phase sought to refine the spacecraft's aerodynamic capabilities during the descent and landing phases.

These ongoing test programs underscored Project Gemini's commitment to rigorous scientific inquiry and technological advancement. Each initiative represented a pivotal step forward in preparing for the challenges and opportunities of manned space exploration, setting the stage for future achievements in aerospace engineering.

Propelling Gemini and Apollo Missions Forward

On December 17th, a pivotal gathering marked the inception of the Scientific Experiments Panel—a transformative initiative poised to shape the future of Gemini and Apollo missions. Convened under the auspices of the Manned Spacecraft Center, this newly formed panel embarked on a mission-critical role: soliciting, evaluating, and implementing scientific experiment proposals destined for space.

At its core, the panel represented a beacon of collaboration and innovation dedicated to expanding the frontiers of scientific discovery beyond Earth's bounds. By fostering a platform for scientific inquiry aboard spacecraft, it sought to harness the unique environment of space to conduct experiments that could unravel mysteries and advance human knowledge.

The panel's mandate extended far beyond mere oversight. It catalyzed interdisciplinary engagement, drawing upon expertise from diverse scientific disciplines to craft experiments tailored for spaceflight conditions. Proposals ranging from astronomy and physics to biology and materials science were welcomed, each promising to unlock new insights into the cosmos and the human experience in space.

On December 19th, Titan II flight N-11 embarked on a critical mission from Cape Canaveral, marking the eighth in a series aimed at advancing this pivotal weapon system. However, this flight carried an additional burden—the challenge of mitigating longitudinal oscillations, famously dubbed as POGO.

POGO, identified as a phenomenon generating significant g-forces during the first stage of operation, posed a formidable risk to astronaut performance. With g-forces reaching up to nine during the initial stages and over three at critical points corresponding to the spacecraft's location, NASA swiftly established 0.25g at 11 cycles per second as the maximum tolerable level for Gemini flights.

Titan II N-11 introduced standpipes within each leg of its stage I oxidizer feed lines to address this pressing issue. This modification aimed to disrupt the coupling between the missile's structural integrity and its propulsion system, presumed to be the root cause of instability.

Despite these efforts, postflight analysis delivered sobering news—rather than mitigating the issue, the POGO fix had inadvertently exacerbated it. Longitudinal oscillations had intensified twofold, highlighting the complexity and urgency of finding a viable solution.

On December 26th, a pivotal initiative unfolded within the Air Force Space Systems Division—establishing the Gemini Launch Vehicle Configuration Control Board. This board was tasked with a critical mission: to develop and enforce rigorous procedures for approving and evaluating specifications and engineering change proposals pertaining to the Gemini launch vehicle.

The Configuration Control Board, which was first convened on March 5, 1963, represented a cornerstone of governance and oversight within the Gemini program. Its mandate extended beyond mere procedural oversight; it aimed to ensure that every facet of the launch vehicle's design, specifications, and modifications adhered to exacting standards of precision and reliability.

At its core, the board served as a nexus of expertise, bringing together stakeholders from diverse technical disciplines to deliberate on critical engineering decisions. Proposals for modifications or enhancements underwent meticulous scrutiny, guided by a commitment to enhancing safety, performance, and mission success.

Air Force Space Systems Division and Aerojet-General negotiated a cost-plus-fixed-fee contract for the first phase of the Gemini launch vehicle engine program, February 14, 1962, through June 30, 1963. The contract required the delivery of one set of engines, with the remaining 14 sets included for planning purposes. The estimated cost of the contract was $13.9 million, with a fixed fee of $917,400 for a total of $14,817,400.

Development and Qualification

January 1963 through December 1963

In January 4, 1963, the Manned Spacecraft Center took a decisive step in the evolution of space exploration, commissioning McDonnell to develop a spacecraft capable of pioneering rendezvous experiments on the upcoming Gemini missions. The mission objectives were clear: to refine techniques essential for future lunar missions by mastering the delicate art of spacecraft rendezvous.

The experimental payload, weighing in at 70 pounds, was meticulously designed. It featured a sophisticated array, including an L-band radar target, a strategically placed flashing light, a robust battery power supply, and intricately engineered antenna systems. These components were pivotal, each playing a crucial role in the success of the ambitious rendezvous experiments planned for the Gemini missions.

The timeline was equally ambitious: the second Gemini flight, slated for a one-day mission, aimed to conduct an open-loop rendezvous experiment. In this scenario, the astronaut would rely on optical observations to guide the spacecraft into position, executing precise maneuvers to achieve rendezvous with the designated target. It was a test of human skill and adaptability in the unforgiving environment of space.

Looking ahead to the third Gemini flight, planned as a seven-day mission, the stakes were raised. Here, the rendezvous experiment was to be conducted in closed-loop mode. This advanced technique allowed the spacecraft's onboard systems to interpret data from its instruments autonomously, translating it into precise maneuvers. It was a pivotal moment in space exploration, marking a transition towards autonomous spacecraft operations and paving the way for future missions to more distant destinations.

These experiments represented not just technical milestones but bold steps toward the ultimate goal of lunar exploration. They underscored the ingenuity and determination of the McDonnell and the Manned Spacecraft Center teams, who pushed the boundaries of what was possible in human spaceflight. Each mission brought new challenges and triumphs, shaping the course of space exploration for generations to come.

Developing the Gemini VII Capsule

McDonnell Aircraft, renowned for its pivotal role in Project Mercury and Gemini, emerged as a leader in space exploration as it embarked on developing the Gemini VII Capsule. Initially competing for the prime Apollo contract against North American Aviation, McDonnell continued its quest to expand the boundaries of spaceflight. Proposals for advanced derivatives of the Gemini program aimed at pioneering cislunar missions and crewed lunar landings, promising cost efficiencies compared to the Apollo program. Despite these ambitions, NASA focused exclusively on Apollo's lunar objectives.

The envisioned applications for Advanced Gemini missions were expansive and ambitious, encompassing military operations, logistics for space stations, and ambitious lunar explorations. Lunar mission proposals ranged from adapting existing docking systems, such as those used with the Agena Target Vehicle, to integrating more powerful upper stages like the Centaur for lunar trajectories. Some concepts even explored radical modifications enabling the Gemini spacecraft to perform lunar landings, offering capabilities such as crewed lunar flybys and emergency rescue missions for Apollo crews.

Design concepts for Advanced Gemini missions varied widely. While some advocated for using "off-the-shelf" Gemini spacecraft with minimal modifications, others proposed enhancements to increase crew capacity, facilitate docking maneuvers with space stations, or support lunar expeditions. Innovations such as adding wings or a parasail to enable horizontal landings reflected the program's forward-thinking approach to reentry and recovery in space environments.

McDonnell Aircraft, already established as the prime contractor for the Project Mercury capsule, was selected by NASA in 1961 to build the Gemini capsule. The resulting spacecraft, delivered from 1963 onward, measured 18 feet 5 inches in length and 10 feet in width, with a launch weight ranging from 7,100 to 8,350 pounds. This marked the beginning of a new era in human space exploration, where McDonnell's expertise and innovation played a crucial role in shaping the capabilities and aspirations of Gemini missions.

The Gemini crew capsule, the Reentry Module, represented a significant evolution from its predecessor, the Mercury capsule. Unlike Mercury, which integrated key systems within the capsule itself, Gemini adopted a modular approach. The retrorockets, electrical power, propulsion systems, oxygen, and water were housed in a detachable Adapter Module positioned behind the Reentry Module. This design ensured that the Adapter Module, designed to burn up during reentry, could be jettisoned after its function was fulfilled, enhancing the capsule's efficiency and safety.

A pivotal advancement in Gemini's design was the modularization of internal spacecraft systems. Each component was housed separately, allowing for independent testing and replacement without disturbing other verified components. This modularity streamlined maintenance procedures and enhanced the spacecraft's reliability during missions.

Access to many components within the capsule was facilitated by small access doors, a practical departure from Mercury's more integrated design. Furthermore, Gemini introduced entirely solid-state electronics, marking a technological leap forward from Mercury's reliance on less advanced systems. This upgrade reduced the spacecraft's weight and improved its overall performance and reliability in the challenging environment of space.

Gemini's modular design and solid-state electronics underscored NASA's commitment to advancing spacecraft technology during the early days of human space exploration. These innovations facilitated smoother operations and maintenance and laid the groundwork for future spacecraft designs, setting a precedent for efficiency, reliability, and adaptability in manned space missions.

Gemini's emergency launch escape system departed from the traditional solid-fuel rocket-powered escape towers used in previous manned missions. Instead, it embraced a more streamlined approach with aircraft-style ejection seats. The decision stemmed from the characteristics of the Titan II booster, which used hypergolic propellants known to ignite immediately upon contact, resulting in a smaller blast and flame compared to cryogenically fueled rockets like the Atlas and Saturn.

The ejection seats were deemed sufficient for safely separating astronauts from a malfunctioning launch vehicle during critical phases of ascent. At higher altitudes where ejection seats were not feasible, the astronauts would rely on the spacecraft itself to detach and return them to Earth.

John Chamberlin, a key advocate for this design choice, had harbored reservations about the Mercury Escape Tower's complexity and weight. He based his advocacy for ejection seats on extensive analysis, including reviewing films of previous Atlas and Titan II ICBM failures. From these assessments, Chamberlin concluded that Titan II's relatively smaller explosion radius made ejection seats a viable safety measure. This approach not only simplified the launch escape system but also contributed to reducing overall spacecraft weight, enhancing mission efficiency and safety.

Gemini's adoption of ejection seats represented a calculated departure from conventional escape tower designs. This decision reflected NASA's evolving understanding of launch vehicle dynamics and commitment to enhancing astronaut safety through innovative engineering solutions.

Maxime Faget, the esteemed designer of the Mercury Launch Escape System (LES), held reservations about the adoption of ejection seats in the Gemini spacecraft. His concerns were multifaceted, focusing primarily on this alternative safety mechanism's potential risks and limitations. Unlike the robust solid-fuel rocket-powered towers he had designed for Mercury, Faget noted several critical issues with the ejection seat setup.

One significant concern was the safety of the astronauts themselves. Faget feared that the powerful ejection seats could cause serious injury to the crew, especially once the booster reached speeds approaching Mach 1, just 40 seconds after liftoff. At these velocities, ejection would become impractical and potentially hazardous.

Moreover, Faget expressed apprehension about the astronauts ejecting through the Titan's exhaust plume during flight. The hot and turbulent environment posed additional risks to the crew's safety in the event of an emergency escape attempt.

Despite these concerns, Faget later acknowledged that Gemini's greatest achievement was its impeccable safety record, which fortunately never required the crew to activate the ejection system in a real-life emergency. This statement underscored the overall success and reliability of the Gemini program in ensuring astronaut safety during its missions.

However, a significant realization came after the tragic Apollo 1 fire in January 1967. The fire, which claimed the lives of astronauts Gus Grissom, Ed White, and Roger B. Chaffee, highlighted the extreme danger of pressurizing a spacecraft with pure oxygen, as was standard practice before launch. The Gemini ejection system had never been tested under these specific conditions of pure oxygen pressurization, a factor that could have exacerbated the risks in an emergency scenario.

General Thomas P. Stafford, a distinguished figure in the history of space exploration, played a pivotal role in advancing safety protocols following his experience during the Gemini 6 mission in December 1965. As commander of Apollo 10, known as the "dress rehearsal" for the first Moon landing, Stafford's contributions extended beyond mission command to encompass significant advancements in astronaut safety and mission preparedness.

Born on September 17, 1930, in Weatherford, Oklahoma, Stafford's career was marked by a commitment to excellence and innovation. He graduated from Weatherford High School before earning a Bachelor of Science degree with honors from the United States Naval Academy in 1952. Stafford's dedication to education and achievement was underscored by numerous honorary degrees, including doctorates from institutions such as Oklahoma State University, the University of Oklahoma, and Embry-Riddle Aeronautical University, among others.

Throughout his career, Stafford's leadership and expertise were instrumental in shaping the course of manned space missions. His reflection on the Gemini 6 near-launch abort highlighted critical safety concerns regarding the pressurization of spacecraft with pure oxygen, a practice that later contributed to the tragic Apollo 1 fire in January 1967. This pivotal event underscored the urgent need for enhanced safety protocols and rigorous testing procedures in subsequent missions.

Gemini represented a significant leap forward in spacecraft technology, notably by including the Gemini Guidance Computer (GGC), also known as the Gemini Spacecraft On-Board Computer (OBC). This onboard computer system, weighing approximately 58.98 pounds (26.75 kg), played a crucial role in managing and executing mission maneuvers. Similar in design to the Saturn Launch Vehicle Digital Computer, the GGC utilized core memory with 4096 addresses, each holding a 39-bit word structured into three 13-bit "syllables". Numeric data, typically represented as 26-bit two's-complement integers, was stored either in the first two syllables of a word or in the accumulator. Instructions were encoded with a 4-bit opcode and 9 bits of operand, capable of being stored in any syllable for flexible operation.

Unlike its predecessor, Mercury, the Gemini spacecraft integrated advanced features borrowed from aviation technology, such as in-flight radar and an artificial horizon. These additions bolstered navigational capabilities, enhancing astronauts' situational awareness and facilitating precise maneuvering during missions.

Similar to Mercury, Gemini retained the use of a joystick for manual control over yaw, pitch, and roll. However, Gemini innovated further by introducing a pair of T-shaped translation control handles for each crew member. These handles precisely controlled the spacecraft's translational movements—forward, backward, up, down, and sideways. This capability was pivotal for performing complex tasks like rendezvous and docking maneuvers and adjusting the spacecraft's trajectory mid-flight. These control mechanisms would later find continuity in the Apollo spacecraft, highlighting their effectiveness and reliability across multiple missions and spacecraft platforms.

Gemini's original design aspirations included a revolutionary landing concept utilizing a Rogallo wing for a solid ground landing with the crew seated upright, actively controlling the craft's forward motion. Unlike previous designs, the airfoil was not just attached to the nose of the capsule but had an additional attachment point near the heat shield for balance, covered by a metal strip running between the twin hatches. However, this ambitious plan was eventually abandoned in favor of the more conventional sea landing using parachutes, akin to the Mercury program. The capsule was oriented closer to horizontal upon splashdown, allowing one side of the heat shield to contact the water first, thus eliminating the need for the landing bag cushion used in Mercury capsules.

The Gemini spacecraft was structured with an Adapter Module consisting of two main components: the Retro module and the Equipment module. The Retro module housed four solid-fuel TE-M-385 Star-13E retrorockets. These retrorockets were spherical except for their nozzles and were structurally attached to beams crossing at right angles in the center of the module. During re-entry, these retrorockets fired sequentially to initiate descent, with abort procedures potentially causing simultaneous firing to separate the Descent module from the Titan rocket quickly.

In contrast, the Equipment module featured the Orbit Attitude and Maneuvering System (OAMS), equipped with sixteen thrusters. These thrusters enabled precise control over translation in all three perpendicular axes—forward/backward, left/right, up/down—and facilitated attitude control for pitch, yaw, and roll adjustments. This capability was crucial for orbital adjustments, including changes in inclination and altitude necessary for rendezvous missions with other spacecraft and docking operations, particularly with the Agena Target

Vehicle (ATV), which had its own rocket engine for significant orbital maneuvers.

Gemini also introduced pioneering technologies such as using fuel cells for electrical power during extended missions, marking the first implementation of such systems in crewed spacecraft. This advancement allowed for longer endurance missions compared to earlier battery-powered missions.

Regarded as a "pilot's spacecraft," Gemini incorporated features reminiscent of jet fighters, influenced significantly by astronaut Gus Grissom. This characterization underscored its agility and versatility in space, enabling milestones like long-duration flights, rendezvous maneuvers, and extravehicular activities (EVAs). These achievements distinguished the US crewed space program from its Soviet counterpart, which faced challenges in developing its crewed lunar program, ultimately leading to its discontinuation due to political and technical obstacles.

The Gemini spacecraft was launched atop the Titan II GLV, an Air Force ICBM converted for crewed spaceflight. Despite early challenges with pogo oscillation and the toxicity of its hypergolic propellants (nitrogen tetroxide and hydrazine), Titan II's reliability and simplicity in design made it a viable choice for NASA's Gemini missions. Its radio guidance system tailored for Cape Kennedy launches ensured precise trajectories essential for successful space missions.

On January 8-9, 1963, a pivotal Design Engineering Inspection was conducted for the advanced trainer of NASA's Paraglider Development Program, Phase II-B(1). This inspection gathered representatives from key NASA centers, including the Manned Spacecraft Center, NASA Headquarters, Flight Research Center, Langley Research Center, and Ames Research Center. The focus of the inspection was to evaluate and advance the development of the advanced trainer designed for the Paraglider Program.

North American Aviation, entrusted with developing this critical trainer, underwent an extensive review process. The inspection resulted in a comprehensive list of 36 alteration requests to refine and enhance the trainer's design. These requests underscored NASA's commitment to meticulous engineering standards and continuous improvement in preparation for the program's next phase.

On January 8-9, the Manned Spacecraft Center (MSC) defined crucial requirements for McDonnell Aircraft Corporation concerning abort scenarios in orbit. These stipulations were pivotal in ensuring future space missions' safety and operational flexibility under the Gemini program. Key requirements outlined during this period included:

Controlled Reentry for All Abort Scenarios: The spacecraft must be capable of executing a controlled reentry in all abort situations, except a failure in the guidance and control systems.

Selection of Emergency Abort Target Areas Onboard: The onboard systems should enable the selection of one of the designated emergency abort target areas. This capability ensured that astronauts could safely abort to predefined landing zones in emergencies.

Navigational Accuracy: The spacecraft's navigational systems must achieve high accuracy, targeting a two-mile radius error at the designated point of impact or landing. This precision was critical for ensuring safe and effective abort maneuvers.

Crew Capability with Paraglider Deployment: In scenarios where the paraglider system was deployed, the crew must be able to eject from the spacecraft safely. This feature was integral to the contingency plans for safe astronaut recovery during abort operations.

On January 9, the Flight Operations Division meticulously defined the operational requirements for the remote stations that would form NASA's worldwide tracking network. This network was crucial for supporting the Gemini missions, ensuring continuous communication and tracking of spacecraft throughout their orbital journeys. The outlined requirements specified for each remote station included:

Gemini System Console: Dedicated to monitoring and controlling Gemini spacecraft operations.

Agena System Console: Specifically for managing Agena target vehicle operations, crucial for rendezvous and docking maneuvers.

Command Console: Responsible for overall mission command and control functions.

Aeromedical Console: Focused on monitoring the health and physiological status of the astronauts during the mission.

Maintenance and Operations Console: Handling station maintenance and operational logistics.

Each Gemini and Agena system console would be equipped with:

42 analog display meters, providing real-time data on various spacecraft and mission parameters.

40 on/off indicators, offering status updates on critical systems and operational modes.

These detailed specifications ensured that each remote tracking station was fully equipped to support the complexities of Gemini missions. The consoles' capabilities allowed for comprehensive monitoring, control, and troubleshooting across different operational domains, from spacecraft systems to astronaut health and safety.

Development and Testing of Silver-Zinc Batteries for the Gemini Spacecraft

On January 10, representatives from the Manned Spacecraft Center (MSC), McDonnell, and the Eagle-Picher Company convened to discuss developing and testing plans for the silver-zinc batteries intended for the Gemini spacecraft. Here are the key points discussed and decisions made during this meeting:

McDonnell had chosen Eagle-Picher Company of Joplin, Missouri, as the vendor for the silver zinc batteries approximately six months earlier.

The primary (main bus) electrical power during launch was intended to be supplied by five silver-zinc batteries.

These batteries were also slated to provide all primary electrical power requirements for one orbit, reentry, and the post-landing phase.

Three additional high-discharge-rate batteries, electrically and mechanically isolated from the main batteries, were designated to power control functions such as relays and solenoids.

Eagle-Picher submitted a test plan proposal by February 9 to outline the testing protocols and procedures for ensuring the batteries' reliability and performance.

McDonnell analyzed the battery power requirements, which led to a decision on February 21. Following this analysis, MSC directed McDonnell to use four batteries instead of five for main bus power on spacecraft Nos. 2 and onwards.

The decision to reduce the number of batteries to four was based on the analysis that closely monitoring a four-battery configuration would meet the spacecraft's power needs, ensuring operational reliability while optimizing weight and space considerations.

On January 11, a significant initiative was proposed by James A. Chamberlin, the Gemini Project Office Manager, to enhance performance among contractor employees working on the Gemini spacecraft. Here are the key details and decisions surrounding this proposal:

The primary goal was to stimulate and encourage better performance from contractor employees involved in the Gemini Project.

James A. Chamberlin suggested organizing astronaut visits to various contractor plants. These visits were intended to boost morale, provide direct feedback from astronauts, and foster a sense of teamwork and dedication among the workforce.

Donald K. Slayton, from the Astronaut Activities Office, supported Chamberlin's proposal. He informed Chamberlin that these visits would commence in February 1963, starting with the Martin Company.

The visits were not merely ceremonial but were designed to provide astronauts with insights into the manufacturing processes, challenges, and innovations at contractor facilities.

By interacting directly with the workers, astronauts could share their perspectives on the importance of their work and its impact on space missions.

Such initiatives aimed to align contractor efforts with NASA's objectives, encourage innovation, and ensure that the Gemini spacecraft met stringent quality and performance standards.

On January 14, the Flight Operations Division's Project Gemini working group identified critical concerns regarding the onboard computer for the Gemini spacecraft. Here are the key points and questions raised:

• There was uncertainty about the specific tasks the on-board computer would perform.

• Clarity was needed on the operational sequence of the computer during missions.

• Understanding what data and commands the onboard computer would require from the ground computer complex and how frequently this interaction would occur.

• How astronauts would interface with and utilize the onboard computer during early missions.

• Defining the essential functions and responsibilities of the onboard computer for the initial Gemini missions.

Addressing these concerns was crucial to ensuring the effective integration and utilization of the onboard computer in the Gemini spacecraft.

Clearing up these uncertainties would enhance mission planning, crew training, and overall mission success.

Manned Spacecraft Center Transition

On January 14, 1963, a pivotal shift in NASA's operations marked the transfer of the Gemini target vehicle program from the Marshall Space Flight Center to the Manned Spacecraft Center (MSC). This transfer followed extensive discussions between MSC and Marshall, culminating in a formal agreement outlined on January 11.

Previously, Marshall Space Flight Center had overseen the Gemini target vehicle program, a critical component of NASA's early space missions. However, with the space program's evolving needs, NASA Headquarters decided on October 12, 1962, to realign responsibilities. This decision aimed to streamline operations, enabling Marshall to focus intensively on developing the Saturn launch vehicle. Concurrently, all other NASA Atlas-Agena programs transitioned to Lewis Research Center, underscoring a strategic effort to consolidate Atlas launch vehicle technology expertise.

Under the agreed-upon terms, Marshall continued to provide advisory support until March 1, after which it assumed a role as technical consultant to MSC upon request. This transitional period ensured continuity and facilitated knowledge transfer, crucial for the seamless integration of responsibilities at MSC.

On January 17, 1963, NASA Administrator James E. Webb and Secretary of Defense Robert S. McNamara signed a landmark agreement reshaping management responsibilities at Cape Canaveral. This agreement marked a pivotal moment in the space exploration collaboration between NASA and the Department of Defense (DOD).

Under the terms of the agreement, the Air Force retained its role as the single manager of the Atlantic Missile Range, overseeing operations across the sprawling 15,000-acre Cape Canaveral launch area. Simultaneously, NASA's Launch Operations Center assumed responsibility as the host agency at the Merritt Island Launch Area, situated north and west of existing DOD installations.

This division of management responsibilities aimed to optimize operational efficiency while delineating clear lines of authority. Both NASA and DOD undertook independent logistics and administration within their respective areas. Crucially, regardless of location, each agency assumed direct responsibility for specific mission functions such as preparation, checkout, launch, and test evaluation.

While DOD retained critical range functions, including scheduling, flight safety, search and rescue operations, and downrange airlift and station operations, the agreement underscored a cooperative framework wherein both agencies worked in tandem to advance aerospace capabilities. This strategic alignment paved the way for enhanced coordination and operational synergy, laying a robust foundation for future space missions and technological advancements.

NASA-DOD Agreement on Project Gemini

On January 21, 1963, a pivotal policy agreement was reached between James E. Webb, NASA administrator, and Robert S. McNamara, Secretary of Defense, defining the roles of NASA and the Department of Defense (DOD) in Project Gemini. This agreement heralded a new phase of collaboration between the two agencies, setting clear guidelines for their joint efforts in advancing manned space exploration.

Central to the agreement was establishing a joint NASA-DOD Gemini Program Planning Board. This board was tasked with planning experiments, conducting flight tests, and analyzing and disseminating results—an initiative to maximize scientific and operational insights from the Gemini missions.

Under the terms of the agreement, NASA retained overall management of Project Gemini. However, DOD assumed significant responsibilities, including involvement in Gemini development, pilot training, preflight checkout, launch, and flight operations. Notably, DOD took on specific roles related to the Titan II launch vehicle and the Atlas-Agena target vehicle, crucial components of the Gemini program's infrastructure.

Financially, DOD committed to contributing funds to support the attainment of Gemini's ambitious objectives, underscoring a shared commitment to the program's success.

On January 22, 1963, during an electrical systems coordination meeting at the Manned Spacecraft Center, significant developments and challenges were reported regarding the operation of fuel cell systems crucial for space missions.

The meeting revealed that the first fuel cell section tested had encountered a critical setback: a fuel cell stack had failed, leading to a fire that burned through the cell's casing. This incident underscored the technical hurdles and risks of pioneering fuel cell technology in space environments. In response, efforts were underway to address these issues by assembling new sections with thicker ion-exchange membranes to enhance durability and reliability.

Despite these challenges, notable successes were reported. One section comprising six fuel cells operated within specified limits for an impressive 707 hours. Even after 875 hours of operation, it remained only five percent below the specified voltage, highlighting its robust performance over an extended period. Similarly, another stack demonstrated consistent performance well within specification after 435 hours of operation.

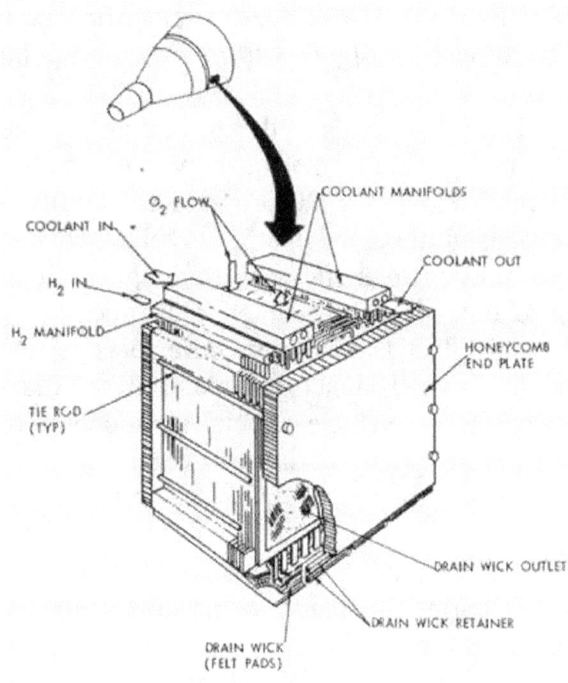

Gemini fuel cell stack.

On January 22, 1963, North American Aviation received a pivotal letter contract marking Phase III, Part I, of the Paraglider Development Program. This contract, Change No. 6 to Contract NAS 9-539, Phase II-B(1), signified a significant milestone in developing advanced landing systems for Project Gemini spacecraft.

Under this agreement, North American Aviation was tasked with producing the Gemini paraglider landing system, a crucial component to facilitate safe and controlled landings for Gemini missions. The paraglider system represented an innovative approach to spacecraft recovery, leveraging aerodynamic principles for precision landing capabilities.

The incorporation of this contract underscored NASA's commitment to enhancing spacecraft recovery technologies, which were essential for ensuring the safety and success of manned space missions. It positioned North American Aviation at the forefront of aerospace engineering, tasked with pioneering solutions that enable astronauts to return from space with unprecedented accuracy and reliability.

Specialization Assignments for New Gemini Astronauts

On January 26, 1963, the Manned Spacecraft Center (MSC) announced specialized roles for nine newly appointed astronauts, marking a strategic allocation of expertise crucial for advancing the Gemini missions.

Neil A. Armstrong was designated for trainers and simulators, underscoring his role in preparing astronauts for the complexities of spaceflight. Frank Borman was assigned to boosters, emphasizing his expertise in the critical propulsion systems vital for launching spacecraft into orbit. Charles Conrad, Jr. took on cockpit layout and systems integration, focusing on the ergonomic design and operational integration of spacecraft controls.

James A. Lovell, Jr. assumed responsibility for recovery systems, ensuring the safe return of astronauts and spacecraft after missions. James A. McDivitt was tasked with guidance and navigation, pivotal for accurate spacecraft maneuvering and orbital adjustments. Elliot M. See, Jr. was assigned to electrical systems, Sequential operations, and mission planning, essential for managing spacecraft operations and mission execution.

Thomas P. Stafford took charge of communications, instrumentation, and range integration, critical for maintaining communication links and monitoring spacecraft performance. Edward H. White II was dedicated to flight control systems, playing a key role in maneuvering spacecraft during critical phases of flight. Finally, John W. Young focused on environmental control systems, personal equipment, and survival gear, ensuring astronaut safety and comfort during missions.

On January 29-30, 1963, during a pivotal launch guidance and control coordination meeting, Aerospace Corporation highlighted significant issues with the General Electric Mod III airborne radio guidance system used in Titan II development flights. This revelation followed three consecutive flight failures attributed to problems within the guidance system, raising concerns about its reliability and performance.

While these failures did not stem from inherent design flaws directly impacting the Gemini program, Aerospace Corporation underscored the need for a more rigorous quality assurance program. Specifically, they criticized General Electric's Mod III (G) quality control practices, citing concerns over poor workmanship as contributing to the system's operational shortcomings.

On January 30, 1963, the Gemini Project Office sought authorization from NASA Headquarters to implement preflight automatic checkout equipment for Project Gemini. This request stemmed from the Mercury Program's experience, where extensive checkout times had posed significant scheduling challenges despite overall mission success.

Recognizing the potential to streamline operations and reduce turnaround times, automatic checkout equipment was proposed to expedite component testing in the Gemini spacecraft. George M. Low, Director of Spacecraft and Flight Missions at the Office of Manned Space Flight, endorsed the initiative and swiftly recommended the provision of four automatic checkout stations for Project Gemini.

Initially approved, the deployment of automatic checkout equipment promised to enhance efficiency and optimize the readiness of Gemini missions. However, as fiscal considerations came into play, the decision to utilize this advanced technology was eventually reversed as an economic measure.

On February 5, 1963, the Crew Systems Division of NASA presented pivotal findings concerning equipment and procedures for extravehicular operations (EVA) during Project Gemini. Representatives outlined strategic initiatives to enhance the spacecraft's capabilities for activities outside the cabin, marking significant progress in space exploration technology.

McDonnell, a key contractor, was tasked with conducting a comprehensive review of current EVA capabilities and initiating a detailed study of operational requirements. The scope of the investigation encompassed several critical areas:

Maneuverability with Closed Hatch: Evaluating the extent to which astronauts could maneuver within the cabin with the hatch closed and cabin pressurized, ensuring safety and functionality during EVA preparations.

Standing in Open Hatches: Assessing requirements to enable astronauts to stand in open hatches without exiting the cabin, facilitating tasks such as observation and maintenance while maintaining crew safety and spacecraft integrity.

Exterior Inspection: Establishing protocols and equipment necessary for astronauts to safely exit the cabin and conduct inspections of the spacecraft's exterior, crucial for maintenance and troubleshooting during missions.

The directive specified that McDonnell should integrate these EVA capabilities into spacecraft starting from No. 2 and onwards, underscoring a proactive approach to integrating advanced operational features into future Gemini missions.

The Gemini Rendezvous and Reentry Panel convened on February 5-6, 1963, highlighting significant delays in acquiring crucial information regarding flight controller procedures for commanding the Agena spacecraft during orbital maneuvers.

The meeting disclosed that efforts to access these essential procedures had been hindered by the stringent security protocols imposed by the Air Force Agena program. These security measures, designed to safeguard sensitive operational details, inadvertently slowed the flow of information critical to Gemini's rendezvous operations.

On February 6, 1963, Titan II development flight N-16 lifted off from Cape Canaveral, marking a significant milestone in aerospace engineering and operational refinement.

This launch represented the eleventh Titan II flight and the third to incorporate increased pressure in the propellant tanks of stage I to mitigate longitudinal oscillations, commonly known as POGO effects. These oscillations posed challenges for spacecraft stability and astronaut comfort during launch.

Titan II flight N-15 was launched from Cape Canaveral on January 10, 1963. It was the tenth in the series of Titan II research and development flights, and the second to achieve significantly reduced levels of longitudinal oscillations by means of propellant tank pressurization. (USAF Photo 33-1, Jan. 10, 1963.)

The efforts reduced POGO levels to approximately 0.5 g, a significant improvement deemed satisfactory from the weapon system's perspective. However, despite this achievement, NASA advocated further reducing POGO levels to 0.25 g—a threshold considered optimal for ensuring safe and comfortable conditions for manned spaceflight.

Conversely, the Air Force, responsible for funding the weapon system aspects of Titan II development, expressed reluctance to allocate additional funds towards achieving the lower POGO threshold sought by NASA. This reluctance stemmed from budgetary considerations and differing priorities between NASA's manned spaceflight objectives and the Air Force's weapon system requirements.

Comprehensive Training of Gemini Astronauts

Among the subjects studied were astronomy, physics of the upper atmosphere and space, global meteorology, selenology, guidance and navigation, computers, fluid mechanics, rocket propulsion systems, aerodynamics, communications, environmental control systems, and medical aspects of space flight. Flight-crew training plans for the rest of the year, which were being formulated during February, called for space science and technology seminars, celestial recognition training, monitoring the Mercury-Atlas 9 flight, weightless flying, pressure suit indoctrination, parachute jumping, survival training, instruction in spacecraft systems and launch support, paraglider flying, centrifuge experience, docking practice, and work with the flight simulator.

On February 7, 1963, a significant milestone was achieved in developing safety systems for manned spaceflight by conducting simulated off-the-pad ejection test No. 8 at the Naval Ordnance Test Station.

Two dummies were ejected during this test, marking the first time a ballute system was incorporated into the ejection procedure. The term "ballute," derived from "balloon" and "parachute," described a novel device intended to stabilize astronauts after ejection at high altitudes.

While the ejection seat and dummy separation proceeded satisfactorily, and the personnel parachute deployed as expected, technical issues arose with the test equipment. These faults prevented the canopy from fully inflating, leading to the failure of the ballute to inflate or release correctly on either dummy.

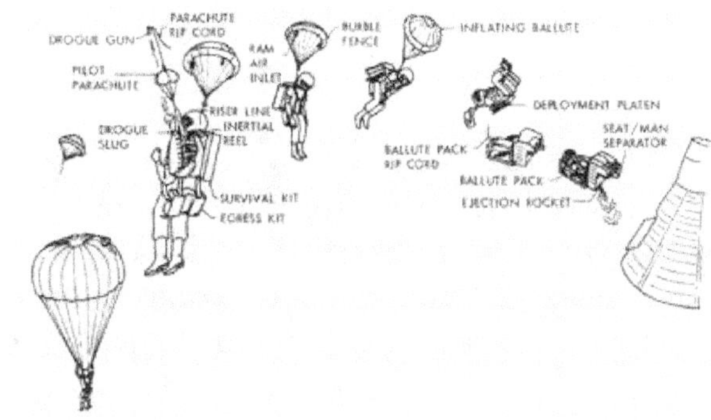

Proposed deployment sequence for the ballute stabilization device. (NASA Photo No. 63-Gemini-12, Jan. 18, 1963.)

In response to these challenges, engineers promptly redesigned the parachute system to ensure more reliable inflation at very low dynamic pressures. This redesigned parachute underwent rigorous testing in a subsequent series of five dummy drops conducted throughout March, all completed successfully.

Department of Defense Objectives in the Gemini Program

On February 8, 1963, Colonel Kenneth W. Schultz of the Headquarters, Air Force Office of Development Planning, articulated the Department of Defense's strategic objectives for the Gemini program during the inaugural meeting of the Gemini Program Planning Board.

He defined three general objectives: conducting orbital experiments related to such possible future missions as the inspection and interception of both cooperative and passive or noncooperative objects in space under a variety of conditions, logistic support of a manned orbiting laboratory, and photo reconnaissance from orbit; gaining military experience and training in all aspects of manned space flight; and assessing the relationship between man and machine in the areas of potential military missions.

On February 8, 1963, Northrop Ventura achieved a significant milestone by completing the first series of 20 drop tests to develop the parachute recovery system for Project Gemini. This crucial phase marked substantial progress in ensuring the safe recovery of spacecraft and astronauts during missions.

During the last two weeks of August 1962, the first four drops used a dummy rendezvous and recovery (R and R) section with the 18-foot drogue parachute to determine the rate of descent of the R and R section. Subsequent drops tested the 84-foot ring-sail main parachute using boilerplate spacecraft No. 1, a steel mock-up of the Gemini spacecraft ballasted to simulate the weight and center of gravity of the flight article. Boilerplate No. 1, manufactured by McDonnell, was delivered to Northrop Ventura on August 1. Drops Nos. 5 and 6 were simple weight drops to determine the structural characteristics of the main parachute. Beginning with drop No. 7, tests were conducted through the entire sequencing of the system from an altitude of 10,000 feet. Through drop No. 13, the main problem was tucking; the edge of the parachute tended to tuck under, hindering full inflation. Drop tests Nos. 5 through 13 were conducted from September through November 1962. The tucking problem was resolved with drop No. 14. Remaining tests in the series demonstrated the structural integrity of the parachute system when deployed at maximum dynamic pressure. Qualification drop tests were expected to begin in April.

In mid-February of 1963, the Gemini Project Office convened its first biweekly Network Coordination Meeting, a crucial forum established to ensure seamless compatibility between ground network equipment configurations, mission requirements, and airborne systems. This initiative underscored the meticulous planning essential for the success of the Gemini missions.

On November 20, 1962, the PCM (Pulse Code Modulation) Working Group concluded that the telemetry system designed for Project Gemini posed no significant compatibility challenges—a pivotal reassurance for the upcoming missions.

Two days later, on February 15th, plans for the checkout of the Agena target vehicle were meticulously presented at a meeting of the Gemini Management Panel. Upon arrival at Cape Canaveral, each Agena vehicle underwent thorough inspection and certification. Mechanical mating and interface assessments with the target docking adapter were completed following these rigorous checks. The subsequent Agena-Gemini spacecraft compatibility tests were conducted with validation and weight verifications. These meticulous procedures were pivotal in ensuring the readiness and reliability of the Agena for joint operations with the Gemini spacecraft.

Later, on February 18th, the Air Force Space Systems Division transmitted the Gemini Launch Vehicle Pilot Safety Program to contractors and organizations involved in the Gemini project. This program underscored the continuity of pilot safety philosophies and procedures from the preceding Mercury-Atlas missions to the upcoming Gemini-Titan missions, highlighting a commitment to astronaut safety amidst technological advancements.

By the end of February, on the 26th and 27th, decisions crucial to spacecraft operations were finalized. The Gemini Project Office determined that spacecraft separation from the launch vehicle for spacecraft Nos. 2 and onward would be managed manually. Notably, this decision obviated the need for a second-stage cutoff signal to the spacecraft, prompting directives to McDonnell to adjust spacecraft hardware accordingly and to Martin to recommend necessary modifications to the launch vehicle hardware. These adjustments were pivotal in refining operational protocols and ensuring mission safety during critical phases of ascent and separation.

On February 28th, pivotal developments marked the progress of Project Gemini, underscoring its rigorous testing and developmental milestones. The Gemini Project Office reassigned spacecraft No. 3 to the Gemini flight program, initially slated for Project Orbit tests—a series of simulated manned orbital flights within the McDonnell vacuum chamber. Static article No. 1 was reclassified in a strategic predesignation as spacecraft No. 3A, supplanting No. 3 in the Project Orbit program. This decision followed a comprehensive review in December 1962, which streamlined the static test program by focusing on articles Nos. 3 and 4 as primary structural test subjects. Craft No. 3 underwent rigorous evaluations simulating launch, reentry, abort scenarios, landings, and parachute deployments, while No. 4 faced tests simulating seat functionality, hatch operations, cabin pressurization, and dynamic responses.

Simultaneously, the Gemini Project Office released a detailed bar chart outlining the preflight checkout process for the Gemini spacecraft within Cape Canaveral's industrial area. Spanning 90 working days and accommodating two full shifts daily—with a partial third shift for testing and maintenance—the chart delineated comprehensive tests across spacecraft segments, the target docking adapter, and the paraglider. This meticulous planning aimed to ensure the spacecraft's readiness from initial receiving inspections to final preparations for transport to the launch pad.

Also, on February 28th, the Gemini Project Office reported a significant breakthrough achieved by Rocketdyne: the successful 270-second steady-state burn-time operation of the orbit attitude and maneuver system's (OAMS) 25-pound thruster. This milestone marked a critical advancement in liquid propulsion system development, aligning with McDonnell's strategic focus on achieving operational longevity and reliability. Although initial beliefs about performance requirements for both the 25-pound and 100-pound thrust chamber assemblies (TCA) required adjustment, Rocketdyne's subsequent shift in focus to enhancing operational efficiency during pulse and steady-state conditions underscored their commitment to overcoming technical challenges.

On March 1, the stage II oxidizer tank from Gemini launch vehicle (GLV) 2 was airlifted from Martin-Denver to Martin-Baltimore to be used in GLV-1. GLV propellant tank and skirt assemblies were manufactured, pressure-tested, and calibrated at Martin-Denver, then shipped to Baltimore, where the GLV was assembled. Martin-Denver had begun major weld fabrication of GLV-1 and GLV-2 tanks in September 1962 and delivered the GLV-1 tanks to Martin-Baltimore on October 10. After extensive testing, the tanks underwent a roll-out inspection on February 14-16, 1963, by Air Force, NASA, Aerospace, and Martin personnel. The inspecting team rejected the stage II oxidizer tank because it was found to be cracked. The rejected tank was returned to Denver and replaced by the GLV-2 stage II oxidizer tank.

In early March of 1963, the Gemini Project Office engaged in pivotal discussions with contractors to establish a strategic framework for the final phase of rendezvous missions—a critical aspect of the Gemini program's objectives. The discussions yielded a set of comprehensive guidelines:

Initiation Timing: Terminal maneuvers would commence when the Agena target vehicle entered the spacecraft's sensor range, typically between its insertion and first apogee.

Guidance Redundancy: Automatic and optical terminal guidance systems would complement each other, with one selected as primary and the other serving as backup for each mission.

Early Mission Flexibility: For early rendezvous missions, the terminal phase could begin as early as the third spacecraft apogee or be delayed until the twelfth due to range radar tracking limitations.

Midcourse Adjustments: To optimize resources, no midcourse corrections would be performed during orbits 4 through 11, barring significant plane or phase errors.

Maneuver Capabilities: In cases of substantial plane or phase discrepancies, the Agena would be maneuvered to align within the spacecraft's operational capability.

Rendezvous Planning: Following major Agena maneuvers, its orbit would be recircularized, followed by two orbits to facilitate spacecraft catchup before initiating the terminal rendezvous plan.

This strategic framework underscored the meticulous planning and operational protocols essential for successful rendezvous missions within the Gemini program.

On March 7th, the Gemini Program Planning Board convened in Washington and made significant decisions to expand the program's scope and objectives. An ad hoc study group was established to compare NASA and Department of Defense (DOD) objectives for the Gemini program, mainly focusing on integrating DOD experiments. Meeting continuously from March 25th to April 26th, the group presented its final report on May 6th. Subsequently, the Board recommended immediate approval of a program for in-flight military experiments and advocated for establishing a dedicated Air Force field office at the Manned Spacecraft Center. This office would oversee DOD's participation in the Gemini program and manage the integration of experimental payloads.

Moreover, the Board emphasized the urgency of addressing longitudinal oscillations in Stage I and combustion instability in Stage II of the Gemini launch vehicle, underscoring their commitment to resolving critical technical challenges.

While the study group proposed additional flights to accommodate experiments beyond the planned framework of the Gemini program, the Board, supported by the Secretary of Defense and NASA Administrator, refrained from recommending such expansions. Instead, they focused on optimizing existing mission parameters and integrating essential military experiments—a pivotal step in advancing civilian and defense-related objectives within the Gemini program. These decisions reflected the program's evolving strategic direction and pivotal role in shaping future space missions and technological advancements.

March 11th marked a critical juncture in the Gemini program's Paraglider Development Program, as setbacks led to significant strategic shifts and reevaluations. Following a series of issues, including the loss of a half-scale test vehicle during a deployment flight test, budget pressures prompted initial considerations of eliminating the paraglider from the Gemini Program as early as October 19, 1962. Despite these challenges, the decision was made to retain the paraglider while reorienting its development plan.

On March 27-28, 1963, NASA and North American representatives met to discuss revised paraglider programs, aiming to redefine its role within the Gemini missions. By March 29th, Andre J. Meyer Jr., of the Gemini Project Office (GPO), announced a decision to delay the paraglider's deployment until the tenth Gemini mission. However, discussions within the Gemini Management Panel on May 2nd indicated optimism that the paraglider might be ready as early as spacecraft No. 7. This sentiment was echoed in GPO's Quarterly Status Report, which projected the paraglider's integration from the seventh flight onward.

Financial considerations also played a pivotal role. By April 9th, North American reported funding constraints for Contracts NAS 9-167 and NAS 9-539, prompting the termination of these contracts and the issuance of a new letter contract, NAS 9-1484, on May 5th. This contract redefined the paraglider effort as the Paraglider Landing System Program, emphasizing its research and development focus.

Simultaneously, on March 12th, North American awarded three major subcontracts for the Paraglider Landing System Program: $461,312 to Northrop for a parachute recovery system, $1,034,003 to Vickers, Inc., for the paraglider control actuation assembly on March 25th, and $708,809 to Minneapolis-Honeywell for the paraglider electronic control system on May 13th. These subcontracts underscored a concerted effort to advance paraglider technology amidst programmatic challenges.

On March 14th, McDonnell presented findings from a study addressing mission "scrub" scenarios, crucial for determining the minimum recycle time between launch attempts—a critical factor for maintaining rendezvous mission schedules within the Gemini program. McDonnell's estimation of a 24.5-hour recycle period prompted further studies at the Manned Spacecraft Center (MSC) to explore compressing this timeframe through concurrent work strategies.

March 19th witnessed significant personnel changes within the Gemini program, marking a pivotal moment in its organizational structure. James A. Chamberlin, previously Manager of Project Gemini, was reassigned to Senior Engineering Advisor to Robert R. Gilruth, Director of Manned Spacecraft Center. Concurrently, Charles W. Mathews transitioned from Chief of the Spacecraft Technology Division to Acting Manager of Project Gemini, assuming leadership responsibilities within the program.

The following day, March 20th, marked the commencement of qualification tests for the production prototype ablation heatshield designed for the Gemini spacecraft. Prior structural and material properties specimen tests had confirmed that the heatshield met or exceeded required design specifications, validating its readiness for further testing and eventual deployment.

On March 21st, a pivotal meeting at the Manned Spacecraft Center established comprehensive guidelines for extravehicular operations (EVA). The meeting affirmed the retention of the pressure suit as a single-wall pressure vessel, open to modifications such as thermal coverings, gloves, and boots as necessary. A specialized tether incorporating 12 nylon-encapsulated communications wires was designed to facilitate astronaut safety and maneuverability during EVAs. This tether, dedicated solely to securing astronauts to the spacecraft during operations, ensured stability while maneuvering was managed through alternative means. These provisions were slated for integration starting with spacecraft No. 4, with a specified operational window of half an hour outside the cabin guiding system design parameters.

Simultaneously, on March 21st, a significant contract amounting to $33,797,565, including fixed fees, was awarded to Philco Corporation of Philadelphia, Pennsylvania. The contract aimed to implement the Integrated Mission Control Center, encompassing all flight information and control display equipment. Philco, selected from a competitive pool of seven qualified bidders, would collaborate with the Manned Spacecraft Center and International Business Machines Corporation (IBM) to construct and maintain the real-time computer complex integral to mission operations. Philco's role would extend to assisting in equipment operation and maintenance for at least one year post-acceptance, underscoring their pivotal role in advancing mission control capabilities within the Gemini program.

On April 1st, the Titan II-Gemini Coordination Committee was established to mitigate longitudinal vibrations (POGO) within the Titan II launch vehicle and enhance engine reliability. This initiative followed extensive briefings by the Air Force Space Systems Division (SSD) and Aerospace to NASA and the Air Force, culminating in a presentation to the Gemini Program Planning Board. The main challenge identified was that the POGO levels, deemed acceptable for military applications, exceeded NASA's stringent standards for the Gemini program. Addressing this required a comprehensive analytical and experimental approach far beyond initial assessments. Endorsing proposals put forth by SSD and Aerospace, the Gemini Program Planning Board formed a high-level committee comprising officials from the Air Force Ballistic Systems Division, SSD, Space Technology Laboratories, and Aerospace to oversee and expedite the development of a viable POGO remedy.

The following day, April 2nd, witnessed significant congressional scrutiny as D. Brainerd Holmes, Director of Manned Space Flight, testified before the Subcommittee on Manned Space Flight of the House Committee on Science and Astronautics. Holmes justified a notable $42.638 million increase in Gemini's 1963 budget, significantly higher than earlier estimates. This increase primarily reflected a $49.9 million augmentation in spacecraft costs, which Holmes attributed to unforeseen challenges and cost escalations encountered during early project phases. McDonnell's initial cost estimates proved inadequate based on minimal deviations from Mercury technology as detailed subsystem specifications were finalized. Key revisions included transitioning from a Mercury-type data transmission system to a more advanced Pulse Code Modulation (PCM) system to meet enhanced data transmission demands while reducing weight and power consumption. Similarly, the design of the rendezvous radar evolved from initial plans based on Bomarc Missile technology to a specialized interferometer-type radar optimized for minimal size, weight, and maximum reliability. The environmental control system, originally planned as two Mercury-type systems, underwent substantial modifications to meet revised performance criteria, highlighting the iterative nature of spacecraft development.

Securing the Future of Manned Space Exploration

On April 2nd, NASA formally announced the signing of a pivotal contract with McDonnell for the Gemini spacecraft, marking a significant milestone in America's ambitious manned space exploration endeavors. Following meticulous negotiations, which concluded on February 27, 1963, the contract was valued at an estimated $428,780,062, inclusive of a fixed fee totaling $27,870,000, resulting in a total estimated cost-plus-fixed-fee of $456,650,062. The contract's terms underscored NASA's commitment to advancing spaceflight capabilities by developing and deploying cutting-edge spacecraft technology.

Initiated in December 1961 under a preliminary letter contract, the comprehensive agreement with McDonnell encompassed the production of 13 flight-rated spacecraft. Among these, 12 were designated for actual space missions, underscoring NASA's ambitious agenda to explore the depths of space. Additionally, McDonnell was tasked with providing essential training resources critical to astronaut preparation, including two mission simulator trainers and a docking simulator trainer. The contract also mandated delivering five boilerplate models for preliminary testing and three static articles dedicated to rigorous vibration and impact ground assessments.

NASA Headquarters dedicated two weeks to thoroughly reviewing the contract's specifics before finalizing the agreement. This meticulous scrutiny ensured that every aspect of the contract aligned with NASA's stringent standards for quality, safety, and technological innovation. The contract aimed to propel the Gemini program forward and laid a solid foundation for future advancements in manned space exploration, positioning America at the forefront of scientific and technological achievement in the space race era.

On April 9th, George M. Low, Director of Spacecraft and Flight Missions at the Office of Manned Space Flight elucidated before the House Subcommittee on Manned Space Flight the rationale behind planning eight rendezvous missions as part of the Gemini program. Central to this strategic initiative was the imperative to develop and refine the capability for orbital rendezvous—a critical precursor to the lunar orbit rendezvous envisioned for the Apollo missions.

Low articulated that the Gemini program aimed to explore various methodologies for conducting rendezvous operations. At one end of the spectrum lay purely manual approaches, where astronauts would visually navigate toward a flashing light on the target vehicle, demonstrating the most straightforward yet propellant-inefficient method. Conversely, at the opposite end stood fully automatic systems employing radar, computers, and stabilized platforms to autonomously execute maneuvers from a considerable distance, offering efficiency but also complexity and potential reliability challenges.

Recognizing the need to balance efficiency with reliability, Low emphasized the importance of gaining practical experience with different rendezvous techniques in the actual space environment. This hands-on approach would allow NASA to validate and refine operational procedures for future lunar missions. The Gemini program aimed to gather crucial data and operational insights to inform and optimize the Apollo program's lunar orbit rendezvous strategies by conducting multiple rendezvous missions with varying degrees of automation and astronaut involvement.

In late April, significant milestones marked the progress of Project Gemini, underscoring the meticulous planning and collaboration essential for its success. On April 22, at Sunnyvale, a pivotal management review convened representatives from the Air Force Space Systems Division (SSD), the Manned Spacecraft Center, and Lockheed. This inaugural meeting focused on the Gemini Agena target vehicle (GATV), a critical program component for advancing manned space exploration.

Modeled after established protocols for satellite and probe programs, the Gemini Target Management Review Meetings provided a monthly platform to assess the GATV's development. This structured approach ensured that all stakeholders remained informed of the program's status, addressing challenges and milestones alike.

The following day, from April 23 to 24, the Gemini Abort Panel convened to address crucial safety considerations. Martin-Baltimore's analysis of recent Titan II flight tests indicated promising outcomes for crew escape scenarios. McDonnell contributed vital data on spacecraft structural capabilities, though uncertainties persisted regarding potential catastrophic failures of the Titan II launch vehicle. Of particular concern was the behavior of the spacecraft's adapter retrosection during and after an abort scenario. Detailed studies, conducted assuming a 70,000-foot altitude, concluded that separation and recontact risks were minimal within the critical 10-second window post-separation.

On April 27, attention shifted to the final design review of complex 14 modifications in Los Angeles, overseen by the Air Force SSD. This review marked a crucial step in preparing the complex for Project Gemini Atlas-Agena launches. All drawings and specifications were formally accepted, setting the stage for SSD's activation of the complex commencing January 1, 1964. Anticipated to take approximately 10 months, this activation phase was essential for ensuring that complex 14 would meet the stringent operational requirements of the Gemini missions.

Chapter 5 - Evolution of the Gemini Flight Program

On April 29th, a pivotal decision reshaped the trajectory of the Gemini flight program, reflecting the adaptive nature of space exploration at NASA Headquarters. Following a comprehensive review by the Gemini Project Office (GPO), significant adjustments were approved to address spacecraft system delivery and integration challenges.

The original flight schedule, heavily influenced by the legacy of Mercury systems in the Gemini spacecraft, underwent substantial revision. The initial unmanned flight, originally slated for December 1963 as a suborbital mission, was now redefined as an orbital test. This adjustment aimed to qualify launch vehicle subsystems and validate spacecraft compatibility without separation or recovery maneuvers.

Originally intended as the program's first manned orbital flight, the subsequent mission transitioned to an unmanned suborbital ballistic flight planned for July 1964. Its primary objective shifted to testing spacecraft reentry capabilities under maximum heating-rate conditions, alongside the comprehensive validation of launch vehicle and spacecraft systems critical for subsequent manned missions.

Initially designed for a manned orbital rendezvous mission, the third flight was repositioned as the program's first manned flight. Scheduled for October 1964, this short-duration mission aimed to evaluate spacecraft systems over a probable three-orbit journey. Subsequent flights were slated at three-month intervals, culminating in January 1967.

Rendezvous terminal maneuvers were planned for the third and fourth missions, contingent upon flight duration allowances. The sixth mission was envisioned as a 14-day long-duration flight, paralleling the objectives of the fourth mission but excluding rendezvous maneuvers with the Atlas-launched Agena D target vehicle.

Splashdowns were planned for the first six flights with water landings aided by parachutes. Beginning with the seventh flight, land landings were to be implemented using paragliders, marking a transition towards enhanced reentry and recovery capabilities as the Gemini program advanced.

April 30th marked a critical juncture for NASA's management stance on Project Gemini, spurred by Secretary of Defense McNamara's fiscal year 1964 budget testimony and subsequent interpretations. In response to concerns over a potential Air Force assumption of Gemini's management, Robert C. Seamans, Jr., NASA's Associate Administrator, emphasized NASA's pivotal leadership role. Acknowledging the Department of Defense's integral involvement in Gemini's management and operations, Seamans asserted NASA's ultimate responsibility for ensuring the program's overall success.

Throughout April, significant technical strides complemented these strategic clarifications. Bell Aerosystems achieved a milestone by successfully firing the Gemini Agena Model 8247 engine at its Buffalo facility. Derived from the proven Model 8096 utilized in NASA and Air Force satellite programs, this new engine represented a critical advancement for Gemini. Unlike its predecessor, the Model 8247 boasted the capability of multiple restarts in orbit, a crucial requirement for Gemini's mission profile.

Key modifications included substituting liquid propellants for solid pyrotechnic "starter cans" to initiate the gas generator, enhancing operational flexibility and reliability. The developmental engine, assembled in March, underwent initial testing at Bell Aerosystems before being transported to the Arnold Engineering Development Center (AEDC) in Tullahoma, Tennessee, in mid-April. By April 12th, test cell preparations at AEDC were finalized, with comprehensive testing slated to commence in May.

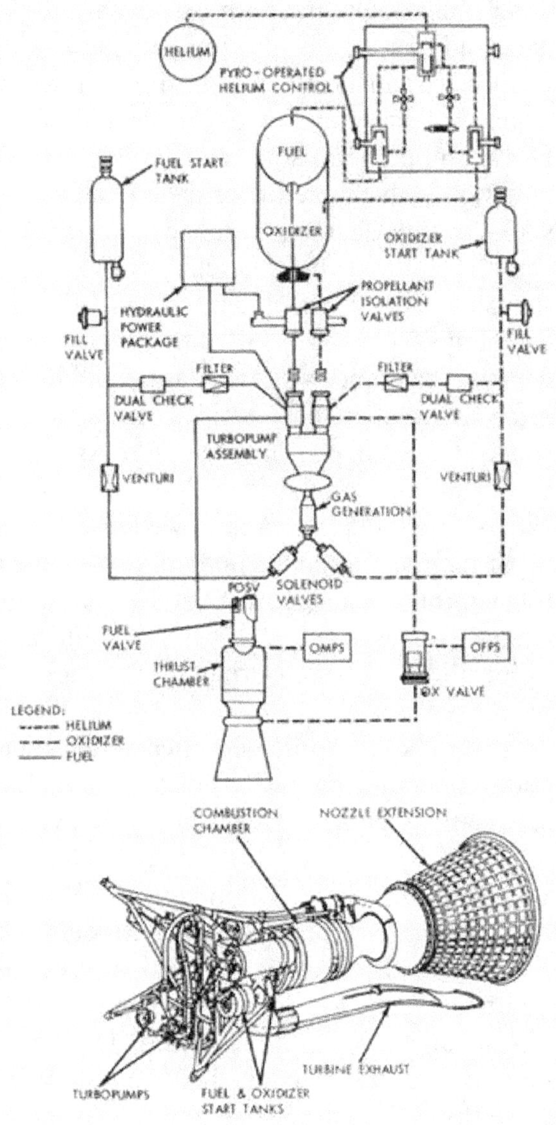

Schematic and drawing of the primary propulsion system of the Gemini Agena target vehicle. (Lockheed, Gemini Agena Target

On May 1st, McDonnell initiated critical tests to validate the Gemini spacecraft's attitude control and maneuver electronics (ACME) system. Following successful development testing, McDonnell focused on qualifying the system's first production prototype, which was received from Minneapolis-Honeywell.

The ACME system was pivotal in Gemini's mission capabilities, enabling precise control and maneuverability essential for orbital operations. These qualification tests aimed to ensure the reliability and functionality of the ACME unit under simulated mission conditions, marking a significant step toward integrating advanced control technologies into the Gemini spacecraft.

On May 2nd, Charles W. Mathews, newly appointed Acting Manager of Project Gemini, conducted a comprehensive review of the program's current status before the Gemini Management Panel. This review encompassed evaluations of spacecraft development, launch vehicles, and ground facilities crucial to the program's success. Modifications of launch complexes 19 and 14 of the tracking network and Atlantic Missile Range checkout facilities were all on schedule, although no margin remained for complex 19 work. The Atlas and Agena presented no problems, but the Gemini launch vehicle schedule was tight; technical problems, notably stage I longitudinal oscillations and stage II engine instability, were compounded by funding difficulties. The Gemini spacecraft, suffering from late deliveries by subcontractors, was being reprogrammed.

On May 3rd, the Arnold Engineering Development Center (AEDC) commenced development testing of the Gemini Agena Model 8247 main engine with an initial instrumentation run. This marked a critical phase in validating the engine's performance capabilities under simulated orbital conditions. After oxidizer contamination resulted in a scrubbed test on May 7, test firing began on May 13. The major objective of AEDC testing was to verify the engine's ability to start at least five times. The AEDC rocket test facility permitted the firing of the engine in an environment simulating orbital temperature and pressure. During the tests, two major problems emerged: turbine overspeed and high-temperature operations of the gas generator valve. At the Atlas/Agena coordination meeting of July 2, Air Force Space Systems Division reported that a turbine overspeed sensing and shutdown circuit had been proposed to resolve the first problem and that solutions to the gas generator problem were being investigated.

Advancing Paraglider Technology for Gemini

On May 5th, NASA awarded Letter Contract NAS 9-1484 to North American Aviation for the Paraglider Landing System Program, marking a significant step in the evolution of landing technology for manned space missions. Valued at $6.7 million, the contract aimed to overhaul and refine the paraglider program, positioning it as a critical component of the Gemini spacecraft's landing strategy.

This contract reflected a reorientation of the paraglider program. Its primary purpose was to develop a complete paraglider landing system and to define all the components of such a system. Among the major tasks this entailed were: (1) completing the design, development, and testing of paraglider subsystems and building and maintaining mock-ups of the vehicle and its subsystems; (2) modifying the paraglider wings procured under earlier contracts to optimize deployment characteristics and designing a prototype wing incorporating aerodynamic improvements; (3) modifying the two full-scale test vehicles produced under Contract NAS 9-167 to incorporate prototype paraglider landing system hardware, modifying the Advanced Paraglider Trainer produced under Contract NAS 9-539 to a tow test vehicle, and fabricating a new, second tow test vehicle; and (4) conducting a flight test program including half-scale tow tests, full-scale boilerplate parachute tests, full-scale deployment tests, and tow test vehicle flight tests. Contract negotiations were completed on July 12, and the final contract was dated September 25, 1963.

On May 6, the Gemini Program Planning Board approved the Air Force Systems Command development plan for the Gemini/Titan II improvement program. The plan covered the development work required to man-rate the Titan II beyond the requirements of the Titan II weapon system. It included three major areas: (1) reducing longitudinal oscillation levels to NASA requirements, (2) reducing the incidence of stage II engine combustion instability, and (3) cleaning up the design of stage I and II engines and augmenting the continuing engine improvement program to enhance engine reliability. The work was to be funded by the Titan Program Office of Air Force Ballistics Systems Division and managed by the Titan II/Gemini Coordination Committee, established on April 1. NASA found the plan satisfactory.

On May 7-17, Aerojet-General delivered the first flight engines for Gemini launch vehicle No. 1 to Martin-Baltimore. Aerojet-General had provided a set of Type "E" dummy engines on March 18. These were installed and used to lay out tubing and wiring while the launch vehicle was being assembled. They were later removed, and flight engines were installed in Stage II on May 7 and Stage I on May 17. Some rework was required because of differences in configuration between the dummy and flight engines, and engine installation was completed on May 21. Wiring and continuity checks followed (May 22-25), and final horizontal tests were completed on May 27.

On May 9, qualification testing of the Gemini parachute recovery system began at E1 Centro, California. Boilerplate spacecraft No. 5, a welded steel mock-up of the spacecraft reentry section, was dropped from a C-130 aircraft at 20,000 feet to duplicate dynamic pressure and altitude at which actual spacecraft recovery would be initiated. Four more land-impact tests followed, the last on June 28; all test objectives were accomplished. The main parachute tucking problem, which had appeared and been resolved during development tests, recurred in drops 4 and 5 (June 17, 28). Although this problem did not affect parachute performance, the Gemini Project Office suspended qualification testing until the condition could be studied and corrected. Northrop Ventura attributed the tucking to excessive fullness of the parachute canopy and resolved the problem by adding control tapes to maintain proper circumference. Four bomb-drop tests during July proved this solution satisfactory, and qualification testing resumed on August 8.

May 15: Simulated off-the-pad ejection seat testing resumed with significant improvements following earlier challenges. McDonnell and Weber Aircraft completely redesigned the ejection seat's blackboard and mechanism linkage to enhance reliability and streamline operation. This redesign aimed to eliminate the complexities that had hindered previous tests, culminating in an unsuccessful test in February. Validated through a series of tests, including a successful preliminary ejection test on April 22nd, tests No. 9 and 9a on May 15th and 25th, respectively, proved completely successful. Subsequent dual ejection tests, No. 10 and 11 on July 2nd and 16th, faced mixed outcomes: while No. 10 was unsuccessful, No. 11 was marred by the failure of a recovery chute, causing major damage upon impact.

May 18: Rocketdyne achieved a successful milestone by testing a 25-pound thrust chamber assembly (TCA) for the reentry control system (RCS), focusing on pulse operation. Earlier efforts had primarily aimed at steady-state performance until it was discovered that this did not guarantee adequate pulse performance. Tests revealed a higher char rate during pulse cycles compared to continuous runs, leading to adjustments such as reducing the oxidizer-to-fuel ratio and increasing ablative chamber wall thickness. Despite these challenges, Rocketdyne continued to grapple with developing a suitable ablative thrust chamber, with delays in delivering TCA hardware to McDonnell affecting overall component schedules.

May 20: The Flight Crew Operations Division reported completing a zero-gravity indoctrination program for nine new crew members at Wright-Patterson Air Force Base, Ohio. Conducted with support from the 6750th Aerospace Medical Research Laboratory, the program utilized modified KC-135 aircraft to simulate 20 zero-gravity parabolas per flight, each lasting 30 seconds. This training was crucial in preparing astronauts for the unique challenges of manned spaceflight aboard Gemini missions.

May 21: The Manned Spacecraft Center initiated a Gemini atmospheric reentry simulation study using a fixed-base simulator with hand controllers and pilot displays. This study aimed to evaluate manual control of the Gemini spacecraft during reentry before commencing a centrifuge program at the Naval Air Development Center. Completed on June 20th, the simulation study provided valuable insights into spacecraft control dynamics under reentry conditions, informing future training and operational protocols.

On May 21, as part of the general revision of the Gemini flight program that NASA Headquarters had approved on April 29, NASA, Air Force Space Systems Division, and Lockheed met to establish basic ground rules for revising Agena development and delivery schedules. The first rendezvous mission using the Agena target vehicle was planned for April 1965, some seven and one-half months later than anticipated in October 1962. Six months would separate the second Agena launch from the first, and subsequent flights would be at three-month, rather than two-month, intervals. The revised schedule was agreed on at the Atlas/Agena coordination meeting on June 6-7, 1963. Among the major features of the new schedule: Agena communications and control subsystem development was to be completed by December 1963 (back six weeks); other Lockheed development work was to be completed by January 1964 (back three and one-half months); assembly and modification of the first target vehicle was to start April 2, 1964, with the vehicle to be accepted and delivered in January 1965; the first Atlas target launch vehicle was to be delivered in December 1964; the schedule for component manufacturing and deliveries was to be so arranged that the second target vehicle could back up the first, given about nine months' notice.

On May 23, the first engineering prototype of the onboard computer completed integration testing with the inertial platform at International Business Machines Corporation (IBM) and was delivered to McDonnell. At McDonnell, the computer underwent further tests. Some trouble developed during the initial test, but IBM technicians corrected the condition, and the computer successfully passed diagnostic test checks.

Testing and Development Milestones in Project Gemini

In late May, North American Aviation initiated testing on the Paraglider Landing System Program's half-scale two-test vehicle (HSTTV). The rigorous series of ground tows from May 27 to July 29 encompassed 121 tests. These tests involved varying wing angles and attachment points to gather essential data for rigging analysis and dynamic tow characteristics. Subsequently, the HSTTV was transported to Edwards Air Force Base, arriving on August 19. The Flight Research Center undertook 133 ground tows starting August 20 and concluding in September. October saw the continuation of 11 helicopter tow tests, primarily aimed at exploring paraglider liftoff behaviors, refining

helicopter tow techniques, and assessing wind-bending effects during high-speed tows.

On May 29, Titan II flight N-20 marked a significant milestone in the Air Force's developmental flight series from Cape Canaveral. Designed to address longitudinal oscillations (POGO) through oxidizer standpipes and fuel accumulators, this mission, unfortunately, encountered failure 55 seconds post-launch. Despite this setback, previous static firings in the spring of 1963 had validated the efficacy of these measures in mitigating POGO instability, attributing it to structural-propulsion system coupling. The subsequent decision by the Air Force Ballistics Systems Division not to continue with the POGO fix on Titan II was later reversed, reinstating it on subsequent flights, including N-25 in November and others after that.

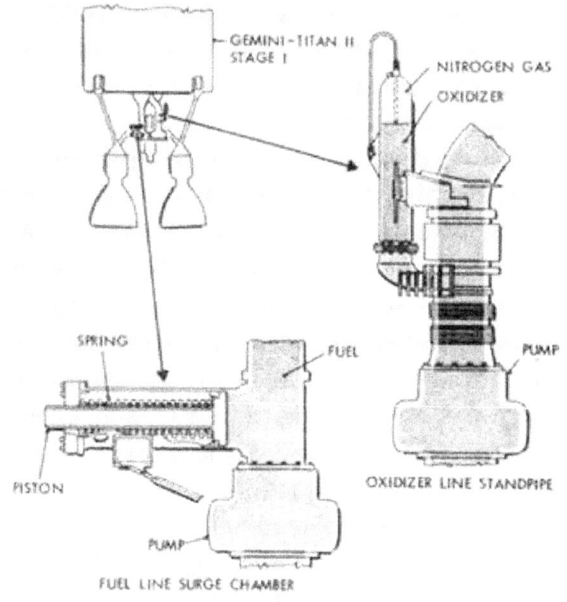

POGO suppression equipment proved out in the Titan II development program. (Martin Photo 8B65766, undated.)

Also, on May 29, activating the Vertical Test Facility (VTF) at Martin-Baltimore marked a pivotal moment in Project Gemini. Comprising a 165-foot tower and an adjoining three-story blockhouse, this facility replicated the testing setup at complex 19. It facilitated comprehensive testing of the fully assembled Gemini launch vehicle, meticulously scrutinizing each subsystem independently before progressing to combined systems tests. The culmination of these evaluations was the Combined Systems Acceptance Test, a critical milestone preceding Air Force approval of the launch vehicle.

Rocketdyne recommenced its rigorous test program for the 100-pound thrust chamber assembly (TCA), integral to the Orbit Attitude and Maneuver System throughout May. Following a hiatus in April due to hardware constraints, Rocketdyne focused on achieving steady-state operation. Encouraging results emerged late in the month, with two tests demonstrating operational durations of 575 and 600 seconds, respectively, showcasing minimal decay in chamber pressure and high-performance efficiencies. These tests underscored the imperative to meet stringent specifications, aiming for sustained operational integrity under mission conditions.

June 2 marked a significant milestone at Martin-Baltimore's Vertical Test Facility with the erection of Stage I of the Gemini launch vehicle 1. This was swiftly followed by the erection of Stage II on June 9, with post-erection inspections concluding by June 12. Concurrently, on June 10, the rigorous Subsystem Functional Verification Tests commenced, setting the stage for comprehensive evaluations of the launch vehicle's operational readiness.

During a pivotal Gemini Abort Panel meeting on June 4-5, McDonnell Aerospace proposed lowering the lower abort limit for Mode 2 to 35,000 to 40,000 feet. They presented computer-modeled data exploring the efficacy of combining Mode 2 and Mode 1 abort procedures, particularly for launch-to-T+10-second abort scenarios where Mode 1 alone might prove inadequate. Gemini's abort modes included Mode 1, utilizing ejection seats from the launch pad to 70,000 feet;

mode 2, employing booster shutdown and retro salvo from 70,000 to approximately 522,000 feet; and Mode 3, employing booster shutdown and normal separation from approximately 522,000 feet until the final seconds of powered flight.

On June 8, a crucial meeting convened representatives from NASA, the Air Force Space Systems Division, Aerospace, McDonnell, and Martin to initiate an investigation into the structural integrity and compatibility of the spacecraft and launch vehicle during the powered phase of missions. This issue had been a concern since the early Mercury-Atlas flights. Contractors were tasked to submit all available structural data to NASA and the Space Systems Division by July 15, 1963, to inform ongoing design refinements.

Gemini launch vehicle 1 undergoing tests in the vertical test facility at Martin's Baltimore plant. (Martin Photo B-58332.)

Training efforts intensified on June 10 as McDonnell's instructors commenced a two-week course on Gemini spacecraft systems for flight controllers at the Manned

Spacecraft Center. This training initiative followed similar sessions conducted for new astronauts in May, with veteran astronauts scheduled for instruction in late June and early July, ensuring comprehensive operational readiness across the Gemini program.

Meanwhile, on June 12, the editorial committee assembled at the Goddard Space Flight Center to outline Gemini Network Operations Directive 63-1. This directive established the overarching framework for the tracking and instrumentation network essential to the Gemini program's operations, building upon the precedents set by the earlier Mercury Network Operations Directive 61-1.

Closing a chapter on June 13, McDonnell's Project Mercury contract officially concluded, signaling a transition towards the Gemini era. McDonnell, having already completed its Mercury activities with spacecraft 15-B delivered to Cape Canaveral, now focused on integrating key personnel and resources into the Gemini program. A termination meeting on June 14 at the Manned Spacecraft Center finalized the disposition of Mercury assets and personnel. McDonnell facilitated the transfer of pertinent resources and expertise to support the burgeoning Gemini missions.

On June 13, Rocketdyne achieved a significant milestone by finalizing the 25-pound thrust chamber assembly (TCA) 's initial design for the Reentry Control System (RCS) and Orbit Attitude and Maneuver System. Just weeks later, on July 5, McDonnell approved a comprehensive redesign of the TCA, slated for installation in spacecraft Nos. 5 and onward. Concurrently, Rocketdyne established a thrust chamber working group to enhance TCA performance. This initiative yielded unexpected success with the rapid design, construction, and successful testing of two 25-pound RCS thrusters under pulse operation conditions. Consequently, this new configuration was swiftly integrated into the manufacturing plan for spacecraft Nos. 2 and beyond. Notably, all TCAs—ranging from 25 to 100 pounds—now shared a uniform design, reflecting streamlined production and operational efficiencies. While the Gemini Project Office attributed this achievement to relaxed test criteria rather than technological breakthroughs, optimizations, including reduced oxidizer-to-fuel ratios and abbreviated firing durations, underscored advancements in thruster performance standards.

Simultaneously, on June 13, the Manned Spacecraft Center's Atlantic Missile Range Operations Office announced the inclusion of the malfunction detection system in upcoming Titan II launches—N-24, N-25, N-29, N-31, and N-32. This initiative, part of the "piggyback program," aimed to enhance launch safety and operational reliability. By June 21, preparations for the maiden flight under this program were completed, including the meticulous installation and validation of all malfunction detection system components. These developments underscored ongoing efforts to integrate advanced safety protocols and operational enhancements into the Gemini program's launch operations, setting the stage for forthcoming missions with heightened confidence and reliability.

Advancing Towards Manned Spaceflight

On June 13, a pivotal moment arrived with the signing of the definitive contract for the Gemini space suit by the David Clark Company. Following negotiations concluded on May 28, the contract was valued at $788,594.80, including a fixed fee of $41,000, marking a significant step forward in ensuring astronaut safety and comfort for upcoming missions.

Two days later, on June 15, the Gemini Project Office (GPO) announced plans for the first manned Gemini mission to encompass three orbits. This decision sparked deliberations regarding the feasibility of conducting the planned rendezvous experiment, with Flight Operations Division's Rendezvous Analysis Branch determining that a three-orbit mission provided ample opportunity for valuable scientific study. McDonnell was promptly directed to explore this operational challenge further.

On June 17, AiResearch began testing a boilerplate spacecraft's Environmental Control System (ECS) developmental unit. Initially utilizing gaseous oxygen due to the unavailability of cryogenic tanks, AiResearch's system development tests continued until September. Concurrently, AiResearch shipped an ECS unit to McDonnell, where boilerplate spacecraft No. 2 installation commenced on July 11 for manned testing.

The following day, June 18, witnessed a flight evaluation test of the Gemini spacecraft's prototype recovery beacon in Galveston Bay. Demonstrating robust capabilities, the beacon achieved a maximum reception range of 123 miles at 10,000-foot altitude during ranging runs conducted by aircraft equipped with receivers.

On June 19, the Cape Gemini/Agena Test Integration Working Group convened to outline "Plan X" test procedures to validate the Gemini spacecraft's command and communication capabilities with the Agena target vehicle. Scheduled at the Merritt Island Launch Area Radar Range Boresight Tower ("Timber Tower"), these tests simulated rendezvous missions to familiarize astronauts with operational protocols.

A milestone in escape system development occurred on June 20 with the successful execution of Sled Test No. 2 at China Lake. This dynamic dual-ejection test validated both seats' ejection and system functionalities despite requiring subsequent reruns due to velocity constraints. After necessary system refinements, subsequent developmental tests, including a redesigned egress kit and soft survival pack, resumed in early 1964.

From June 20-21, McDonnell hosted a comprehensive design review meeting attended by experienced NASA personnel, yielding 76 requests for design modifications. This meeting, which included a crew station mock-up review, underscored collaborative efforts to enhance spacecraft functionality and astronaut safety.

Wrapping up the month on June 24, the Arnold Engineering Development Center conducted a retrorocket abort test, meeting test objectives despite nozzle assembly and cone failures. These setbacks prompted redesign efforts, with plans for further tests in October 1963 to validate revised configurations, ensuring mission-critical systems operated flawlessly under various scenarios.

Instrumented mannequin being lowered into a boilerplate Gemini spacecraft in preparation for a dynamic sled test of the Gemini ejection system. Notice the rocket motors at the rear of the sled that propelled it along the track. (NASA Photo 63-Gemini-60, released Sept. 30, 1963.)

On June 24, North American initiated a crucial Paraglider Landing System Program phase with a series of five drop tests using a boilerplate test vehicle. These tests aimed to qualify the parachute recovery system for the full-scale test vehicle. Earlier successes in the reoriented paraglider program included two effective parachute recovery system bomb-drop tests on May 22 and June 3. However, the initial boilerplate drop test on June 24 resulted in minor damage to both the main parachute and the boilerplate itself.

Subsequent drop tests showed significant progress: drops No. 2 (July 2), No. 3 (July 12), and No. 4 (July 18) were successful, demonstrating improved performance and reliability. Unfortunately, the fifth drop test on July 30 encountered multiple malfunctions, leading to a complete failure of the recovery system and the destruction of the test vehicle upon impact.

Despite this setback, North American considered the overall objectives of the flight qualification program for the parachute system to have been achieved. They proposed ending the parachute program, citing irreparable damage to the boilerplate vehicle. However, the Manned Spacecraft Center rejected this request and issued Change Notice No. 3 to contract NAS 9-1484, directing North American to collaborate with McDonnell on conducting two additional drop tests.

Further investigations, including wind tunnel tests using a 1/20-scale spacecraft model, pinpointed the root causes of the malfunctions. With insights gained from these tests, a modified parachute recovery system was successfully integrated and tested with a new boilerplate vehicle on November 12. Subsequent validation through a second drop test on December 3 confirmed the system's reliability and effectiveness.

Ultimately, these efforts culminated in successfully qualifying the parachute recovery system for the full-scale test vehicle, marking a critical advancement in the readiness of the Paraglider Landing System Program for upcoming missions.

On June 25, Martin-Baltimore received the stage II fuel tank for Gemini launch vehicle 2 from Martin-Denver. This new tank replaced one that was rejected due to cracks found during heat treatment. Additional components, including the stage II oxidizer tank and stage I fuel and oxidizer tanks, were received on July 12 after undergoing a roll-out inspection at Martin-Denver from July 1-3.

On June 27, Charles W. Mathews, Acting Manager of the Gemini Project Office, reported to the Gemini Management Panel that the launching azimuth for the first Gemini mission had been adjusted from 90 to 72.5 degrees. This change, aligning with the azimuth used for Mercury orbital launches, aimed to optimize tracking network coverage. The spacecraft slated for this mission would feature a complete production shell with shingles and heatshield. They would carry simulated computer systems, an inertial measuring unit, and environmental control systems in the reentry module. Simulated equipment would also be deployed in the adapter section. Instruments onboard would record various data points, including pressures, vibrations, temperatures, and accelerations.

On June 28, during a spacecraft operations meeting, McDonnell presented a "scrub" recycle schedule as part of ongoing investigations into the feasibility of delayed Gemini launches meeting successive launch windows, particularly crucial for rendezvous missions. Initially estimated at 24.5 hours, the recycle time was refined to 48 hours under normal conditions. The Gemini Project Office aimed to reduce this to 24 hours further and ultimately below 19 hours, potentially by substituting fuel cells with batteries specifically for rendezvous missions. This adjustment was intended to enhance operational flexibility and mission readiness.

On July 5, McDonnell began the first phase of Spacecraft Systems Tests (SST) on the instrumentation pallets to be installed in spacecraft No. 1. Numerous troubles brought a halt to SST on July 21 for two weeks of corrective action, including the return of one telemetry transmitter and the C-band beacon to the vendors for out-of-specification performance. Phase I of SST resumed on

August 5 and was completed well within test specifications on August 21.

Simultaneously, on July 5, the first engineering prototype inertial guidance system underwent integration and compatibility testing with a complete guidance and control system at McDonnell. All spacecraft wiring was compatible with the computer, and the components operated with complete accuracy.

On July 8, McDonnell warned the Gemini Project Office that the capacity of the spacecraft computer was in danger of being exceeded. The original function of the computer had been limited to providing rendezvous and reentry guidance. Other functions were subsequently added, and the computer's spare capacity no longer appeared adequate to handle them. McDonnell requested an immediate review of computer requirements. In the meantime, it advised International Business Machines to delete one of the added functions, orbital navigation, from computers for spacecraft Nos. 2 and 3.

The reentry control system unit for Gemini spacecraft No. 1 at the McDonnell plant. (NASA Photo #124, June 1963.)

The following day, the Gemini Phase I Centrifuge Program began at Naval Air Development Center, using the Aviation Medical Acceleration Laboratory centrifuge equipped to simulate the command pilot's position in the Gemini spacecraft. The program had two parts: an engineering evaluation of command pilot controls and displays required for the launch and reentry phases of the Gemini mission, including evaluation of prototype Gemini seat contours, pressure suit operation under acceleration, and the restraint system; and pilot familiarization with Gemini launch, reentry, and selected abort reentry acceleration profiles. The engineering evaluation was completed on August 2. Pilot familiarization was conducted between July 16 and August 17. The participating astronauts were generally satisfied with the design and operation of displays and controls, though they recommended minor operational changes. They could not cope with the reentry tasks without undue difficulty, even under the high acceleration of extreme abort conditions.

Suit Redesign Delays Gemini Project Progress

On July 9th, 1963, a critical evaluation of the G2C Gemini pressure suit underscored significant setbacks during the engineering mock-up phase at McDonnell Aircraft Corporation. Designed by the David Clark Company and recently delivered, the suit's torso exhibited unexpected deformation, rendering it ill-fitted for the rigorous demands of space.

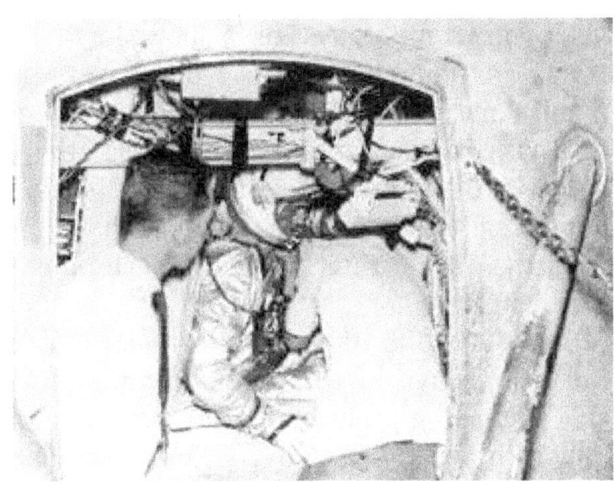

Dr. Howard A. Minners observes Astronaut Donald K. Slayton being readied for a run in the centrifuge at Aviation Medical Acceleration Laboratory, Johnsville, Pennsylvania. (NASA Photo S-63-11195, July 1963.)

Within the confines of the Gemini spacecraft mock-up, engineers discovered that the helmet visor guard exacerbated issues by increasing the helmet's height, leading to troublesome interference with the spacecraft hatch mechanism. Recognizing the urgent need for modifications, McDonnell promptly returned the suit to David Clark, outlining specific adjustments required to rectify these critical flaws.

The ensuing redesign process, focused primarily on eliminating the fixed visor guard and correcting the suit's compromised torso fit, proved more time-consuming than initially anticipated. As a result, the Gemini Project faced an unexpected delay of approximately six weeks before the prototype suit could be reevaluated and production could commence in earnest.

On July 11th, 1963, Walter C. Williams, Deputy Director for Mission Requirements and Flight Operations at the Manned Spacecraft Center (MSC), addressed Major General Leighton I. Davis, the DOD Representative for Project Gemini Operations, outlining critical safety concerns within the Gemini program. Collaboratively identified by MSC's Range Safety Office, NASA, and contractors, these concerns primarily revolved around potential launch vehicle failures. The ramifications of a catastrophic launch vehicle failure were of particular concern, highlighting unresolved issues crucial to mission safety.

In response, Aerojet-General initiated a rigorous test program in September, focusing on comparing the performance of cryogenic and hypergolic propellants. These tests crucially demonstrated that hypergolic propellants burn rather than explode upon tank rupture—an important distinction that could mitigate potential catastrophic scenarios.

On July 12th, 1963, the Gemini Project Office (GPO) at Ames Research Center completed a comprehensive test program using a centrifuge to evaluate the effects of longitudinal oscillations (known as POGO) on pilot performance within the Gemini launch vehicle. These oscillations, ranging from 0 to ±3g superimposed on a steady-state load of 3.5g, significantly impacted pilot capabilities.

Findings indicated that beyond ±0.25g, pilots experienced substantial performance degradation, impaired vision, reduced eye scan rates, compromised sensory perception, and diminished kinesthetic cues, affecting speech and overall operational effectiveness. As a result, GPO reaffirmed the critical need to limit POGO oscillations to a maximum of 0.25g to safeguard pilot performance and mission success.

Preparations and Challenges in Early Gemini Missions

On July 12th, 1963, acting Manager Charles W. Mathews of the Gemini Project Office at the Manned Spacecraft Center (MSC) informed senior staff of a strategic contingency plan for the Gemini program. In response to uncertainties surrounding the first Gemini flight (GT-1), discussions centered on preparing a backup mission designated Gemini-Titan 1A (GT 1A). NASA Headquarters approved this additional flight article in August, allocating an estimated budget of $1.5 to $2 million.

The payload for GT 1A included a boilerplate reentry module originally intended for flotation tests at MSC. Local contractors had manufactured it, with subsequent modifications by MSC upon delivery in September. Additionally, an adapter identical to that of Spacecraft No. 1 was prepared by McDonnell and slated for direct shipment to Cape Canaveral. Telemetry equipment and wiring harnesses were to be integrated into the boilerplate at the Cape.

Should GT-1 fail to achieve its objectives, GT 1A would replicate its mission goals, providing a critical contingency measure to maintain program momentum.

Development tests of the Agena Model 8247 main engine at the Arnold Engineering Development Center encountered a significant setback on July 15th. Testing ceased abruptly when a latch-type gas generator valve failed, necessitating an emergency shutdown that inadvertently led to turbine overspeed and complete failure of the engine's turbine pump assembly. In response, the valve underwent a redesign process. Recognizing the uncertainty of its success, engineers initiated a parallel program to develop an alternative solenoid-operated valve configuration to enhance engine reliability and simplify its operation.

Despite setbacks, intensive development efforts culminated in the selection of the solenoid valve type during a pivotal meeting at Bell Aerosystems on November 15th. This decision marked a significant advancement in engine technology, albeit at the cost of delaying Preliminary Flight Rating Tests initially scheduled for September 1963.

At Ames Research Center, wind tunnel tests of a half-scale paraglider test vehicle commenced on July 18th, supporting the Paraglider Landing System Program. These tests evaluated longitudinal and lateral aerodynamic characteristics and the static deployment properties of a new low-lobe wing configuration jointly developed by North American Aviation and NASA. Initial findings indicated potential lateral stability issues, prompting meticulous data collection until the tests concluded on August 8th.

July 20: Fuel Cell Development Challenges

On July 20th, the Gemini Project Office reported delays in the development of fuel cell technology attributed to vendor part rejections, extended delivery schedules, and delayed determination of production procedures. While the full extent of the delay remained under assessment, these setbacks highlighted the intricate logistical challenges inherent in advancing critical spacecraft technologies.

Beginning July 31st, Electronic-Electrical Interference (EEI) tests of the Gemini launch vehicle (GLV) 1 commenced at Martin-Baltimore's vertical test facility. These tests aimed to identify and mitigate interference issues between GLV electrical and electronic systems. Initial tests revealed unacceptable interference from five systems, persisting until modifications were implemented, culminating in resolving all issues by September 5th by adding filters and grounds to both aerospace ground equipment and airborne circuits.

On August 1st, 1963, a pivotal Design Engineering Inspection marked a significant milestone for the Paraglider Landing System Program at North American's Space and Information Systems Division. This inspection focused on the full-scale test vehicle (FSTV) and its associated wing and hardware, a crucial component of contract NAS 9-1484 aimed at advancing paraglider technology.

The paraglider full-scale test vehicle in the Design Engineering Inspection briefing room at North American. (NASA Photo S-63-20931, undated.)

Originally intended to serve dual purposes—developing wing deployment systems and evaluating flight performance and control characteristics—the FSTV underwent critical revisions during the inspection. To streamline operations and reduce costs, the decision was made to prioritize system development over comprehensive flight evaluations. As a result, the objective of assessing glide performance was phased out.

The inspection yielded 30 alteration requests, predominantly mandatory in nature. These underlined the rigorous standards and meticulous attention to detail required to refine and enhance the paraglider's operational capabilities.

On August 1st, 1963, a pivotal Design Engineering Inspection marked a significant milestone for the Paraglider Landing System Program at North American's Space and Information Systems Division. This inspection focused on the full-scale test vehicle (FSTV) and its associated wing and hardware, a crucial component of contract NAS 9-1484 aimed at advancing paraglider technology.

Originally intended to serve dual purposes—developing wing deployment systems and evaluating flight performance and control characteristics—the FSTV underwent critical revisions during the inspection. To streamline operations and reduce costs, the decision was to prioritize system development over comprehensive

flight evaluations. As a result, the objective of assessing glide performance was phased out.

The inspection yielded 30 mandatory alteration requests, underscoring the rigorous standards and meticulous attention to detail required to refine and enhance the paraglider's operational capabilities.

Desert Survival Training

On August 5th, 1963, a comprehensive desert survival course commenced at Stead Air Force Base, Nevada, heralding a critical phase in the preparation of new flight crew members and two Mercury astronauts for upcoming Gemini missions. Tailored specifically for the unique challenges of Gemini missions, the course unfolded over five intensive days, structured into three distinct phases.

Astronauts after a training session in desert near Stead Air Force Base, Nevada. Front row, left to right: Frank Borman, James A. Lovell, Jr., John W. Young, Charles Conrad, Jr., James A. McDivitt, Edward H. White II. Back row, left to right: Raymond G. Zedekar (Astronaut Training Officer), Thomas P. Stafford, Donald K. Slayton, Neil A. Armstrong, and Elliot M. See, Jr. (NASA Photo No. 63-Astronauts-135, released Aug. 16, 1963.)

The initial phase spanned one and a half days of academic sessions, focusing on the characteristics of global desert environments and essential survival techniques. This theoretical groundwork was followed by a full day of practical field demonstrations, during which participants learned hands-on skills in utilizing survival equipment and harnessing parachutes to construct clothing, shelters, and distress signals.

The culmination of the course involved two days of rigorous remote site training.

Here, two-man teams were isolated in the desert, tasked with applying their acquired knowledge and skills under real-world conditions—a critical test of their readiness to handle unforeseen challenges in the harsh desert terrain.

Following a month-long delay in resolving technical issues with the parachute tucking problem, qualification testing of the Gemini parachute recovery system resumed on August 8th over the Salton Sea Range, California. This marked the sixth test in the series, with the subsequent test on August 20th differing from previous land-impact tests by focusing on water-impact scenarios.

These tests successfully validated the system's ability to achieve low accelerations during water landings, ensuring astronauts' safety. However, further qualification testing was halted on September 3rd following a decision to incorporate a high-altitude stabilization parachute into the recovery system, underscoring ongoing efforts to enhance safety and reliability in preparation for manned Gemini missions.

On August 9th, 1963, representatives from the Manned Spacecraft Center (MSC), Arnold Engineering Development Center, McDonnell, and Thiokol convened to address critical issues with the retrorocket abort system for the Gemini spacecraft. Recent tests at Arnold had revealed failures in components such as retrorocket nozzle exit cones and their mounting structures, primarily due to deficiencies in their design for joining and retaining the nozzle throat and exit cones.

In response, MSC and McDonnell opted to halt development testing of the current nozzle assembly and initiate a comprehensive redesign effort. Thiokol conducted preliminary tests on the redesigned assembly from September 18-20. Subsequent full-scale tests at Arnold on October 4th validated the structural integrity of the redesigned assembly, confirming its operational reliability without any malfunctions.

Water impact test of the Gemini parachute recovery system in the Salton Sea, California. (Northrop Ventura Photo 0748-65-33328, undated.)

Beginning August 20th, Rocketdyne embarked on a rigorous series of tests to validate its new thrust chamber assembly (TCA) design, integral to both the reentry control system (RCS) and the orbit attitude and maneuver system (OAMS) of the Gemini spacecraft. The test plan verified each type of TCA—25-pound RCS, 25-pound OAMS, 85-pound OAMS, and 100-pound OAMS—across mission duty cycles, steady-state life, limited environmental exposure, and performance metrics.

Rocketdyne submitted its design verification test schedule to McDonnell and the Gemini Project Office on August 27th, optimistically projecting that all 16 tests would conclude by September 10th. However, due to unforeseen complexities, the design verification testing extended beyond schedule, reaching completion only in October.

On August 21st, 1963, the Titan II development flight N-24 launched from the Atlantic Missile Range, marking the inaugural test in the Gemini malfunction detection system (MDS) piggyback series. However, just 81 seconds after liftoff, all MDS parameters were lost due to a short circuit, highlighting an early setback. Subsequent flights in the series—N-25 on November 1st, N-29 on December 12th, N-31 on January 15th, 1964, and N-33 on March 23rd, 1964—proved more successful, verifying the Gemini MDS's performance under actual flight conditions and engine operations, despite minor instrumentation issues during N-25.

The Manned Spacecraft Center (MSC) released a work statement on August 21st to procure eight Atlas launch vehicles crucial for the Gemini program. Following a defense purchase request on August 28th, initial obligations amounted to $1.4 million, with a projected final cost of $40 million. The Atlas launch vehicles, like others in the Gemini program, were acquired through the Air Force Space Systems Division, underscoring collaborative efforts in advancing manned space exploration.

McDonnell reported on August 24th that spacecraft No. 2 faced delays of approximately one month, primarily due to late deliveries of critical onboard systems from vendors. Components such as the orbit attitude and maneuver system, reentry control system, fuel cells, and cryogenic storage tanks had also encountered issues during vibration qualification, necessitating modifications. Despite these setbacks, the Development Engineering Inspection of spacecraft No. 2 was rescheduled for February 1964 after initial delays pushed it from October 1963.

On August 25th, McDonnell completed the fabrication and assembly of spacecraft No. 1, marking a significant milestone in the Gemini program. Phase II of Spacecraft Systems Tests (SST) commenced promptly, encompassing comprehensive tests on the complete launch configuration, including adapters. This phase alternated with final manufacturing cleanup over subsequent weeks, culminating in vibration testing from September 17-20, Altitude Chamber Tests from September 21-23, and an Integrated Systems Test on September 30th. The spacecraft passed its final roll-out inspection on October 1st before being shipped to the Atlantic Missile Range on October 4th.

On August 31st, the Gemini Project Office (GPO) explored innovative landing systems for the spacecraft, specifically investigating a parasail and landing rocket system. This system aimed to enable controlled land landings, utilizing components such as the parasail for gliding and turning and landing rockets to adjust descent rates to between 8 and 11 feet per second before touchdown. Despite initial studies by McDonnell and ongoing research and development by the Landing and Impact System Section, NASA Headquarters opted not to proceed further with the parasail system following a briefing on September 6th.

Also, on August 31st, GPO reported plans to resume systems testing for the orbit attitude and maneuver system (OAMS) and reentry control system (RCS) in early October. Initially commenced in August 1962 but halted due to thrust chamber unavailability, these tests were critical in validating various operational parameters and performance under extreme conditions. However, despite plans, actual resumption of these tests did not occur until May 1964, underscoring ongoing challenges in system readiness and integration.

Throughout August, the Gemini Pyrotechnic Ad Hoc Committee finalized its report, prompted by a spacecraft design review in June. The committee, led by Russell E. Clickner of the Mercury Project Office

(MPO), recommended significant modifications to enhance pyrotechnic circuitry, redundancy, system design, and qualification testing. These enhancements aimed to bolster safety and reliability in critical spacecraft operations, reflecting ongoing efforts to refine Gemini's technological capabilities amidst the demands of early space exploration.

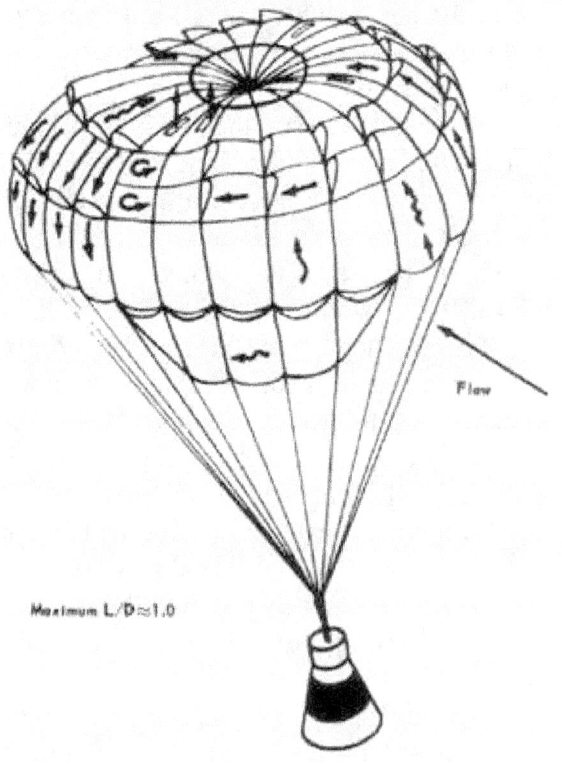

Sketch of the parasail landing system proposed for the Gemini spacecraft. (NASA Photo S-64-481, undated.)

On September 3rd, 1963, the Gemini Project Office (GPO) established a Mission Planning Coordination Group to oversee monthly activities related to operations, network guidance and control, and trajectories and orbits for the Gemini missions. This group aimed to ensure coordination among various elements at the Manned Spacecraft Center actively involved in mission planning. Their inaugural meeting on September 9th focused on reviewing Gemini mission planning documentation, the mission plan for Gemini-Titan (GT) 1,

MISTRAM (missile tracking and measurement system) requirements utilizing the J-1 computer and outlining mission objectives and tests for GT-2 and GT-3.

Also on September 3rd, GPO decided to suspend qualification testing of the parachute recovery system to incorporate a drogue parachute. This addition aimed to stabilize the spacecraft during the final reentry phase, specifically at altitudes ranging from 50,000 to 10,000 feet. Initially intended for the reentry control system (RCS), which was experiencing development challenges, the revised design would also facilitate dumping RCS propellants before deploying the main recovery parachute.

GPO outlined a three-phase drop test program to develop and qualify the revised recovery system:

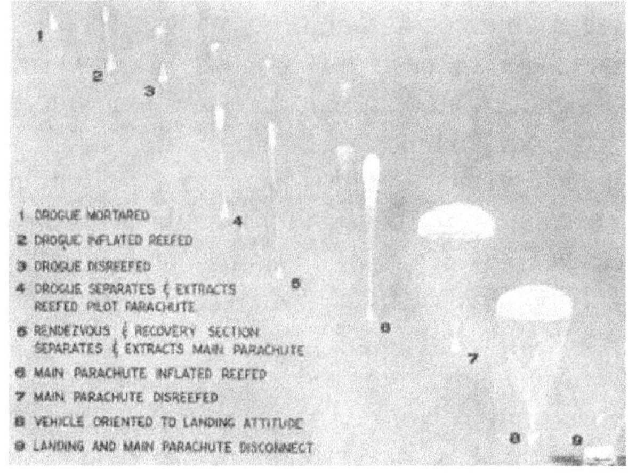

The sequence of events in the operation of the Gemini parachute recovery system incorporating the drogue chute. (Northrop Ventura Photo 0748-94-38242, undated.)

Phase I (January-February 1964): Utilizing boilerplate No. 5 as a test vehicle, this phase would focus on deploying the pilot parachute through the stabilization chute. The sequence involved deploying the stabilization chute at 50,000 feet, releasing it at 10,600 feet, and using a lanyard to deploy the pilot chute 2.5 seconds later. The rendezvous and recovery (R&R) section would then separate from the spacecraft to allow deployment of the main chute.

Phase II (March-August 1964): Using a parachute test vehicle (an instrumented weight bomb), this phase aimed to complete the development of the stabilization chute.

Phase III (June-October 1964): This phase would qualify the entire recovery system using static article No. 7, a boilerplate pressure vessel and heatshield equipped with production RCS and R&R sections. The qualification tests were scheduled to conclude before the third Gemini mission, ensuring system readiness for operational deployment.

Additionally, to maintain continuity, the existing parachute recovery system would be qualified, with three more drops scheduled for February and March 1964, using static article No. 7 as the test vehicle before transitioning to Phase III testing.

On September 4, Manned Spacecraft Center's Instrumentation and Electronics Systems Division representatives convened with McDonnell to synchronize efforts for the Gemini radar program. With the Gemini Project Office pushing for expedited progress, the focus was on achieving operational status for the rendezvous radar system.

On September 5, Lockheed's Gemini Agena target vehicle (GATV) contract underwent revisions. Initially scheduled for launch within seven-and-a-half months, the deadline relaxation prompted Lockheed to opt for the enhanced AD-62 block of Agena vehicles instead of the AD-13. Subsequently, a shift to the AD-71 block was mandated in January 1964, causing delays—two months for the AD-62 and an additional five weeks for the AD-71. These adjustments consumed much of the allotted contingency time, intensifying concerns over potential launch delays.

On September 6, the Department of Defense sanctioned the Titan II Augmented Engine Improvement Program, marking a pivotal development. Aerojet-General secured a contract on November 15 to overhaul engine components, aiming to rectify inherent design weaknesses and enhance overall reliability through comprehensive redesign rather than incremental fixes. Although the primary objective wasn't fully realized, the initiative yielded secondary benefits such as addressing minor design flaws, refining welding techniques, and optimizing assembly procedures.

On the same day, the inaugural Combined Systems Acceptance Test (CSAT) of Gemini launch vehicle No. 1 occurred at Martin-Baltimore's vertical test facility. Preceded by preliminary dry runs and Electronic-Electrical Interference (EEI) Tests in August, the formal CSAT on September 6 simulated a complete launch countdown, engine start, liftoff, and flight up to stage II engine shutdown. Both primary and backup guidance and control systems were rigorously evaluated. Martin engineers meticulously analyzed telemetry and ground equipment data, presenting the vehicle for final acceptance to the Air Force Space Systems Division/Aerospace Vehicle Acceptance Team on September 11. This milestone underscored a critical step forward in ensuring the readiness and reliability of the Gemini launch vehicle.

On September 8, sixteen astronauts embarked on intensive training in water and land parachute landing techniques, marking a pivotal phase in preparing for potential emergencies during Project Gemini missions. This specialized training was crucial, particularly for scenarios where a low-level abort (below 70,000 feet) might necessitate the pilot's ejection from the spacecraft, followed by a personal parachute descent.

To simulate these critical procedures, astronauts underwent rigorous exercises employing a towed 24-foot diameter parasail. This equipment lifted them to altitudes reaching up to 400 feet before releasing the towline, allowing the astronaut to glide down and execute a controlled landing. These training efforts were essential to ensure the crew's preparedness and safety during mission contingencies during the Gemini program.

During September 11-12, responding to the Gemini Project Office's directive to achieve operational status for the Gemini rendezvous radar system, personnel from the Manned Spacecraft Center's Instrumentation and Electronics System Division met with Westinghouse in Baltimore for a critical review of the test program. Despite Westinghouse completing its radio frequency anechoic chamber test, anomalies persisted that couldn't definitively be traced to the radar system due to potential chamber reflections. To resolve this, plans were made for an outdoor range test to assess the chamber's suitability for radar testing under more realistic conditions.

From September 11 to 20, the vehicle acceptance team overseeing Gemini launch vehicle (GLV) 1 undertook an exhaustive inspection, scrutinizing its manufacturing and testing history with a focus on the results of the Combined Systems Acceptance Test (CSAT) conducted on September 6. Regrettably, GLV-1 was deemed unacceptable primarily due to severe contamination of its electrical connectors. Moreover, the proper documentation of several major components' qualifications had not been adequately maintained. Subsequently, from September 21 to 29, Martin engineers meticulously inspected all 350 electrical connectors on GLV-1, identifying 180 that required cleaning or replacement. Similar inspections and maintenance were also conducted on GLV-2. This extensive reevaluation necessitated the rerunning of subsystem tests and CSAT. The preliminary CSAT was successfully completed on October 2, followed by the final CSAT on October 4.

On September 14, the Gemini Project Office reported an approximately three-week setback in the battery qualification program. McDonnell dispatched a team to investigate issues related to high-porosity welds in titanium battery cases, a critical concern uncovered during prequalification vibration tests. Although the batteries didn't fail electrically, they exhibited excessive vibration, potentially attributable to a low amplification factor that could be mitigated through potting measures. These challenges underscored the meticulous testing and troubleshooting required to ensure the reliability and performance of critical systems in the Gemini spacecraft.

On September 23, a comprehensive technical development plan for Department of Defense experiments slated for Gemini missions was issued. The plan outline 13 Air Force and nine Navy experiments at an estimated cost of $22 million. The Manned Spacecraft Center meticulously reviewed these experiments for feasibility during the planning phase. However, their inclusion in Gemini flights remained tentative pending further technical refinement of the experiments and clarification of spacecraft weight and volume constraints.

On September 27, Electro-Mechanical Research achieved a significant milestone by successfully testing the compatibility of airborne and ground station PCM (pulse code modulated) telemetry equipment. These tests confirmed that the Gemini spacecraft and Agena used telemetry formats fully compatible with NASA's ground stations, ensuring seamless data transmission and recording capabilities during missions.

That same day, a Development Engineering Inspection was conducted at North American's Space and Information Systems Division for the tow test vehicles (TTVs) as part of the Paraglider Landing System Program. These manned vehicles were equipped with predeployed wings to evaluate flight performance and control, focusing on landing maneuvers. The inspection identified 33 areas requiring alterations, with 24 deemed mandatory to enhance vehicle functionality and safety.

On the same day, North American discontinued efforts to retrofit Gemini prototype paraglider deployment hardware onto the full-scale test vehicle (FSTV). Initially part of the Paraglider Landing System Program's contract, the decision to halt retrofitting was reached during discussions between North American and NASA, culminating in its formal removal from contractual obligations on November 7 via Change Notice No. 5 to Contract NAS 9-1484.

On September 30, the Manned Spacecraft Center awarded its inaugural incentive-type contract to Ling-Temco-Vought, Inc. of Dallas, Texas. This contract aimed at fabricating a trainer integral to the Gemini launch vehicle training program. Structured as a fixed-price-incentive-fee contract with a target cost of $90,000 and a ceiling of $105,000, it included a profit-sharing arrangement. Ling-Temco-Vought would either pay 20% of savings under the target cost or receive 20% of cost savings, aligning financial incentives with efficient project execution.

Additionally, on September 30, the Air Force Space Systems Division contracted Aerojet-General to develop a backup solution for the injectors of the Gemini launch vehicle's second-stage engine. Driven by concerns over incipient combustion instability observed during Titan II development flights, the Gemini Stability Improvement Program focused on redesigning the stage II engine injector for enhanced reliability. Commencing as a contingency plan, this 18-month effort evolved to maximize mission success probability, culminating in a thoroughly redesigned injector system by December 24, 1963.

On October 1, the Gemini Project Office (GPO) initiated a design study tasking McDonnell with assessing the feasibility and configuration requirements for utilizing batteries instead of fuel cells in spacecraft designated for two-day rendezvous missions. This decision followed GPO personnel's visit to General Electric to evaluate experimental results to determine the theoretical operational lifespan of fuel cells for powering Gemini spacecraft. Initial tests indicated a lifespan of approximately 600 hours. However, modifications to the spacecraft's coolant system inadvertently raised operating temperatures, reducing fuel cell longevity to between 150 and 250 hours—a significant setback. Until methods for enhancing fuel cell durability could be implemented, this issue remained a critical challenge in the development program.

On the same day, the Gemini Project Office compiled an abstract outlining flight qualification requirement for experimental equipment slated for Gemini missions. This document provided a succinct overview of essential environmental criteria impacting the design, fabrication, and installation of experimental gear destined for spacecraft deployment.

Installation of right ballast seat and instrumentation pallet in Gemini spacecraft No. 1. (NASA-USAF Photo 63-13025, Dec. 7, 1963.)

On October 4, Gemini spacecraft No. 1 arrived at the Atlantic Missile Range and was transferred to Hangar AF. Following a receiving inspection on October 7 and a Voltage Standing Wave Ratio Test on

October 8, its instrument pallets were removed for laboratory testing and checkout procedures commencing on October 9. The spacecraft performed meticulous weight, balance, and overall functionality checks during this phase. The instrument pallets were reinstalled on November 26, initiating further assessments of individual and integrated systems, including communications, instrumentation, and environmental controls. The final phase of industrial area testing culminated in a confidence level test conducted on February 12, 1964, marking a critical milestone in affirming the readiness and reliability of Gemini spacecraft No. 1 for upcoming missions.

On October 8, Martin-Baltimore completed its evaluation of the data from the second Combined Systems Acceptance Test of Gemini Launch Vehicle (GLV) 1, deeming it acceptable. The GLV-1 Vehicle Acceptance Team (VAT) inspected the vehicle and decided to ship it to the Atlantic Missile Range (AMR) on October 12. Despite lacking flight-qualified components, relocating GLV-1 to AMR allowed for early compatibility checks and the final acceptance of complex 19. The vehicle underwent additional tests, including leak checks and inspections, before being formally accepted by the Air Force Space Systems Division on October 25 and airlifted to AMR the next day.

Simultaneously, on October 14, North American completed the first full-scale prototype paraglider wing for the Paraglider Landing System Program. The wing was sent to Ames Research Center for wind tunnel tests to assess its aerodynamic characteristics, structural integrity, and limits. Initial testing concluded on October 28, providing limited data. A subsequent round of testing from December 4 to 9 proved more fruitful, successfully meeting all test objectives.

During the same period, on October 14, the Mission Planning Coordination Group deliberated on the feasibility of conducting rendezvous maneuvers at the first apogee. Proposed by Richard R. Carley of the Gemini Project Office, this capability was considered as a test objective for all rendezvous flights moving forward.

On October 15, a critical meeting convened in Los Angeles involving personnel from the Air Force Space Systems Division (SSD), Air Force Ballistic Systems Division (BSD), and Titan II contractors. Discussions centered on integrating Gemini launch vehicle fixes into Titan II development flights. BSD initially paused these integrations due to the limited remaining flights to qualify the missile. However, under the directive of General Bernard A. Schriever, Commander of Air Force Systems Command, efforts resumed on November 1 with the flight of Titan II N-25, marking a renewed push to address launch vehicle issues effectively.

Chapter 6 - New Astronauts and Technical Challenges

On October 18, officials from the Manned Spacecraft Center (MSC) in Houston proudly introduced fourteen new astronauts, expanding NASA's cadre to 30 individuals dedicated to training for space exploration. This diverse group included seven Air Force volunteers: Major Edwin E. Aldrin, Jr.; Captains William A. Anders, Donn F. Eisele, Charles A. Bassett II, Theodore C. Freeman, David R. Scott, and Michael Collins. From the Navy, Lieutenant Commander Richard F. Gordon, Jr., and Lieutenants Eugene A. Cernan, Alan L. Bean, and Roger B. Chaffee joined alongside Marine Captain Clifton C. Williams, Jr. The civilian contingent comprised R. Walter Cunningham and Russell L. Schweickart. Selected from a pool of approximately 500 military and 225 civilian applicants, these new astronauts began training at MSC on February 2, 1964.

On October 21, Rocketdyne achieved a significant milestone three days later by successfully test-firing an orbit attitude and maneuver system (OAMS) 85-pound thruster to meet a new mission duty cycle requirement. This development prompted McDonnell to reassess OAMS mission duty cycles, aiming to extend the operational life of thruster components. Concurrently, Rocketdyne paused testing to analyze small ablative rocket engine performance, while McDonnell revised duty cycle specifications. By January 1964, NASA, McDonnell, and Rocketdyne collaboratively clarified enhanced life requirements for OAMS engines, setting new standards for performance under rigorous mission conditions.

On October 25, North American completed modifications on the Advanced Paraglider Trainer, transforming it into a full-scale tow test vehicle (TTV) for the Paraglider Landing System Program. This vehicle underwent ground tow tests at Edwards Air Force Base starting December 28, with initial testing concluding on January 14, 1964. A second TTV was prepared and delivered to Edwards by February 14, followed by extensive ground tow tests through June and the installation of flight-worthy control systems beginning in April.

October 26 marked a pivotal moment as Gemini launch vehicle 1 arrived at the Atlantic Missile Range and was assembled at complex 19. Stage I and Stage II were erected and cabled together in preparation for the Sequence Compatibility Firing scheduled for mid-December. Electronic-Electrical Interference Tests were completed by early November, with full power applied to the vehicle by November 13.

Closing the month, on October 30, a critical meeting addressed challenges with the ejection seat system, focusing on improving the ballute—a device crucial for stabilizing astronauts during ejection and descent. Wind tunnel data highlighted issues with supersonic performance and deployment at subsonic speeds, prompting design enhancements. Additionally, discussions led to recommendations for automatic separation mechanisms between the seat backboard and egress kit, a proposal under review by McDonnell for implementation.

On November 1, a pivotal milestone was reached in developing the Titan II rocket at the Atlantic Missile Range. Known as flight N-25, this mission carried crucial innovations to mitigate longitudinal vibrations that had plagued earlier flights. Once exceeding 0.25g, these vibrations posed a significant challenge to NASA's ambitions for manned space flight. However, with meticulous engineering, flight N-25 achieved a groundbreaking reduction to just 0.22g, marking the first instance where these vibrations fell within acceptable limits.

The key to this success lay in integrating an oxidizer surge chamber and fuel accumulator kit. Although tested initially on a single flight, the Gemini Project Office swiftly gained confidence in its efficacy. By November 6, their decision was resolute: several more kits were ordered for installation in subsequent Gemini launch vehicles. This confidence was reaffirmed by later Titan II development flights, notably N-29 on December 12, 1963, and N-31 on January 15, 1964, and the historic flight of Gemini-Titan 1.

On November 5, McDonnell Aerospace undertook a critical review of advancements in beryllium shingle technology, aimed at safeguarding the Gemini spacecraft's reentry control system and rendezvous and recovery structures from intense reentry heat. Earlier setbacks, compounded by manufacturing complexities, had hindered progress in shingle testing throughout the year.

The primary obstacles in manufacturing the cross-roll beryllium shingles included the final product's flaking, lamination, and cracking flaws. To address these challenges decisively, a pivotal meeting convened at Pioneer Astro Industries in Chicago, Illinois, on November 14, 1963. Engineers and project leaders deliberated strategies to enhance manufacturing reliability during this session.

A strategic decision emerged from the meeting: where feasible, chemical etching would replace traditional machine tooling methods, aiming to mitigate the inherent flaws associated with mechanical processes. Additionally, when machine tooling remained indispensable, lighter cuts would be implemented to minimize the risk of manufacturing defects.

On November 7, Major General Leighton I. Davis, the Department of Defense (DOD) Representative for Project Gemini Support Operations, unveiled DOD's comprehensive plan for facilitating Gemini missions. In his pivotal role as the liaison between the DOD and NASA, General Davis assumed responsibility for coordinating the DOD's support across various critical areas.

Central to his mandate was ensuring seamless collaboration with NASA to meet operational requirements. This encompassed launch logistics provisions, extensive tracking network management, planned and emergency recovery operations, robust communications infrastructure, public affairs management, and medical support.

General Davis's strategic oversight marked a pivotal phase in integrating military resources and expertise into the civilian-led Gemini program. His directive aimed to enhance mission success and underscore the cooperative spirit between NASA and the DOD, which was essential for navigating the complexities of manned space exploration during the Gemini era.

On November 12, concerns over delays in the fuel cell development program prompted the Gemini Project Office to take decisive action. They directed McDonnell to modify the electrical system for spacecraft No. 3, allowing for flexibility in power sources post-delivery to Cape Canaveral. This strategic decision aimed to ensure that fuel cells or a silver-zinc battery power system could seamlessly integrate as needed.

Recognizing the urgency of this adaptation, a contract change reflecting these modifications was swiftly issued on January 20, 1964. This adjustment safeguarded against potential setbacks caused by the ongoing fuel cell challenges and underscored the Gemini Project's commitment to maintaining schedule integrity and mission readiness.

On November 13, the Gemini Management Panel convened to assess the progress of spacecraft and launch vehicle readiness, leading to a critical decision to reevaluate the Gemini launch schedules. Aiming to optimize testing procedures at Cape Canaveral and enhance confidence in mission success, the panel directed Gemini Project Manager Charles W. Mathews and Colonel Richard C. Dineen, Chief of the

Gemini Launch Vehicle at the Air Force Space Systems Division, to form an ad hoc group. This group was tasked with conducting an intensive 30-day study of work plans and schedules with a clear objective: achieving manned flight in 1964.

The following day, on November 24, key stakeholders from NASA, the Air Force, and industry convened at Cape Canaveral to outline study areas and establish a collaborative framework. Over subsequent 10-day intervals, they meticulously developed ground rules, reviewed progress, and coordinated efforts to streamline operations.

Mathews subsequently presented the study's outcomes at the panel meeting on December 13. He emphasized crucial ground rules to advance Gemini-Titan (GT) 3, slated as the first manned flight in 1964. Central to these efforts was a strategic focus on reducing redundant tests conducted at Cape Canaveral that had already been completed at McDonnell while integrating the overall testing regimen.

The resultant master schedule, derived from this rigorous study, outlined the launch timeline: GT-1 on March 17, 1964; GT-2 on August 11; and GT-3 on November 6. Additionally, GT-1A was designated strictly as a backup, to be deployed only in the event of GT-1's failure.

On November 14, the Manned Spacecraft Center (MSC) initiated a pioneering drop-test program over Galveston Bay, employing a helicopter-towed paraglider half-scale tow test vehicle. This ambitious endeavor aimed to meticulously study trim conditions and stability characteristics across various deployment configurations.

The inaugural drop successfully validated the U-shaped deployment configuration, marking a promising start to the experimental series. However, challenges emerged swiftly. The second test, conducted on November 19, was regrettably abortive, resulting in minor damage. Undeterred, MSC pressed forward with the third test on November 26, which, unfortunately, resulted in irreparable damage to the wing upon impact.

In response, MSC swiftly secured a replacement wing from North American and resumed testing. The fourth test, conducted on December 19, achieved partial success despite earlier setbacks. MSC concluded the experimental phase after this test, opting not to pursue further trials.

On November 15, a significant milestone was achieved in the Gemini program by delivering the first production version of the inertial guidance system to McDonnell. This advanced system represented a critical advancement in spacecraft navigation, essential for ensuring precise control and orientation during missions.

The development phase culminated in the delivery of the production unit, marking a pivotal step towards integrating cutting-edge technology into the Gemini spacecraft. Specialized tests on the configuration test unit, utilizing guidance and control equipment from spacecraft No. 2, were slated for completion by January 1964. These tests aimed to validate the system's reliability and performance under simulated mission conditions.

On November 16, a pivotal agreement was reached between the Flight Crew Support Division and the Flight Operations Division concerning the flight profile and a rendezvous evaluation experiment for the upcoming Gemini-Titan 4 mission. The primary objective of this experiment was to simulate a standard Agena/Gemini rendezvous procedure and to replicate part of the maneuver using a loss of signal/manual technique.

Central to the mission's strategy was the adoption of circular phasing and catch-up orbits, proposed by the Flight Crew Support Division. These techniques aimed to refine and optimize rendezvous procedures crucial for future space missions.

As preparations advanced, detailed assessments of exact fuel requirements and ground tracking logistics were underway within the Flight Operations Division. These studies were essential for ensuring the feasibility and precision of the planned maneuvers, laying the groundwork for successfully executing the Gemini-Titan 4 mission.

Starting on November 17, Douglas Aircraft Corporation, based in Tulsa, Oklahoma, embarked on a crucial series of tests to validate the structural integrity of the Gemini Target Docking Adapter (TDA) during shroud separation. This critical component, integral to the successful docking maneuvers of the Gemini spacecraft with the Agena target vehicle, underwent rigorous evaluation under simulated altitude conditions.

Jump test of the 36-inch ballute with dual suspension at the Naval Parachute Facility, El Centro, California. The second figure is a free-falling photographer with a camera mounted in his helmet. A second observer jumped later and took this picture. (NASA Photo 64-Gemini-120, released Dec. 18, 1963.)

The primary focus of these tests was to ensure the flawless operation of pyrotechnic devices responsible for shroud separation and to verify adequate clearance between the shroud and the TDA. The testing regimen, spanning several days, culminated successfully on November 21.

Results from these tests unequivocally demonstrated the TDA's compatibility with the shroud system under operational conditions. Importantly, there were no indications of structural damage or failure within the TDA, affirming its robustness and reliability for upcoming Gemini missions.

On November 22, a pivotal phase began in developing the Gemini escape system with the initiation of a series of 24 test drops aimed at refining the ballute stabilization system.

These tests commenced with a live jump over El Centro, marking the beginning of rigorous evaluations crucial for enhancing astronaut safety during emergency egress scenarios.

Initially employing a three-foot diameter ballute, subsequent live jumps, and dummy drops—totaling five and four, respectively, culminating on January 9, 1964—revealed excessive rotation rates. Engineers progressively increased the ballute diameter in response and transitioned from single-point to two-point suspension configurations.

Between January 14 and February 4, 14 tests—comprising 12 human and two dummy drops—were executed at altitudes ranging from 12,500 to 35,000 feet. These tests proved instrumental in establishing the 48-inch diameter ballute as optimal for stability and performance within the Gemini context.

Armed with conclusive findings, the Gemini Project Office promptly directed McDonnell to adopt the 48-inch diameter ballute for the upcoming qualification drop test program. This comprehensive program included structural integrity tests in the wind tunnel at Arnold Engineering Development Center, ensuring the ballute's readiness for real-world emergency scenarios.

On November 25, the Manned Spacecraft Center (MSC) received proposals for developing the Gemini Extravehicular Life Support Package, marking a crucial step in enhancing astronaut safety and operational capabilities during extravehicular activities (EVAs). This initiative, prompted by requests for proposals issued in October, aimed to procure a robust life support system capable of sustaining astronauts during initial EVAs using a hardline tether for 10 to 15 minutes.

The proposed system included essential components such as a high-pressure gaseous oxygen supply bottle and regulators and valves designed to control oxygen flow within an open-loop system. This setup was crucial for maintaining vital life support functions in the challenging space environment.

MSC expedited the evaluation process, aiming to complete assessments by the end of December to advance toward system integration and testing swiftly. In January 1964, following a thorough evaluation, the Garrett Corporation was awarded the contract to develop and manufacture the Gemini Extravehicular Life Support Package.

This proactive approach underscored the Gemini program's commitment to advancing technological solutions tailored for the unique demands of space exploration. The development of this life support system represented a critical milestone in enabling safe and effective extravehicular activities, setting the stage for future missions, and enhancing the overall capabilities of manned spaceflight during the Gemini era.

By the end of November, the Gemini Project Office (GPO) faced significant challenges in testing and developing the Orbit Attitude and Maneuver System (OAMS) at Rocketdyne. The research and development phase of testing OAMS components was expected to extend into 1964, primarily due to ongoing issues with the thrust chamber assembly (TCA). The development of a reliable TCA remained a major hurdle, compounded by uncertainties in hardware availability and complications arising from McDonnell's revision of mission duty cycles.

Moreover, delays in hardware delivery were impeding system testing, which was now projected to conclude no earlier than the second quarter of 1964. These persistent setbacks were, in turn, causing serious delays in the qualification test program, crucial for meeting the planned manned Gemini launch schedule for 1964.

To mitigate these delays, GPO contemplated the unconventional step of commencing qualification tests before the completion of development testing, highlighting the urgency and strategic flexibility required to keep the Gemini program on track amidst technical challenges.

November (Throughout the Month): Gemini Agena Target Vehicle (GATV) Schedule

Throughout November, Lockheed incorporated a milestone schedule for the Gemini Agena Target Vehicle (GATV) in its monthly progress reports. This schedule adjustment followed the revised Gemini flight program announced on April 29, necessitating corresponding modifications to the Agena program.

Key milestones for the first GATV were now slated to occur five to six months later than originally scheduled in January 1963. Engineering development completion was rescheduled for May 15, 1964, with modification and final assembly planned for June 12 instead of January 10, 1964. Preliminary vehicle systems testing was shifted to September 11, 1964, from the earlier April 10 date.

Special tests, including a Radio Frequency Interference Test, were now scheduled to conclude on November 20, moving from May 22, 1964. Final Vehicle Systems Tests were adjusted to finish by December 18, 1964, with shipment expected on January 6, 1965, instead of June 30, 1964. Consequently, the launch of the first GATV was rescheduled to April 15, 1965, reflecting a delay of seven and a half months from the previously targeted date of September 1, 1964.

On December 3, the Gemini Program Planning Board issued a pivotal memorandum outlining plans to rectify several critical deficiencies identified in the Titan II launch vehicle, which was crucial for ensuring the success and safety of the Gemini missions.

A primary focus of the corrective measures was addressing longitudinal oscillations, commonly known as POGO (Pronounced Oscillation in Gravity Optimization), which had been observed during earlier Titan II flights. NASA and the Air Force collaborated on a comprehensive program to mitigate this effect, involving ground-proof tests of all subsystem modifications to control oscillations. Flight tests to validate these solutions were slated to precede their application to the Gemini launch vehicle, ensuring reliability during manned missions.

Additionally, efforts were directed toward enhancing the combustion stability of the Titan II engines. This included rigorous testing using artificially induced disturbances to demonstrate dynamic stability. Engine performance in flight conditions would be finalized on unmanned vehicles before human-rated missions, affirming their readiness and safety.

Furthermore, the program encompassed a thorough initiative to rectify design deficiencies that emerged during the Titan II development flights. This comprehensive engine improvement program aimed to bolster overall reliability by addressing underlying design flaws.

On December 9, McDonnell Aerospace delivered Gemini boilerplate No. 201 to Houston, marking a significant step in astronaut training for the Gemini program. An egress trainer boilerplate was essential for simulating emergency exit procedures from the spacecraft in various conditions.

With the delivery of boilerplate No. 201, preparations promptly commenced for egress tests scheduled to take place in January 1964. These tests were slated to occur in a specially designed water tank at Ellington Air Force Base, Texas. The tank provided a controlled environment mimicking the conditions astronauts would face during water landings, ensuring they were well-prepared to execute safe and efficient egress procedures.

On December 10, Aerojet-General achieved a significant milestone by delivering the Stage II engine for Gemini Launch Vehicle (GLV) 2 to Martin-Baltimore. This engine delivery marked a crucial step in the assembly and testing process for GLV-2.

The installation of the Stage II engine was completed swiftly on December 31. Shortly before, on December 29, an interim Stage I engine was also received and installed on January 9, 1964. This Stage I engine was designated for testing purposes at the Martin plant, serving as a precursor to installing a flight engine before GLV-2 was transported to Cape Canaveral.

GLV-2's horizontal testing concluded successfully by January 17, paving the way for further preparations. Before erecting GLV-2 in the vertical test facility, additional measures were taken to enhance its performance and safety. Specifically, a longitudinal oscillation (POGO) kit was installed in Stage I. This kit included an oxidizer standpipe and a fuel surge chamber meticulously designed to mitigate pressure pulses within the propellant feed lines. This reduced POGO oscillations to levels deemed safe for manned flight operations.

Crew Selection and Rearrangements in the Gemini Program

In late 1963, Crew Operations Director Deke Slayton made initial crew selections for upcoming Gemini missions. Shepard and Stafford were chosen for Gemini 3, McDivitt and White for Gemini 4, and Schirra and Young for Gemini 5, intended as the first Agena rendezvous mission. Grissom and Borman were assigned as backup crew for Gemini 3 and slated for Gemini 6, which was planned as the first long-duration mission. Conrad and Lovell were designated as backup crew for Gemini 4.

The crew assignments underwent several rearrangements due to health and compatibility concerns. Shepard's Ménière's disease led to Grissom assuming command of Gemini 3. Slayton then reassigned Stafford and Young, feeling Young was a better match for Grissom. Cooper was appointed to command Gemini 5, with Conrad moving to pilot and Borman to backup command. Armstrong and Elliot See were assigned as backup crew for Gemini 5.

Further changes occurred when Slayton reassigned See to command Gemini 9 due to concerns about his readiness for EVA. Scott was appointed pilot of Gemini 8, with Charles Bassett slated as pilot of Gemini 9. The final rearrangement followed the tragic deaths of See and Bassett in a jet crash. Stafford and Cernan were promoted to the prime crew of Gemini 9A, with Lovell and Aldrin moving to the backup crew.

These adjustments shaped the Gemini missions and influenced crew selections for the Apollo program. The deaths of Grissom, White, and Chaffee in the Apollo 1 fire further impacted crew assignments, paving the way for the first lunar landing missions.

Announced by the Manned Spacecraft Center, Elliot M. See, Jr., and Charles A. Bassett II were initially selected as command pilot and pilot for Gemini IX. Following their tragic accident, Stafford and Cernan were appointed as the new prime crew for Gemini 9A, with Lovell and Aldrin shifting to the backup crew, setting a course that would eventually lead Lovell and Aldrin to become the prime crew for Gemini 12.

Gemini Launch Vehicle GLV-3 Assembly

On December 13, Martin-Baltimore received a pivotal shipment: the propellant tanks for Gemini Launch Vehicle (GLV) 3, marking a significant milestone in the launch vehicle's assembly process.

The fabrication of these propellant tanks commenced at Martin-Denver in June, underscoring the meticulous planning and manufacturing timeline required for such critical components. Splicing of the oxidizer and fuel tanks for each stage was completed by April 17, 1964, setting the stage for further integration efforts.

Subsequent milestones included the arrival of flight engines from Aerojet-General on May 10 and their installation by June 6. This phase represented a critical step in preparing GLV-3 for comprehensive testing and validation.

Final horizontal tests of the fully assembled launch vehicle commenced on June 1 and concluded on June 17, culminating in an extensive Air Force inspection. This thorough evaluation ensured that GLV-3 met rigorous standards before its final phase: erection in the vertical test facility.

On December 17-18, a critical evaluation phase unfolded for the G2C training and qualification pressure suit, concurrent with a mock-up review of the spacecraft crew station at McDonnell. This comprehensive assessment ensured the pressure suit met rigorous standards for crew comfort, functionality, and spacecraft compatibility.

Overall, the G2C suit garnered positive feedback from the crew, affirming its acceptability and seamless integration with the spacecraft's operational environment. Notably, adjustments to the helmet design had been successfully implemented, resolving previous concerns without introducing new design issues.

Looking ahead, plans were set to deliver eleven G2C suits by the end of February 1964, including five tailored explicitly for astronauts. The remaining 23 suits were slated for delivery in March 1964, marking the commencement of qualification and reliability testing under the management of the Crew Systems Division at the Manned Spacecraft Center.

On December 20, McDonnell achieved a significant milestone by shipping its Gemini Mission Simulator No. 1 portion to Cape Kennedy. This simulator represented a critical tool in preparing astronauts for the complexities of Gemini missions, enhancing training through realistic mission scenarios.

The shipment included essential components of the simulator, laying the groundwork for its assembly and integration at Cape Kennedy. Anticipation was high for the arrival of the computers required for the training device, expected to be delivered by mid-January 1964. These computers would serve as the computational backbone of the simulator, simulating various mission parameters and scenarios to ensure astronauts were thoroughly prepared for spaceflight challenges.

On December 21, the Gemini Project Office (GPO) made significant decisions regarding the technological systems onboard the spacecraft:

GPO announced that spacecraft No. 3 would integrate a silver-zinc battery power system instead of the originally planned fuel cell system. This adjustment was necessitated by delays in qualifying the fuel cell system in time for the mission. By late January 1964, McDonnell reviewed the status of the fuel cell program and proposed integrating an improved fuel cell design into spacecraft No. 5, while opting to use the battery power system for spacecraft Nos. 3 and 4.

Following a flammability test conducted by McDonnell, the Gemini Project Office reported that Teflon-insulated wiring would be implemented throughout the spacecraft. This modification aimed to enhance overall safety by reducing fire risks and ensuring robust electrical integrity in the spacecraft's systems.

On December 23, persistent challenges in developing engines for the Gemini orbit attitude and maneuver system prompted a comprehensive review by the management of the Manned Spacecraft Center. Following extensive discussion, three strategic decisions were made to address these issues:

A directive was issued to explore the feasibility of further reducing the oxidizer-to-fuel ratio, currently set at 1.3:1. This investigation aimed to maintain stable combustion and starting characteristics while lowering operating temperatures. Such optimization was expected to enhance engine longevity, crucial for sustained mission operations.

An investigation was initiated to evaluate the possibility of realigning the lateral-firing thrusters more closely with the spacecraft's center of gravity. This adjustment was intended to reduce the strain on the 25-pound thrusters, which had not yet demonstrated a complete operational duty cycle without failure. The efficiency of maintaining spacecraft attitude during lateral maneuvers could be significantly enhanced by improving alignment.

Integration of Ablation Material Laminates: A decision was made to construct an engine billet incorporating ablation material laminates oriented approximately parallel to the motor housing. Initial tests of this parallel laminate material had shown promising results in addressing the thrusters' operational duty cycle challenges—this innovative approach aimed to optimize performance and reliability, crucial for the success of future Gemini missions.

Testing and Preparation for Launch

On December 31, Gemini Launch Vehicle 1 reached a significant milestone as its two stages, positioned side by side at Complex 19, completed the Combined Systems Test (CST). Initially scheduled for December 13, the CST was delayed due to the late completion of complex support systems needed to ensure operational compatibility with the launch vehicle. This comprehensive test validated the integration and functionality of both stages in preparation for the next critical phase, the Sequence Compatibility Firing (SCF).

The Wet Mock Simulated Flight for SCF was completed on January 7, 1964, marking another step towards readiness. However, the SCF planned for January 10 faced a setback and was halted at T-20, necessitating a reschedule to January 14 due to adverse cold weather conditions. Finally, overcoming these challenges, the SCF—a static firing involving both stage I and stage II engines—was successfully executed on January 21. This pivotal test confirmed the readiness and performance of the launch vehicle under simulated flight conditions.

Chapter 7 - Development and Qualification

January 1964 through December 1964

Adjustments in the Gemini Program Schedule

On January 1, 1964, NASA's headquarters issued a directive to the Gemini Project Office, significantly altering the trajectory of the Gemini-Titan (GT) missions 3 and 4. Specifically, initially slated for inclusion in these missions, the radar and rendezvous evaluation pod were now earmarked for later deployment. This decision was pivotal, as it paved the way for GT-4 to become a pioneering battery-powered long-duration flight within the Gemini program.

The reevaluation placed GT-5 in the spotlight, where the radar and rendezvous evaluation pod would find its new home. This adjustment inevitably introduced a ripple effect across the mission schedule, potentially delaying the first planned Agena flight. This strategic shift underscored NASA's meticulous planning and adaptability in navigating the complexities of space exploration during the Gemini era.

Development and Testing of Gemini Pressure Suits

In early January 1964, a crucial meeting convened between NASA's Crew Systems Division (CSD) representatives and the David Clark Company. The agenda: a meticulous review of the design for the G2C training and qualification pressure suit. This meeting marked a pivotal moment in the Gemini program's preparation phase, where several critical components awaited final approval before integration into the G3C flight suit configuration.

By January 17, the Crew Systems Division had diligently compiled a comprehensive statement of work for procuring the G3C flight suits. The procurement process, anticipated to commence in March, signaled NASA's forward momentum in outfitting astronauts with advanced gear tailored for the rigorous demands of space missions.

Simultaneously, plans were underway to initiate rigorous qualification and reliability tests for the G2C pressure suit, also scheduled to commence in March. These tests were pivotal in ensuring the utmost safety and functionality of the suits destined to accompany astronauts into space.

On January 13, 1964, a pivotal phase began in the Gemini program as spacecraft No. 2 commenced its Spacecraft Systems Tests (SST) at McDonnell. This marked the start of Phase I, focusing on module tests crucial for ensuring the spacecraft's operational readiness.

Since the successful completion of SST for spacecraft No. 1, NASA's approach to checkout procedures had undergone significant evolution. All test activities, from manufacturing to module testing, now fell under the purview of a dedicated Launch Preparations Group (LPG). This group, overseen by the NASA-MSC Florida Operations Assistant Manager for Gemini, was a collaborative effort involving McDonnell, NASA operators, and quality control personnel stationed at Cape Kennedy.

Initially based in St. Louis, the LPG reviewed and approved test procedures while conducting various tests on spacecraft Nos. 2 and 3. This arrangement aimed to streamline operations and enhance scheduling efficiency by eliminating redundant testing practices.

However, the modular SST of spacecraft No. 2 encountered numerous challenges, necessitating troubleshooting, equipment modifications, and structural adjustments. These issues prompted delayed progress to Phase II, where mated SST was slated to begin. Ultimately, Phase II mated SST for spacecraft No. 2 was postponed until July, reflecting the rigorous standards and meticulous attention to detail required to achieve operational readiness for manned space missions.

On January 15, 1964, a pivotal phase began in developing the drogue stabilization parachute for the Gemini parachute recovery system. Phase I started with the successful test drop of boilerplate spacecraft No. 5 at El Centro, marking a crucial milestone in NASA's quest to perfect its recovery capabilities.

Sequence Compatibility Firing of the two stages of Gemini launch vehicle 1 at pad 19, Jan. 21, 1964. (KSC Photo 64P-7, Jan. 21, 1964.)

Phase I aimed to evaluate the drogue chute's deployment dynamics, specifically focusing on the effects of deploying the pilot chute via a lanyard attached to the drogue chute. Initial tests on January 15 and 28 were deemed successful, demonstrating promising outcomes for the evolving parachute system.

However, the third test on February 6 encountered a setback when the cables connecting the drogue-and-pilot-chute combination to the rendezvous and recovery (R&R) section of the boilerplate failed during pilot-chute deployment. Despite the main chute deploying adequately and achieving a routine landing for the boilerplate, the R&R section sustained significant damage upon impact.

In response to the failure, testing was temporarily halted while McDonnell conducted a thorough analysis to identify the root cause. After necessary adjustments and improvements, testing resumed on April 10 with the fourth drop test. Phase I concluded successfully on April 21 with the fifth and final drop test, validating the refined design and deployment mechanisms.

Following Phase I, boilerplate No. 5 was repurposed into static article No. 4A by September 18, preparing it for Phase III tests. This phase would further refine and validate the parachute system's performance under more rigorous conditions, marking another step forward in NASA's meticulous preparation for manned space missions.

On January 20, 1964, Martin-Baltimore conducted a critical static test to evaluate the spacecraft/launch vehicle interface structure under extreme conditions. This test, designed to fail, provided essential data regarding the structural integrity of components crucial for the upcoming Gemini-Titan 1 mission.

The test results were highly encouraging, revealing a significant margin of structural strength. It demonstrated a robust minimum structural margin of 23 percent above the ultimate conditions anticipated during the transonic buffet phase of launch. These findings underscored the reliability and resilience of the interface structure under simulated launch stresses.

Buoyed by the successful test outcomes, plans were swiftly implemented for a comprehensive structures meeting scheduled in Houston from March 17 to 19, 1964. This meeting was slated to review all load conditions, stress distribution analyses, and safety margins across the spacecraft and launch vehicle interface. Such meticulous preparations ensured that all structural aspects were optimized and thoroughly vetted before the Gemini-Titan 1 mission.

On January 22, 1964, North American Aviation embarked on a series of pivotal flights for the Paraglider Landing System Program, demonstrating the deployment and maneuverability of a full-scale wing from the rendezvous and recovery can. This initiative marked a significant stride in NASA's pursuit of advanced landing technologies for manned space missions.

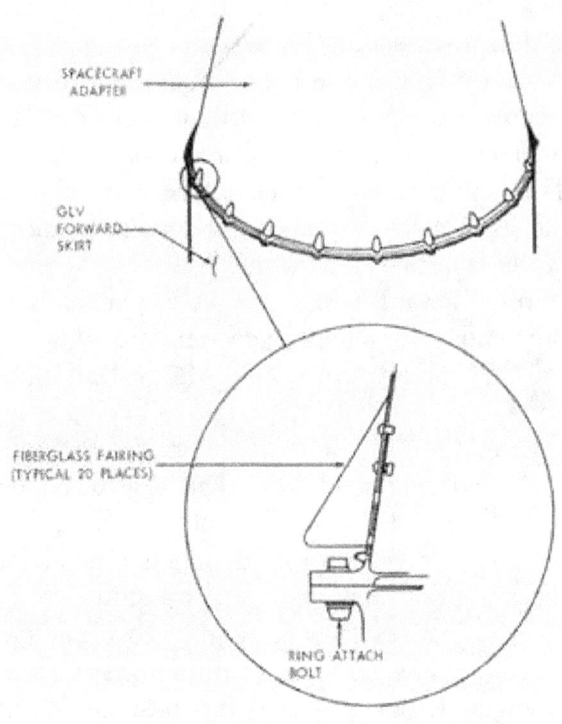

The interface between Gemini launch vehicle and spacecraft.
(NASA Photo S-64-3065, undated.)

Initially planned for 20 tests, the program conducted 25 deployment flight tests. The early flights, including those on February 18, March 6, April 10, and April 22, showed promising progress but faced challenges in achieving consistent success.

Flight No. 6 on April 30 marked a breakthrough, completing the entire sequence successfully, followed by another successful flight on May 28 (Flight No. 7). However, subsequent flights encountered intermittent difficulties, with flights on June 12, June 29, July 15, and July 23 experiencing setbacks despite generally satisfactory deployment sequences.

Following Flight No. 12 on July 29, a series of flights throughout August, September, and October encountered various issues, primarily related to achieving stable glides post-deployment. Despite these challenges, the program persisted in refining its approach.

The final flights on October 23, November 6, and December 1 demonstrated significant improvement. They successfully executed the complete test sequence without incident. These flights validated the effectiveness of the Paraglider Landing System, highlighting NASA's perseverance and incremental advancements in pioneering new technologies for safe spacecraft landings.

On January 25, 1964, Rocketdyne achieved a significant milestone in developing the Gemini spacecraft's orbit attitude and maneuver system (OAMS). They successfully tested a 100-pound thrust chamber assembly (TCA) to the rigorous 757-second mission duty cycle without encountering any failures. This TCA incorporated an innovative modified injector design that utilized boundary-layer cooling, spraying approximately 25 percent of the fuel down the chamber wall before combustion. This technique effectively reduced heat stress, with the ablative material charring to a depth of just 0.5 to 0.55 inches, well within operational tolerances. Encouraged by these results, the Gemini Project Office (GPO) expressed confidence that the boundary-layer cooling method resolved durability concerns for the 100-pound TCAs.

Efforts to apply the same technique to the 25-pound TCA initially faced challenges but saw success with a modified injector design featuring a rounded-edge splash plate. Tested under a mixture ratio of 0.7:1, this configuration demonstrated improved performance, with only 0.18 inches of uncharred material remaining after the rigorous 570-second mission duty cycle. Previous attempts with similar mix ratios had failed after 380 seconds. As a result, GPO anticipated readiness for both 25- and 100-pound TCAs to integrate into spacecraft 5 and beyond.

Also, on January 25, the Gemini Project Office reported on Ames Research Center's successful completion of a visual reentry control simulator program. This initiative aimed to assess the feasibility of spacecraft attitude control during reentry using the horizon as the sole visual reference. The simulation confirmed earlier analytical studies, affirming that astronauts could effectively maintain control of reentry attitudes with the horizon view alone. This capability was deemed well within astronaut capabilities, further enhancing confidence in Gemini's reentry procedures.

In January, the Gemini Project Office published the program plan for extravehicular operations (EVA), outlining ambitious objectives to evaluate human capabilities in performing tasks in space. The plan aimed to augment spacecraft capabilities and assess advanced EVA equipment to support future manned space endeavors. Flight Crew Operations Directorate initiated detailed planning for EVA activities across multiple Gemini-Titan missions:

• GT-4: Depressurizing the cabin, opening the hatch, and standing up.

• GT-5: Performing complete egress and ingress maneuvers.

• GT-6: Egressing and accessing the equipment adapter interior to retrieve data packages.

• GT-7 and GT-8: Evaluating maneuvering capabilities using tethers and handholds.

• GT-9: Testing astronaut maneuvering units.

• GT-10 through GT-12: Evaluating advanced EVA equipment and procedures.

The Crew Systems Division concurrently conducted ground tests of EVA equipment, including simulated zero-gravity egress and ingress exercises. These preparations underscored NASA's proactive approach to advancing EVA capabilities crucial for future space missions.

On February 1, 1964, McDonnell initiated critical tests of the spacecraft pyrotechnic hatch firing system using boilerplate spacecraft No. 3A. The initial test involved firing a single hatch, which successfully opened and locked into place. However, the opening time of 350 milliseconds exceeded the allowable limit by 50 milliseconds, indicating a need for optimization to meet stringent operational requirements.

Following adjustments, McDonnell proceeded with a dual-hatch firing test on February 10, which yielded satisfactory results. This test confirmed the system's capability to deploy both hatches within acceptable operational parameters, showcasing significant improvement over the initial single-hatch test.

In preparation for further testing and qualification, boilerplate spacecraft No. 3A was readied for shipment to Weber Aircraft. It would undergo comprehensive evaluation as part of the ejection seat system qualification program there. This rigorous testing was essential to ensure the system's reliability and safety for astronauts during potential emergency scenarios.

On February 2, 1964, manufacturing of the heatshield for spacecraft No. 3 reached a significant milestone with its completion. This heatshield represented the first production article to feature the full thickness of 1.0 inch, marking a substantial advancement in spacecraft thermal protection technology.

Previous heatshields for spacecraft Nos. 1 and 2 were approximately half as thick, highlighting the ongoing refinement and enhancement of thermal management capabilities for spacecraft reentry and atmospheric entry conditions. The thicker heatshield of spacecraft No. 3 aimed to provide enhanced protection and stability during the intense heat and pressures encountered during reentry into Earth's atmosphere.

Completing the 1.0-inch-thick heatshield underscored NASA's commitment to advancing materials science and engineering to support the Gemini program's manned space missions. This milestone represented a crucial step forward in ensuring the safety and reliability of spacecraft components, crucial for the success of upcoming missions.

Gemini boilerplate 3A in the production area at the McDonnell plant before being shipped to Weber Aircraft. (NASA Photo 1053, Feb. 18, 1964.)

On February 3, 1964, NASA awarded a significant contract to the Garrett Corporation's AiResearch Manufacturing Division to develop the extravehicular pressurization and ventilation system. This cost-plus-incentive-fee contract, totaling $133,358, marked a crucial step in advancing capabilities essential for extravehicular activities (EVA) during the Gemini program.

The initial phase of the contract focused on conducting a detailed study to define the configuration of the pressurization and ventilation system. This involved comprehensive analysis and planning to establish the system's functional requirements, design parameters, and integration considerations within the Gemini spacecraft.

The awarding of this contract underscored NASA's commitment to enhancing astronaut safety and mission effectiveness through advanced life support technologies. The pressurization and ventilation system would be pivotal in enabling astronauts to perform critical tasks outside the spacecraft, such as spacewalks and other EVA operations planned for future Gemini missions.

By engaging the expertise of the Garrett Corporation's AiResearch Manufacturing Division, NASA aimed to leverage cutting-edge engineering and innovation to develop a robust and reliable system capable of withstanding the challenges of space exploration. This initiative represented a significant investment in technological advancement crucial for expanding human presence and capabilities in space during the pioneering days of manned spaceflight.

On February 5, 1964, significant progress was made in the Gemini program as Gemini launched vehicle 2, including stage I and interstage components, which were erected in the vertical test facility at Martin Baltimore. This marked a crucial step toward the comprehensive testing and readiness assessment of the launch vehicle's subsystems.

Two days later, on February 7, stage II of the Gemini launch vehicle was also erected in the vertical test facility, further advancing preparations for upcoming subsystem functional verification tests.

Subsystem Functional Verification Tests commenced on February 21. They aimed to rigorously evaluate the operational functionality and integration of critical systems within the Gemini launch vehicle, which was essential to ensure the vehicle's readiness for upcoming missions.

Simultaneously, Bell Aerosystems initiated Preliminary Flight Rating Tests (PFRT) of the Agena primary propulsion system (PPS) on February 5. The testing, initially scheduled for completion by April 24, encountered unforeseen challenges that extended the timeline until late June.

Testing progressed smoothly until early April, after which a significant setback occurred when the start tanks of the test unit's fuel and oxidizer system failed due to visible longitudinal cracks in the outer shell. This issue was traced to intergranular corrosion of the stainless-steel tanks, necessitating replacement with heat-treated shells delivered by April 24.

Following the replacement of defective tanks, PFRT resumed in early May, marking a critical phase in ensuring the reliability and performance of the Agena propulsion system for Gemini missions.

During a Gemini Management Panel meeting on February 7, 1964, Bernhard A. Hohmann of Aerospace expressed significant concerns regarding the ongoing increase in spacecraft weight. His apprehensions were echoed by Major General Ben I. Funk of the Air Force Space Systems Division, who highlighted potential implications for the Department of Defense (DoD) experiments program.

The issue of spacecraft weight growth had become a persistent challenge, raising alarms among stakeholders involved in the Gemini program. Major General Funk emphasized the need for a detailed study to assess the problem's scope and impact thoroughly. This study aimed to identify potential solutions crucial for maintaining mission objectives and accommodating vital DoD experiments.

By September 29 of the same year, George M. Low, NASA's Deputy Associate Administrator for Manned Space Flight, noted a concerning trend: spacecraft No. 8 had been steadily increasing in weight by an average of 35 pounds per month since early 1963. This steady growth posed logistical and operational challenges, potentially affecting mission planning and spacecraft capabilities.

The Management Panel's focus on spacecraft weight growth underscored the program's commitment to addressing technical and operational hurdles. Future meetings were slated to discuss possible strategies and adaptations necessary to mitigate weight increases without compromising mission objectives or essential experiments.

On February 8, 1964, the Manned Spacecraft Center (MSC) made a significant decision regarding the processing and reduction of data for postlaunch analysis. It was decided that MSC facilities would be utilized for this critical task, rather than outsourcing to Lockheed facilities.

Before this decision, MSC had conducted an assessment exploring the feasibility of utilizing Lockheed's facilities for the postlaunch data analysis. However, after careful evaluation and cost analysis, MSC determined that leveraging its own facilities would be more cost-effective and efficient for the Gemini program.

The decision to use MSC facilities was driven by financial considerations, with estimates indicating potential savings of approximately $300,000. This financial advantage was pivotal in the decision-making process, aligning with NASA's commitment to optimizing resources while ensuring rigorous data analysis capabilities.

By centralizing postlaunch data processing and reduction at MSC, NASA aimed to streamline operations, enhance control over data management, and expedite the analysis of critical mission data. This strategic move underscored MSC's role as a hub for mission support and data operations within the Gemini program, reinforcing its capabilities in managing complex data-intensive tasks.

The Gemini Project Office announced on February 15, 1964, that the developmental test program for the Gemini spacecraft retrorockets had reached completion at Thiokol. This marked a significant milestone in the spacecraft's development, setting the stage for the next phase: qualification tests scheduled to commence in March 1964.

Thiokol's completion of the developmental test program signified the successful evaluation and validation of the retrorockets' design and performance under simulated operational conditions. These retrorockets were crucial components essential for the precise maneuvering and reentry phases of the Gemini missions.

With developmental testing concluded, the focus shifted to rigorous qualification tests scheduled to begin the following month. These tests aimed to verify and certify the retrorockets' reliability, functionality, and safety for actual mission deployment, ensuring they met stringent operational standards required for manned spaceflight.

Simultaneously, on February 15, the Manned Spacecraft Center's Flight Operations Division completed a series of simulated Gemini rendezvous missions. These simulations evaluated the adequacy and sequential usage of trajectory and real-time control displays planned for use during actual rendezvous operations.

The simulated rendezvous missions provided critical insights into the operational readiness of the trajectory and control display systems. They assessed their effectiveness in supporting complex orbital maneuvers and rendezvous procedures essential for upcoming Gemini missions.

On February 16, 1964, Bell Aerosystems achieved a significant milestone by delivering the first Gemini Agena Model 8247 main engine to Lockheed. This engine marked a crucial component in the propulsion system destined for testing at Lockheed's Santa Cruz Test Base.

The delivered engine was designated for integration into the Propulsion Test Vehicle Assembly (PTVA). This specialized unit was specifically designed to conduct comprehensive tests on the Gemini Agena spacecraft's primary and secondary propulsion systems.

Following the delivery of the main engine, Bell Aerosystems continued their contribution by delivering two secondary propulsion system modules for the PTVA. These modules were crucial for ensuring comprehensive testing and validation of the entire propulsion system assembly.

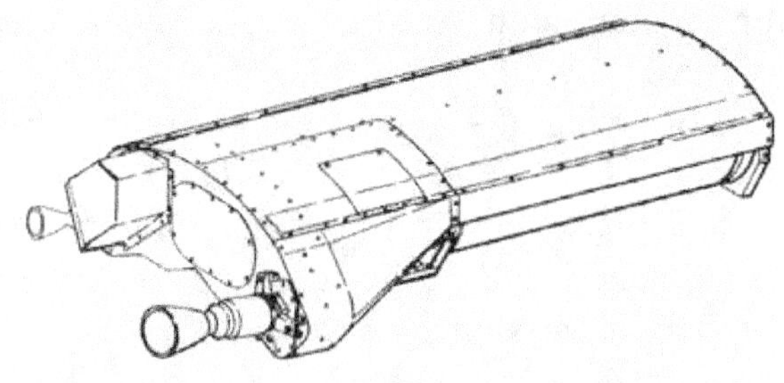

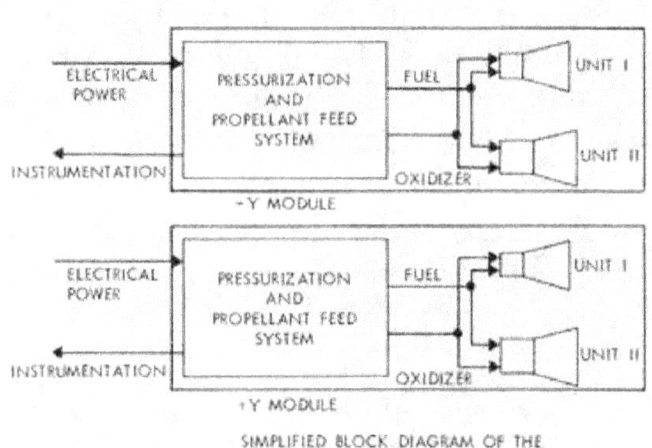

The Agena secondary propulsion system. (Lockheed, "Gemini Agena Target Vehicle Familiarization Handbook," LMSC A602521, Apr. 1, 1964, pp. 4-1, 4-3.)

Delivery Timeline:

Main Engine: February 16

Secondary Propulsion Modules: March 6 and 14

Assembly and Testing Timeline

The integration and assembly of the PTVA were swiftly executed, with the installation of the secondary propulsion system modules completed by March 26.

Subsequently, the fully assembled PTVA was delivered to Lockheed's Santa Cruz Test Base, where it would undergo extensive testing and validation procedures.

Preliminary Flight Rating Tests of Gemini Agena Secondary Propulsion System

Bell Aerosystems initiated Preliminary Flight Rating Tests (PFRT) for the Agena secondary propulsion system (SPS) on February 17, 1964. These tests were integral to evaluating the performance and reliability of the SPS under simulated operational conditions.

The PFRT proceeded smoothly through its initial phases, including acceleration and vibration tests, without significant issues. The SPS began calibration firings in early April to fine-tune its operational parameters.

However, the testing phase faced setbacks. A propellant valve failure caused a minor delay in Unit I of the SPS. The more serious issue emerged in late April during high-temperature firings of Unit II's 200-pound thrust chamber. The chamber's wall near the injector face burned through after accumulating 354 seconds of firing time, below the specified limit of 400 seconds but well above the maximum orbital useful time of 200 seconds.

To address the burn-through problem, Bell Aerosystems promptly replaced the damaged thrust chamber and resumed testing. Despite the setback, the PFRT, originally scheduled to conclude on June 19, faced delays, ultimately concluding in mid-August after several adjustments and rescheduling efforts.

Bell Aerosystems initiated a dedicated test program in September to investigate the root cause of the thrust chamber's burn-through issue. This investigative effort aimed to implement corrective measures and ensure the reliability and safety of the SPS for future Gemini Agena missions.

Gemini Spacecraft No. 1 Preparing for Launch

In February 18-19, the Gemini Program Office at Cape Kennedy conducted a critical preflight readiness review of Gemini spacecraft No. 1. This meticulous assessment followed the comprehensive Spacecraft Systems Tests conducted earlier in the industrial area at Cape Kennedy on February 12.

Each spacecraft component was scrutinized during the review for any outstanding issues, deviations from specifications, and qualification status. Despite several open items identified during the review process, none posed constraints on the spacecraft's integration with its launch vehicle. Importantly, none of these items indicated any potential delays in the scheduled launch.

After completing these assessments, the spacecraft was transferred to Complex 19 on March 3rd. It was carefully positioned within the spacecraft erector support assembly in the designated erector white room. On March 4th, the premate Spacecraft Systems Test was successfully executed, marking another crucial milestone in the meticulous preparation process.

The Shift Away from Paragliders

On February 20, George E. Mueller, NASA's Associate Administrator for Manned Space Flight, delivered a pivotal announcement to the Gemini Project Office (GPO) staff. He confirmed that all upcoming 12 Gemini missions would conclude with water landings. This decision marked a significant departure from earlier plans outlined in Project Gemini Quarterly Report No. 8, which as of February 29, 1964, still detailed considerations for using the paraglider system in the final three Gemini missions.

By April 29, during a GPO staff meeting, a consensus emerged to scale back activities related to the paraglider program and initiate its phased withdrawal from Gemini operations. This shift was formally solidified during discussions between NASA and North American representatives on May 4, emphasizing a refocus on the essential flight testing aspects of the program. Consequently, on June 12, Gemini Project Manager Charles W. Mathews officially notified the Gemini Procurement Office of the program's decision to remove the paraglider recovery system requirement from its specifications. He urged expedited adjustments to the McDonnell contract to reflect this change.

On February 21, at Complex 19, the Gemini launch vehicle 1 underwent critical Subsystem Functional Verification Tests (SSFVT) to ensure launch readiness. These tests were a crucial step in the meticulous preparation process, mirroring previous SSFVT conducted at Martin-Baltimore's vertical test facility. The primary objective was to rigorously verify the readiness of the launch vehicle's subsystems for the upcoming systems tests.

Throughout the testing period, which concluded on March 3, engineers and technicians meticulously assessed each subsystem's functionality and integration. These tests validated the performance of critical systems and ensured that any potential issues were identified and resolved before proceeding with further testing phases.

Planning scientific endeavors in space, on February 26, George M. Low, NASA's Deputy Associate Administrator for Manned Space Flight, conveyed pivotal information to Gemini Project Manager Charles W. Mathews regarding the approved experiments slated for the initial five Gemini missions. This directive followed NASA Associate Administrator Robert C. Seamans, Jr.'s endorsement of recommendations put forth by the Manned Space Flight Experiments Board. These recommendations were contingent upon completing feasibility studies conducted by the Gemini Project Office (GPO).

The approved experiments were categorized as Category B, emphasizing their prioritization without compromising scheduled launch timelines. Notably, these experiments did not initially include those essential for gathering critical design insights for both Gemini and the forthcoming Apollo missions. However, the GPO retained the authority to incorporate these high-priority items into the mission agenda.

On February 28, Gemini Project Manager Charles W. Mathews addressed senior staff at the Manned Spacecraft Center, outlining initiatives to enhance control over the weight and configuration of Gemini spacecraft. Mathews appointed Lewis R. Fisher from his office to spearhead these efforts to lead a newly established Systems Integration Office within the Gemini Project Office.

The primary mandate of the Systems Integration Office was to oversee and document key aspects of spacecraft integration meticulously. This included maintaining precise records of spacecraft weight, monitoring interface interactions between the spacecraft and its launch vehicle, and coordinating interactions with the Agena target vehicle.

On February 29, in response to ongoing challenges with the Gemini spacecraft's orbit attitude and maneuver system thrusters, the Gemini Project Office (GPO) initiated backup engine programs as contingency measures. These efforts addressed development issues and potential additional requirements affecting spacecraft operations.

At Marshall Space Flight Center, efforts focused on developing a robust 100-pound engine. This engine targeted applications for the Gemini spacecraft and the Saturn S-IVB launch vehicle, demonstrating its versatility and potential dual-purpose capabilities.

Simultaneously, the Manned Spacecraft Center developed a specialized 25-pound radiation-cooled engine. This engine was designed to meet specific operational needs, particularly when radiation exposure could impact conventional engine performance.

On February 29, the GPO disclosed the results of a comprehensive test program evaluating the effects of cracked throats or liners on the orbit attitude and maneuver system thrusters. Due to inherent challenges in the manufacturing process, nearly all thrust chamber assemblies (TCA) exhibited such cracks, rendering them unsuitable for delivery initially.

However, tests revealed promising outcomes: there was no discernible degradation in engine performance or lifespan attributable to these cracks. Rocketdyne, the engine manufacturer, confirmed that none of their five space engine programs had experienced failures due to cracked throats. Consequently, with specific operational limitations, cracked throats were deemed acceptable for use.

Mating and Testing of Gemini Launch Vehicle and Spacecraft No. 1

On March 5, a significant milestone was achieved at Complex 19 as Gemini Launch Vehicle (GLV) 1 and Spacecraft No. 1 were mechanically mated. This critical step marked the physical integration of the spacecraft with its launch vehicle, setting the stage for subsequent testing phases.

Before proceeding to electrical integration, a rigorous Combined Systems Test (CST) was conducted on March 10 to verify the launch vehicle's operational status. This comprehensive assessment ensured that all integrated systems functioned seamlessly, paving the way for the next testing phase.

Gemini-Titan 1 during Electronic-Electrical Interference Tests with the launch vehicle erector lowered. (NASA Photo No. 64-Gemini 1-44.)

From March 12 to March 25, a series of specialized Electronic-Electrical Interference (EEI) Tests were conducted. These tests evaluated and mitigated potential electromagnetic interference between the spacecraft and its electrical systems. Despite encountering persistent anomalies during testing, the overall objectives of the EEI tests were successfully achieved.

Following the EEI testing phase, a successful post-EEI systems reverification CST was executed on March 27. This final verification step confirmed the integrity and readiness of the integrated systems after addressing any identified issues during the EEI testing period.

On March 6, Martin-Baltimore reached a significant milestone in preparing Gemini launch vehicle 4 (GLV-4) by receiving the propellant tanks from Martin-Denver. These tanks, whose fabrication commenced in November 1963, were crucial components for the vehicle's propulsion system.

Splicing the tanks was completed by July 21, marking a pivotal step in ensuring structural integrity and readiness for subsequent integration phases. Concurrently, Aerojet-General contributed significantly by delivering the stage II flight engine on June 26, followed by the stage I engine on July 28. These engine installations were finalized by

September 4, consolidating the vehicle's propulsion capabilities.

After completing final horizontal tests by October 26, Martin-Baltimore received authorization to erect the vehicle in the vertical test facility. This comprehensive testing and integration process highlighted meticulous planning and execution, essential for ensuring the reliability and functionality of GLV-4 for its upcoming mission.

Addressing unforeseen challenges, from March 17 to 19, the structures panel convened to meticulously review and resolve all outstanding issues concerning the structural integrity of the interface between the spacecraft adapter section and the launch vehicle upper skirt. This critical meeting aimed to ensure that all components were robust and could withstand the rigorous demands of spaceflight.

During the review process, a significant discrepancy arose when Aerospace's analysis indicated load factors were approximately ten times higher than initially predicted by McDonnell. This unexpected finding prompted further analysis by McDonnell, reaffirming the validity of its original estimates.

The resolution of this discrepancy underscored the meticulous scrutiny applied to every aspect of the Gemini spacecraft's design and integration. Such thorough review processes were crucial in identifying and addressing potential vulnerabilities early in the development phase, ensuring the spacecraft's overall structural integrity and mission readiness.

Shift in Focus: Gemini Program and Titan II

Refocusing priorities, on March 19, a significant decision was made regarding the Titan II launch vehicle integral to the Gemini program. George E. Mueller, NASA's Associate Administrator for Manned Space Flight, communicated to Associate Administrator Robert C. Seamans, Jr., that the Titan II launch vehicle was no longer considered the pacing item in the Gemini program. This marked a notable shift in perspective and priorities within NASA's strategic planning.

The decision to discontinue the Air Force Systems Command weekly report, initiated in September 1963 to monitor and address Titan II development challenges, reflected the program's progress and increased confidence in the launch vehicle's reliability.

Mueller's assessment signaled a broader acknowledgment within NASA that other aspects of the Gemini program were now critical paths to mission success, potentially superseding previous concerns about the Titan II's development pace. This strategic realignment underscored NASA's adaptive approach to managing and advancing the complexities of manned spaceflight, ensuring that resources and attention were directed where they were most needed to achieve program goals effectively.

Ensuring mission preparedness, on March 20, the Manned Spacecraft Center (MSC) approved recommendations from the Air Force Space Systems Division (SSD) for a comprehensive test program aimed at bolstering confidence in 16 critical electronic and electrical components essential to the Gemini Agena target vehicle. This decision marked a significant step in ensuring the reliability and readiness of key mission components.

The approved test program encompassed a range of rigorous assessments, including complete electromagnetic interference (EMI) testing explicitly tailored to the unique requirements of the Gemini mission. Elevated stress and extended life tests were included to evaluate the durability and performance under simulated operational conditions.

While SSD initially recommended subsystem-level and component-level EMI testing, MSC did not approve this aspect of the program. Despite this, SSD directed Lockheed to commence the approved test program on March 23.

Scheduled to conclude by July 1, the EMI tests aimed to identify and mitigate any potential electromagnetic interference issues that could impact mission-critical systems. Concurrently, stress and life tests, slated for completion by September 1, 1964, focused on validating component longevity and robustness in varying operational scenarios.

Optimizing mission resilience, during a meeting on March 25, the Trajectories and Orbits Panel of the Gemini Project Office convened to discuss mission plans for the first Agena rendezvous flight. Representatives from the Flight Operations Division presented two primary mission plans currently under consideration, each proposing distinct orbital strategies to maximize mission effectiveness and resilience.

The first plan, championed by Howard W. Tindall, Jr., and James T. Rose from their earlier tenure in the Space Task Group, advocated for tangential orbits of both the Agena target vehicle and the Gemini spacecraft. This approach optimized rendezvous and docking procedures by leveraging precise orbital mechanics.

In contrast, the second plan, proposed by Edwin E. Aldrin, Jr., of the Air Force Space Systems Division, suggested concentric orbits rather than tangential ones. This alternative strategy offered a significant advantage in maximizing onboard backup techniques. Specifically, it was designed to effectively utilize remaining onboard systems in the event of failures in critical components such as the inertial guidance system platform, computer, or radar.

Assessing impact characteristics, on March 26, a significant milestone was reached in the testing phase of the Gemini program as Boilerplate Spacecraft No. 4 underwent its inaugural drop from a test rig. This test was pivotal in gathering essential data on the spacecraft's landing dynamics and impact characteristics.

During the drop, Boilerplate No. 4 achieved a horizontal velocity of 60 feet per second and a vertical velocity of approximately 40 feet per second upon impact with the water surface. These velocities were carefully monitored to assess the spacecraft's behavior under various speeds and attitudes during landing scenarios.

The test's primary objective was to collect precise data on landing accelerations, crucial for validating and refining the spacecraft's design and ensuring the safety of future manned missions. This comprehensive evaluation highlighted NASA's commitment to meticulous testing and preparation, essential for successfully executing Gemini spacecraft landings.

Overcoming technical challenges, the Propulsion Test Vehicle Assembly (PTVA) arrived at Santa Cruz Test Base on March 26, marking a significant step in the Gemini program's propulsion system testing.

The PTVA comprised a basic Agena structure equipped with propellant pressurization, feed-and-load systems, the Primary Propulsion System (PPS), and two Secondary Propulsion System (SPS) modules attached to the aft rack. This assembly was crucial for conducting comprehensive tests to validate the adequacy of propellant loading systems, ground equipment, and overall system operation.

The test program was designed to include loading operations and hot firings of the PPS and SPS modules. These tests demonstrated the functionality of propulsion systems under operational conditions, gathered engineering data on system performance, and evaluated the environmental impact of propulsion operations.

However, the start of testing faced delays due to unforeseen issues with the PPS start tank identified during Preliminary Flight Rating Tests conducted at Bell Aerosystems in April. Subsequently, Lockheed returned the defective PTVA main engine start tanks to Bell for inspection and replacement, causing a delay in testing activities.

By mid-May, new start tanks were ready, but additional minor issues persisted, prolonging the delay until June 16, when hot-firing finally commenced. These setbacks underscored the challenges inherent in complex aerospace testing programs and highlighted NASA's meticulous approach to ensuring the reliability and readiness of propulsion systems crucial for manned space missions.

On March 28, the Gemini Project Office disclosed the findings from the potability test conducted on water from the fuel cells intended for use on spacecraft No. 2. The test results revealed that while the water was slightly acidic, it was deemed safe and suitable for drinking purposes.

This critical assessment ensured that the water sourced from the spacecraft's fuel cells met stringent safety standards necessary for manned missions. Despite its slightly acidic nature, the water's overall quality met the criteria required for human consumption, confirming its suitability for astronauts' use during their missions.

On March 30, Director Robert R. Gilruth announced a significant reorganization of the Florida unit of the Manned Spacecraft Center (MSC), renaming it to MSC-Florida Operations. This restructuring was part of NASA's efforts to streamline operations and enhance efficiency in preparation for upcoming missions within the Gemini program.

G. Merritt Preston, who had overseen MSC activities at the Cape since 1961, was appointed to lead MSC-Florida Operations. The primary objective of this reorganization was to optimize the Florida unit's workflow and responsibilities, aligning them closely with Project Mercury's operational needs.

One notable change introduced by this reorganization was the integration of Florida personnel into spacecraft testing activities at McDonnell. This strategic shift aimed to reduce duplicate testing efforts conducted at the Cape by ensuring that flight-ready spacecraft were delivered directly to the launch site.

By eliminating redundant testing processes and leveraging expertise across different facilities, NASA sought to enhance efficiency and accelerate spacecraft readiness for upcoming Gemini missions.

Preparing Gemini Launch Vehicle 1:

On March 31, the Gemini launch vehicle (GLV) 1 underwent significant progress towards readiness for its upcoming mission. Electrical and mechanical modifications to its airborne components were completed, marking a crucial step in the vehicle's preparation process.

Initially shipped to the Cape with certain ground test-only items, GLV-1 began its transformation on January 31. These components were systematically replaced with flight-ready units, ensuring the vehicle was equipped with the necessary systems and components for its upcoming mission.

A key milestone in the preparation phase was the GLV-1 Wet Mock Simulated Launch, conducted on April 2. This exercise involved a complete countdown simulation, including propellant loading, to validate operational procedures and readiness under simulated launch conditions.

Subsequent testing culminated on April 5 with the completion of a Simulated Flight Test. This comprehensive simulation evaluated the vehicle's performance across all critical phases of flight, ensuring that all systems operated as intended and preparing personnel for actual mission scenarios.

On April 1, 1964, astronauts from NASA's Project Gemini embarked on a critical visit to St. Louis, marking a pivotal moment in the program's rigorous operational evaluations. Their mission: to meticulously assess the translation and docking trainer—a key component in preparing for manned space missions.

The astronauts, seasoned pioneers of space exploration, arrived in St. Louis with a focus that mirrored the precision demanded by their training. They meticulously scrutinized every aspect of the trainer, identifying minor discrepancies that McDonnell, the aerospace company tasked with the trainer's development, promptly addressed. This attention to detail was crucial, ensuring the equipment met the stringent standards required for human spaceflight.

After the astronauts' comprehensive evaluation, McDonnell completed engineering evaluation tests by April 6, 1964. The thoroughness of these tests underscored the commitment to safety and functionality that defined Project Gemini—a program dedicated to advancing the frontiers of human capability in space.

With the evaluations successfully concluded, the translation and docking trainer was carefully disassembled for transport to the Manned Spacecraft Center in Houston. This journey marked a logistical milestone and a symbolic step towards realizing NASA's ambitious goals for manned space exploration.

On April 2, 1964, NASA's Project Gemini embarked on a challenging 36-hour open-sea qualification test in the waters of Galveston Bay, utilizing static article No. 5. This pivotal test aimed to simulate the conditions astronauts would face during water recovery scenarios, a critical aspect of space missions.

The test, however, encountered unforeseen challenges early on. Within just two hours, the test subjects—integral to assessing human response and equipment functionality—succumbed to seasickness, prompting the premature end of the exercise.

Despite the abbreviated duration, valuable technical insights were gleaned.

Among the technical issues encountered during this brief exposure was the failure of a suit ventilation fan, which highlighted vulnerabilities in life support systems crucial for sustaining astronauts in hostile environments. Additionally, structural integrity issues with the high-frequency whip antenna underscored the importance of robust equipment design capable of withstanding rigorous operational demands.

This test, though abbreviated, served as a pivotal learning experience for the Project Gemini team. It underscored the importance of comprehensive preparation and resilient equipment for future manned space missions. Each setback provided valuable data, informing subsequent refinements and enhancements that would ultimately contribute to the program's success.

Gemini Spacecraft and Subsystems

The Gemini spacecraft was a cone-shaped capsule with two components: a reentry module and an adaptor module. The adaptor module made up the base of the spacecraft. It was a truncated cone 228.6 cm high, 304.8 cm in diameter at the base, and 228.6 cm at the upper end, attached to the reentry module's base. The re-entry module consisted of a truncated cone which decreased in diameter from 228.6 cm at the base to 98.2 cm, topped by a short cylinder of the same diameter and then another truncated cone decreasing to a diameter of 74.6 cm at the flat top. The reentry module was 345.0 cm high, giving the Gemini spacecraft a total height of 573.6 cm.

The adaptor module was an externally skinned, stringer-framed structure with magnesium stringers and an aluminum alloy frame. It comprised two parts: an equipment section at the base and a retrorocket section at the top. The equipment section held fuel and propulsion systems and was isolated from the retrorocket section by a fiber-glass sandwich honeycomb blast shield. The retrorocket section held the capsule's re-entry rockets.

The reentry module consisted mainly of the pressurized cabin holding the two Gemini astronauts. Separating the reentry module from the retrorocket section of the adaptor at its base was a curved silicone elastomer ablative heat shield. The module was composed predominantly of titanium and nickel alloy with beryllium shingles. At the narrow top of the module was the cylindrical reentry control system section, and above this was the rendezvous and recovery section, which holds the reentry parachutes. The cabin had two seats equipped with emergency ejection devices, instrument panels, life support equipment, and equipment stowage compartments in a total pressurized volume of about 2.25 cubic meters. Two large hatches with small windows could be opened outward, one positioned above each seat.

Attitude control was effected by two translation-maneuver hand controllers, an attitude controller, redundant horizon sensor systems, and reentry control electronics, with guidance provided via an inertial measuring unit and radar system. The orbital attitude and maneuver system used a hypergolic propellant combination of monomethylhydrazine and nitrogen tetroxide supplied to the engines by a helium system pressurized at 2800 psi. Two 95 lb translation thrusters and eight 23 lb attitude thrusters were mounted along the bottom rim of the adaptor, and two 79 lb and 4 95 lb thrusters were mounted at the front of the adaptor. Power was supplied by three silver-zinc batteries to a 22- to 30-volt DC two-wire system. During reentry and post-landing, power was supplied by four 45 amp-hr silver-zinc batteries.

Voice communications were performed at 296.9 MHz with an output power of 3 W. A backup transmitter-receiver at 15.016 MHz with an output power of 5 W was also available. Two antenna systems consisting of quarter-wave monopoles were used. Telemetry was transmitted via three systems, one for real-time telemetry, one for recorder playback, and a spare. Each system was frequency-modulated with a minimum power of 2 W. Spacecraft tracking consisted of two C-band radar transponders and an acquisition-aid beacon. One transponder was mounted in the adaptor with a peak power output of 600 W to a slot antenna on the bottom of the adaptor. The other was in the reentry section, delivering 1000 W to three helical antennas mounted at 120-degree intervals just forward of the hatches. The acquisition-aid beacon was mounted on the adaptor and had a power of 250 mW.

During reentry, the spacecraft would be maneuvered to the appropriate orientation, and the equipment adaptor section would be detached and jettisoned, exposing the retrorocket module. The retrorockets consisted of four spherical-case polysulfide ammonium perchlorate solid-propellant motors mounted near the center of the reentry adaptor module, each with 11,070 N thrust. They would fire to initiate the spacecraft's reentry into the atmosphere, with attitude being maintained by a reentry control system of 16 engines, each with 5.2 N thrust. The retrorocket module would then be jettisoned, exposing the heat shield at the base of the reentry module. Along with the ablative heat shield, thermal protection during reentry was provided by thin Rene 41 radiative shingles at the base of the module and beryllium shingles at the top. Beneath the shingles was a layer of MIN-K insulation and thermoflex blankets. At roughly 15,000 meters, the astronauts would deploy a 2.4-meter drogue chute from the rendezvous and recovery section. At 3230 meters altitude, the crew releases the drogue, which extracts the 5.5-meter pilot parachute. The rendezvous and recovery section was released 2.5 seconds later, deploying the 25.6-meter main ring-sail parachute stored at the bottom of the section. The spacecraft was then rotated from a nose-up to a 35-degree angle for water landing. At this point, a recovery beacon was activated, transmitting via an HF whip antenna mounted near the front of the reentry module.

Gemini-Titan 1: Testing the Foundations of Space Exploration

On April 8, 1964, at 11:00 a.m. e.s.t., the inaugural mission of NASA's Gemini program, Gemini-Titan 1 (GT-1), roared to life from Complex 19 at Cape Kennedy. This mission marked a crucial milestone in American space history, utilizing the first operational Gemini spacecraft paired with the newly developed Gemini Launch Vehicle (GLV).

GT-1 was an unmanned mission meticulously crafted to validate the structural robustness of both the GLV and its spacecraft counterpart. It aimed to demonstrate the GLV's capability to position the spacecraft precisely in a predefined Earth orbit. The mission parameters dictated that the spacecraft would remain coupled to Stage II of the GLV throughout its orbital journey.

Six minutes after liftoff, GT-1 achieved orbital insertion, placing the GLV and spacecraft as a unified entity in orbit around Earth. The planned mission encompassed three orbits, culminating approximately 4 hours and 50 minutes after launch with a final pass over Cape Kennedy. Unlike subsequent Gemini missions, GT-1 did not incorporate a recovery phase. Instead, tracking efforts led by the Goddard Space Flight Center meticulously monitored the spacecraft's progress until its eventual reentry on the 64th orbit over the southern Atlantic Ocean on April 12, where it gracefully disintegrated upon atmospheric reentry.

The success of GT-1 solidified the GLV's technical proficiency and validated the structural integrity of the Gemini spacecraft, paving the way for subsequent manned missions that would push the boundaries of human space exploration.

On April 9, Phase II of the program to integrate a drogue stabilization chute into the parachute recovery system commenced at El Centro. This phase aimed to refine the stabilization chute design and establish optimal reefing parameters crucial for safe spacecraft recovery.

Parachute test vehicle after drop test on July 16, 1964. (NASA Photo No. 64-H-2451, July 16, 1964.)

The inaugural test in the series encountered significant setbacks. Using a weighted, instrumented bomb-shaped Parachute Test Vehicle (PTV), the test experienced multiple malfunctions, culminating in the loss of all parachutes and the destruction of the PTV upon impact with the ground. Despite extensive post-test analysis, investigators could not pinpoint the exact cause of these malfunctions.

The second drop, conducted on May 5, faced its own complications when an emergency drag chute prematurely deployed, disrupting the intended test conditions and yielding no actionable data. However, subsequent tests demonstrated marked improvement, with Phase II concluding on November 19 after completing the 15th drop in the PTV series.

With the conclusion of Phase II, developmental testing of the parachute recovery system's drogue configuration was finalized. This paved the way for qualification tests, which commenced on December 17, to certify the system's readiness for operational use.

On April 9, a significant milestone was achieved in developing the ballute stabilization system at the Arnold Engineering Development Center. The completion of structural qualification testing marked a crucial step forward in validating the effectiveness and reliability of this innovative stabilization device.

The testing regimen included a series of evaluations conducted in both subsonic and supersonic wind tunnel environments. Specifically, two subsonic runs and four supersonic runs were performed under design conditions, assessing the ballute's performance across varying dynamic pressures. Additionally, two ultimate runs were executed at 150 percent of the design's maximum dynamic pressure, further scrutinizing the device's structural integrity under extreme conditions.

These rigorous tests confirmed the four-foot ballute as fully satisfactory as a stabilization device. The ballute demonstrated robustness and stability across the range of conditions tested, affirming its capability to stabilize spacecraft during the descent and recovery phases effectively.

Following the successful structural qualification, the ballute system's final qualification continued as part of a broader personnel parachute and high-altitude drop test program initiated in January 1965. This comprehensive program aimed to finalize the ballute's integration into operational systems, ensuring its readiness for deployment in actual space missions.

On April 9, Flight Crew Support Division (FCSD) members visited McDonnell to address and strategize solutions for Gemini cockpit stowage challenges. This initiative was prompted by concerns over the adequate storage of essential equipment and supplies within the spacecraft, particularly in anticipation of upcoming missions.

The primary objective of the visit was to conduct a thorough review and discussion on Gemini cockpit stowage issues. FCSD members brought a mock-up of various equipment items, including the 16-millimeter camera window mount, flight medical kit, defecation gloves, and star chart with its holder. These items were integral to assessing and determining the optimal stowage requirements for mission success.

FCSD highlighted the potential criticality of stowage arrangements, especially considering the significant volume of camera equipment expected for upcoming missions. This foresight underscored the importance of efficiently utilizing space within the Gemini spacecraft to ensure accessibility and functionality during critical mission phases.

The collaboration between FCSD and McDonnell aimed to address and mitigate any potential issues related to cockpit stowage. By leveraging mock-ups and conducting detailed discussions, the team sought to optimize the spacecraft's interior layout to accommodate essential equipment while maintaining operational efficiency and crew comfort.

The insights gathered during this review session provided a foundation for future planning and enhancements in spacecraft design and functionality. Addressing cockpit stowage challenges early on was crucial in preparing for the demands of upcoming Gemini missions, reinforcing NASA's commitment to meticulous preparation and operational readiness.

On April 9, the Arnold Engineering Development Center initiated a test program to evaluate the thermal conditions on the base of the Gemini spacecraft during potential abort scenarios. This critical evaluation focused on assessing heat levels generated by the retrorockets when fired from altitudes exceeding 150,000 feet.

The primary objective of the test program was to simulate abort conditions and measure the thermal effects on the spacecraft's base during retrorocket firings. This assessment was crucial to ensure that excessive heat did not pose a risk to the spacecraft's structural integrity or operational systems.

Initial evaluations following the tests indicated that no significant heating issues were observed on the base of the Gemini spacecraft under simulated abort conditions. This preliminary conclusion provided reassurance that the spacecraft's design and thermal management systems were effectively mitigating potential heat-related concerns during critical phases of flight.

The findings from this test program laid a foundation for further refinement and validation of the spacecraft's thermal protection systems. They also informed ongoing engineering efforts to enhance the spacecraft's resilience and safety in abort scenarios.

On April 9-10, the Crew Systems Division of NASA convened a crucial design review focused on the food, water, and waste management systems intended for use aboard the Gemini spacecraft. This review session was pivotal in evaluating the functionality and readiness of several key production prototypes essential for supporting astronauts during their missions.

The primary objective of the design review was to assess the performance, reliability, and integration of various systems critical to sustaining crew members in space. Specifically, the review scrutinized production prototypes of the following components:

• Urine Transport System: Designed to manage and transport astronaut urine efficiently within the spacecraft.

• Water Dispenser: Responsible for delivering potable water to astronauts during their missions.

• Feeder Bag: A component involved in food and water distribution mechanisms.

• First-Day Urine Collection Bag: Used for initial urine collection during the early phases of the mission.

• Sampling Device: Instrumentation essential for collecting and analyzing samples in space conditions.

During the review, the urine transport system and water dispenser designs received official approval, indicating that they met stringent criteria for performance, safety, and integration within the spacecraft environment. These approvals marked significant milestones in the development process, ensuring these vital systems were ready for further testing and implementation.

While the urine transport system and water dispenser designs were approved, the remaining components—the feeder bag, first-day urine collection bag, and sampling device—were acknowledged as promising but requiring additional refinement. This recognition prompted further development efforts to optimize these systems to meet exacting standards for spaceflight operations.

The successful design review underscored NASA's commitment to meticulous planning and developing crew support systems for Gemini missions. By addressing feedback and iteratively improving system designs, NASA aimed to enhance the overall reliability and functionality of spacecraft operations in orbit.

Announcement of Gemini Prime and Backup Crews

On April 13, Director Robert R. Gilruth of NASA's Manned Spacecraft Center made a significant announcement, unveiling the prime and backup crews for the upcoming first manned Gemini flight. This pivotal moment marked a crucial step forward in NASA's ambitious human spaceflight program.

Virgil I. "Gus" Grissom, a seasoned astronaut renowned for his previous Mercury missions, was selected as the commander for the inaugural manned Gemini flight. Joining him as the pilot was John W. Young, an accomplished astronaut known for his technical expertise and contributions to space exploration.

Standing ready as the backup crew was Walter M. "Wally" Schirra, Jr., a veteran astronaut who had flown on Mercury-Atlas 8 and Gemini 6A missions, and Thomas P. Stafford, an experienced astronaut noted for his role in Gemini program preparations.

The announcement of these crews underscored NASA's meticulous crew selection process and marked a crucial milestone in preparing for the first manned Gemini mission. Virgil Grissom and John Young and their backup counterparts were tasked with pioneering new capabilities and demonstrating extended spaceflight operations as part of NASA's ongoing efforts to advance human exploration beyond Earth's orbit.

With the prime and backup crews in place, NASA moved forward with rigorous training and simulation exercises to prepare the astronauts for the challenges of spaceflight aboard the Gemini spacecraft. The announcement fueled excitement and anticipation among space enthusiasts worldwide as it heralded the imminent launch of manned missions that would push the boundaries of human achievement in space.

The selection of Virgil I. Grissom and John W. Young as the prime crew for the first manned Gemini flight, supported by Walter

M. Schirra, Jr., and Thomas P. Stafford as backups, marked a significant achievement in NASA's quest for space exploration milestones. Their dedication, expertise, and bravery paved the way for subsequent Gemini missions and contributed to the foundation of knowledge essential for future manned space endeavors.

Left to right are astronauts John W. Young, Virgil I. Grissom, Walter M. Schirra Jr. and Thomas P. Stafford. Gemini III crew assignments are as follows: Grissom, command pilot; Young, pilot, on the prime crew, with Schirra (command pilot) and Stafford (pilot) serving as alternates.
NASA

On April 13, the Air Force Space Systems Division (SSD) recommended a Gemini Agena launch without rendezvous to enhance confidence in the target vehicle's performance before attempting a rendezvous mission. However, the Gemini Project Office (GPO) deemed this plan impractical due to schedule, launch sequence, and cost constraints.

Instead of the proposed non-rendezvous mission, the GPO accepted SSD's alternate recommendation to designate one target vehicle as a Development Test Vehicle (DTV). This decision allowed for more extensive subsystems and systems testing, malfunction studies, and necessary modifications at Lockheed's facilities.

Gemini Agena Target Vehicle (GATV) 5001 was selected as the DTV for the Gemini program. Despite its designation as a DTV, GPO insisted on maintaining GATV 5001 in flight status until further authorization from the program office. All originally planned tests remained essential to demonstrate the vehicle's performance as a flight-ready component.

The designation of GATV 5001 as the DTV represented a strategic shift in testing and validation procedures within the Gemini program. By prioritizing comprehensive testing and modifications, NASA and its partners aimed to ensure the reliability and functionality of the Gemini Agena target vehicles for future missions, including critical rendezvous operations.

On April 14, the Electrical-Electronic Interference Tests commenced on Gemini Launch Vehicle (GLV) 2 at the vertical test facility located at Martin-Baltimore. These tests were crucial to ensuring the reliability and compatibility of electrical systems between GLV and Aerospace Ground Equipment (AGE).

Oscillograph recorders were employed to monitor 20 circuits involving GLV and AGE. During the tests, anomalies were detected in five circuits. Specifically, two hydraulic switchover circuits exhibited voltage transients that exceeded established failure criteria.

Upon identifying the anomalies, a special test was conducted to address the issues. The root cause was pinpointed to the AGE rather than the GLV itself. This discovery enabled technicians to implement corrective measures directly within the ground equipment, effectively resolving the voltage transient issues.

The Electrical-Electronic Interference Tests were pivotal in ensuring the robustness of electrical systems crucial to the Gemini launch vehicles. By meticulously monitoring and addressing anomalies during these tests, engineers and technicians upheld stringent quality standards and readiness criteria for upcoming Gemini missions.

Following the successful evaluation of Gemini-Titan (GT) 1, the Gemini Management Panel expressed optimism regarding the timeline for upcoming manned missions within the Gemini program. Their deliberations on April 15, 1964, highlighted key milestones and adjustments necessary for the next missions.

Based on the reviewed results, the panel projected that GT-2 could potentially launch on August 24, 1964, followed by GT-3 on November 16, 1964. These dates included a buffer for up to four-week delays per mission, ensuring flexibility in unforeseen challenges.

Special attention was directed towards GT-2, where the spacecraft itself had become the critical path item, a role previously held by the launch vehicle during GT-1. Despite a one-month delay in starting systems tests for Spacecraft No. 2, progress was reported to be satisfactory. However, concerns were noted regarding the tight schedule for commencing systems tests for Spacecraft No. 3, scheduled to begin on June 1.

The panel's optimistic outlook underscored their confidence in achieving manned flight capabilities by the end of 1964, building upon the success and lessons learned from GT-1. This readiness was a testament to NASA's meticulous planning and systematic approach to advancing human space exploration during the Gemini program.

On April 22, 1964, the formal Combined Systems Acceptance Test (CSAT) for Gemini Launch Vehicle (GLV) 2 was successfully concluded at the vertical test facility in Martin-Baltimore. This milestone marked a critical step in preparing GLV-2 for upcoming missions within the Gemini program.

Before the formal CSAT on April 22, three preliminary CSATs were conducted from April 17 to April 20. During these preliminary tests, any identified anomalies were diligently addressed and resolved to ensure the launch vehicle's readiness.

Following the completion of the preliminary tests, three additional nonscheduled tests were performed on GLV-2 before it was finally removed from the test facility:

Radio Frequency Susceptibility Test (April 26): This test aimed to demonstrate the GLV-2 ordnance's ability to withstand electromagnetic fields with intensities up to 100 watts per square meter. The test was conducted with live ordnance items connected in flight configuration to simulate real-world conditions.

Electrical-Electronic Interference Test (May 1): This test assessed the interference levels across the interface between the GLV and a spacecraft simulator. It aimed to ensure that electronic systems on the GLV and the spacecraft simulator could operate without mutual interference, which was crucial for mission success.

Rate Switch Package Replacement: The rate switch package was damaged during the CSAT on April 17. Following the formal CSAT on April 22, the damaged package was replaced, necessitating a retest to verify its functionality and integration with the GLV systems.

Gemini Launch Vehicle (GLV) 2 Acceptance and Preparation

Between April 27 and May 2, 1964, significant progress was made in preparing Gemini Launch Vehicle (GLV) 2 for its upcoming missions within the Gemini program.

From April 27 to May 1, the GLV-2 Vehicle Acceptance Team (VAT) convened at Martin Baltimore to conduct a comprehensive inspection. This inspection ensured that GLV-2 met all operational and safety standards required for its role in the Gemini missions.

On April 30, 1964, significant developments related to the Gemini Agena Target Vehicle (GATV) 5001 occurred, marking its acceptance and subsequent modifications for the Gemini program.

Air Force Space Systems Division (SSD) Acceptance: SSD accepted the Gemini program's first Agena D (AD-71). This Agena D vehicle was procured from Lockheed via routine procedures for NASA's use in the Gemini missions.

Retrofit Operations: Following its acceptance, minor retrofit operations were conducted on AD-71 to prepare it for its role as GATV 5001.

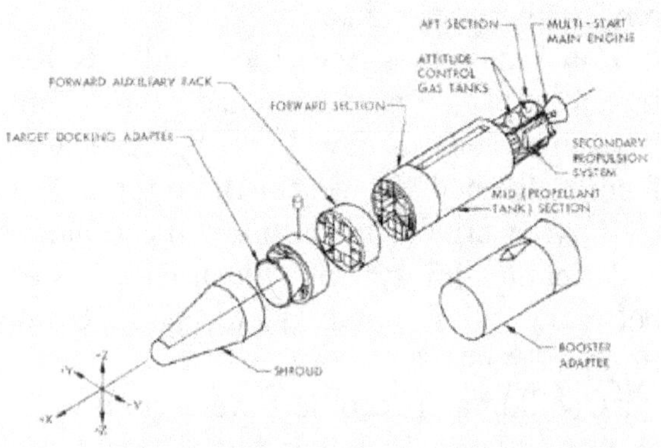

Configuration of the Gemini Agena target vehicle. (Lockheed, "Gemini Agena Target Vehicle Familiarization Handbook," LMSC A602521, Apr. 1, 1964, p. 1-6.)

Conversion to GATV 5001: On May 14, AD-71 entered the manufacturing final assembly area at the Lockheed plant for conversion into GATV 5001. This conversion process involved several significant modifications to adapt the Agena D into a specifically configured target vehicle for Gemini rendezvous missions.

Key Modifications Included:

• Installing a target docking adapter supplied by McDonnell was essential for the docking maneuvers planned during Gemini missions.

• An auxiliary equipment rack was added to accommodate extra systems and instrumentation needed for Gemini mission objectives.

• Integration of external status displays to provide visual feedback and critical information about the vehicle's operational status during missions.

• A secondary propulsion system to support maneuvering and docking operations during rendezvous with the Gemini spacecraft.

• Adding an L-band tracking radar enhanced tracking and navigation capabilities, crucial for rendezvous and docking maneuvers in space.

The conversion of AD-71 into GATV 5001 represented a pivotal step in preparing the Gemini program for rendezvous missions. These modifications ensured that GATV 5001 could meet the stringent requirements of the Gemini missions, including rendezvous and docking operations in Earth orbit. This process demonstrated the collaborative efforts between NASA, Lockheed, and SSD in advancing space exploration capabilities during the Gemini era.

The VAT inspection was completed on May 1, and GLV-2 was found acceptable after thorough scrutiny and assessment of its components and systems.

All Hands on Deck Preparing for the Gemini Flight

On May 1, 1964, pivotal advancements unfolded in the meticulous preparation for the Gemini spacecraft, heralding critical steps towards upcoming missions. The day commenced with successfully completing the Predelivery Acceptance Tests (PDA) for the Spacecraft Computer Formal Qualification Unit. This crucial milestone affirmed that the computer unit had met all designated specifications, underscoring its readiness for seamless integration into the Gemini spacecraft. With this achievement, the unit was promptly conveyed to McDonnell, where it would soon become an integral part of the ongoing Gemini spacecraft assembly process.

Simultaneously, another significant event transpired as the Spacecraft No. 2 flight unit arrived early in the month. This delivery encompassed a myriad of essential components and systems vital for the operational integrity of Spacecraft No. 2 during its upcoming missions.

The following day, May 2, witnessed the relocation of GLV-2 from the vertical test facility to the assembly area, marking the commencement of the next crucial phase in preparing the launch vehicle for its forthcoming mission. This transition underscored the meticulous planning and execution required as NASA and its contractors meticulously prepared for the challenges and triumphs in the Gemini program.

On May 5, a critical phase in perfecting the Gemini parachute recovery system for spacecraft No. 2 commenced with the first of a series of three tests using static article No. 7 at El Centro. This initial configuration omitted the drogue stabilization chute intended for spacecraft Nos. 3 and beyond. However, the test encountered setbacks as several failures were observed during the first drop. These failures necessitated immediate action from McDonnell, prompting redesign efforts to reinforce the brackets securing the parachute container to the rendezvous and recovery section, alongside adjustments to the sequencing circuitry.

Further refinements were imperative following the second test on May 28, where although the brackets buckled, they did not fail outright. Addressing these issues was paramount, leading to subsequent modifications and strengthening measures to enhance durability and reliability.

The culmination arrived on June 18 with the successful completion of the third and final test, marking the qualification milestone for the parachute system. Subsequently, static article No. 7 underwent modifications to facilitate Phase III testing to validate an updated parachute system incorporating the drogue chute. This iterative process, initiated on December 17, underscored NASA's commitment to rigorous testing and continuous improvement as they navigated towards ensuring the safety and efficacy of the Gemini missions.

From May 5 to May 7, the Manned Spacecraft Center's Landing and Recovery Division undertook rigorous rough water suitability tests using Gemini boilerplate spacecraft in the Gulf of Mexico. These tests unfolded amidst challenging sea conditions characterized by waves ranging from 4 to 8 feet and surface winds blowing at 20 to 25 knots.

Central to these trials was evaluating the spacecraft's flotation collar, initially air-dropped into the water. Crew members executed egress procedures from the spacecraft while on the water, validating essential survival kit recovery beacon operations. The efficacy of the dye marker, designed to produce a visible water pattern for aerial identification, was scrutinized and found partially unsatisfactory during the tests.

Despite the formidable conditions, the flotation collar demonstrated robust endurance, showcasing its reliability in mitigating the impact of rough seas. These trials underscored the meticulous preparation and dedication of the Manned Spacecraft Center's team as they refined critical procedures essential for ensuring the safety and recovery of astronauts during the Gemini missions.

On May 8, Langley Research Center concluded a series of tests involving a model of the Gemini launch vehicle, focusing on assessing both static and dynamic loads exerted by ground winds on the vehicle and its erector. These comprehensive tests spanned from April 15 and culminated in evaluations under simulated wind velocities ranging from 5 to 52 miles per hour.

The findings from these tests were reassuring, as they revealed that the wind velocities tested did not generate significant loads, which raised concerns. This validation underscored the robust design and structural integrity of the Gemini launch vehicle and its erector system, affirming their capability to withstand varying wind conditions without compromising operational safety or performance.

From May 11 to June 13, a pivotal operation unfolded as engineers replaced the interim Stage I engine on GLV-2 with a new flight engine. This meticulous process guaranteed the launch vehicle's utmost reliability and optimal performance for its upcoming missions.

The meticulousness involved in this engine replacement underscores NASA's commitment to ensuring every aspect of the Gemini program was thoroughly vetted and prepared for the challenges ahead.

On May 11, sea trials commenced for the tracking ship Rose Knot in Chesapeake Bay, marking a significant effort to analyze the impact of shock vibrations on Gemini equipment. Specific vibration issues were identified during these trials, particularly affecting the pulse-code-modulation system.

In parallel, simulations of Gemini-Agena systems were conducted using an instrumented Lockheed Super Constellation aircraft. This approach allowed for comprehensive testing and validation of equipment under conditions akin to those expected during operational missions.

Crew Station Evaluation: Gemini-Titan 3

On May 11-12, the primary and backup crews for Gemini-Titan 3 meticulously inspected the spacecraft No. 3 crew station mock-up at McDonnell. This critical evaluation aimed to ensure that all elements of the crew station met rigorous standards ahead of the mission.

The inspection yielded positive results, with both crews deeming the significant aspects of the crew station acceptable. While a few minor issues were noted, they were identified as non-critical and would not impact the mission's launch schedule.

On May 13, the Flight Operations Division presented Plan No. 3 of the Gemini Program Office's proposed mission to the Trajectories and Orbits Panel. This plan focused on the first Agena rendezvous flight and outlined preliminary details for achieving rendezvous at the first apogee during what was envisioned as a nominally perfect mission.

Though incomplete, the proposal aimed to establish a framework for the mission's trajectory and operational objectives. It underscored the meticulous planning and coordination required to orchestrate a successful rendezvous in space, marking a pivotal step forward in advancing the capabilities and ambitions of the Gemini program.

On May 19, a significant development unfolded. The Manned Spacecraft Center formally requested McDonnell submit a proposal for converting the Gemini spacecraft contract to a cost-plus-incentive-fee type. This transition aligned financial incentives with project performance and milestones, fostering efficiency and accountability in the program's execution.

The initiative stemmed from earlier efforts, starting on April 6, 1964, when Gemini Program Manager Charles W. Mathews appointed a committee led by Deputy Manager Kenneth S. Kleinknecht. This committee was tasked with preparing the proposal request, which included defining performance and schedule criteria critical to the contract's modification. By April 19, the Gemini Program Office had completed and reviewed these criteria, laying the groundwork for NASA Headquarters to approve the request during the week of May 3.

Later, on May 22, a complete inertial guidance system formal integration PDA was completed for spacecraft No. 2. This involved rigorous testing and verification of the inertial guidance system to ensure its proper guidance system to ensure its proper integration and functionality within spacecraft No. 2.

On May 29, Gemini spacecraft No. 3 embarked on Phase I of the modular Spacecraft Systems Tests (SST) at McDonnell, overseen by the Launch Preparation Group. This phase marked the initiation of comprehensive testing to validate the spacecraft's integrated systems and readiness for upcoming missions.

Following Phase I, the spacecraft's Development Engineering Inspection occurred on June 9-10, ensuring all systems met rigorous standards and specifications. Subsequent efforts focused on integrating the new rendezvous and recovery section, which included the high-altitude drogue parachute. This installation and checkout process spanned through July and August, crucially enhancing the spacecraft's capabilities for rendezvous and recovery scenarios.

By September 12, the modular SST and Phase II mated SST preparations were completed.

On May 31, the Manned Spacecraft Center (MSC) announced the development of various devices to prepare flight crews for scheduled extravehicular tests during the Gemini missions. These preparations were comprehensive and included several specialized training tools:

• Data Simulator: Designed to replicate the mechanical effects of a zero-gravity (zero-g) environment, providing crews with realistic training scenarios.

• Gemini Boilerplate No. 2: Utilized in a vacuum chamber to simulate space conditions, crucial for testing and familiarizing crews with spacecraft operations in a space-like environment.

• KC-135 Aircraft: Used for training in zero-g environments by performing parabolic flight maneuvers, allowing crews to practice ingress and egress procedures under weightless conditions.

• Gemini Mission Simulator: Used for procedural training and practicing vehicle control maneuvers while wearing pressurized suits, simulating mission-specific scenarios.

• Crew Procedures Development Trainer: A dedicated simulator for developing and refining crew procedures specific to Gemini missions.

• Flight Spacecraft: This involved direct training on the actual spacecraft, enabling crews to familiarize themselves with onboard systems and operations.

MSC anticipated that developing these devices and preliminary procedures would enable the commencement of a structured training program starting in August 1964. These comprehensive preparations highlighted NASA's proactive approach in ensuring that astronauts were thoroughly trained and prepared for the unique challenges of extravehicular activities during the Gemini missions.

By May 31, the Gemini Program Office (GPO) reported significant progress in developing and testing the orbit attitude and maneuver system thrust chamber assemblies (TCAs). Encouraged by a series of highly successful tests, the GPO confirmed the freezing of all TCA designs, marking a critical milestone in ensuring their reliability and performance for upcoming missions.

Key findings from the tests included:

• 25-pound TCA: Tested up to 2100 seconds under mission duty cycle conditions, with a maximum skin temperature of 375 degrees F, well within the 600 degrees F allowable limit.

• 85-pound TCA: Successfully operated for 3050 seconds under mission duty cycle conditions, with skin temperatures not exceeding 320 degrees F.

• 100-pound TCA: Two tests were conducted:

The first test ran for 757 seconds, achieving skin temperatures of 230 to 250 degrees F before termination.

The second test lasted 1950 seconds until fuel exhaustion, reaching a maximum skin temperature of 600 degrees F.

GPO attributed the success of these tests to improved injector screening techniques, reorientation of ablation material laminates, and the implementation of boundary-layer cooling techniques suggested by Rocketdyne. These innovations significantly enhanced the thermal management and performance capabilities of the TCAs.

In May, Rocketdyne finalized the design for long-duration TCAs, which were slated for production and installation starting with spacecraft No. 5. This upgrade included outfitting spacecraft No. 5 with 100-pound aft-firing thrusters and all 25-pound thrusters equipped with long-life TCAs. A complete set of long-life TCAs was scheduled for installation in spacecraft No. 6, underscoring NASA's commitment to incorporating advanced technologies to enhance mission reliability and success in the Gemini program.

On June 3, in collaboration with the Air Force and NASA, Lockheed launched the Gemini Extra Care Program to minimize equipment failures and discrepancies arising from poor or careless workmanship during the modification and assembly of the Agena target vehicle. This initiative introduced a range of measures to improve manufacturing practices and foster a robust team spirit across all operational levels.

Key components of the Gemini Extra Care Program included:

• Increased Inspection: Heightened scrutiny and quality checks throughout the assembly process.

• Exhortation: Encouragement and motivation initiatives to emphasize the importance of meticulous workmanship.

• Morale Boosters: Activities designed to maintain and enhance team morale and cohesion.

• Special Awards: Recognition for exemplary performance and contributions to quality improvement.

• Other Activities: Various supplementary efforts aimed at reinforcing a culture of precision and excellence.

The program's effectiveness was swiftly validated by a substantial reduction in Failed Equipment and Discrepancy Reports (FEDRs) documented in the Gemini final manufacturing area across successive vehicles. This outcome underscored the program's pivotal role in elevating manufacturing standards and ensuring the reliability of components critical to the Gemini missions.

On June 4, dynamic qualification testing of the Gemini ejection seat commenced with sled test No. 6 at China Lake. This initial test aimed to validate the functionality of hatches and hatch actuators under abort conditions, simulating scenarios without actual ejection attempts. The test proved successful, paving the way for formal qualification testing starting from July 1 with test No. 7.

Test No. 7 simulated high dynamic pressure conditions following an abort from the powered phase of Gemini flight, replicating the vehicle's orientation heatshield forward as during reentry. Both seats were ejected, and all systems operated flawlessly as designed. However, subsequent sled testing was delayed due to slow delivery of pyrotechnics, with test No. 8 postponed until November 5.

During test No. 8, a critical structural deficiency was identified when one dummy's feet came out of the stirrups, causing the seat to pitch and yaw to the left. This overload led to the left side panel breaking off, interrupting the ejection sequence and resulting in the destruction of both the seat and dummy upon impact.

In response to this failure, representatives from the Manned Spacecraft Center and McDonnell convened during the week of November 15 to reassess the test program. They agreed to proceed with test No. 9 under conditions approximating the most severe for which the ejection system was designed. This test, conducted on December 11, successfully demonstrated the entire ejection sequence, confirming the effectiveness of the structural redesign.

With the completion of test No. 9, the qualification sled test program concluded, marking a critical phase in ensuring the reliability and safety of the Gemini ejection seat system under demanding operational conditions.

On June 6, the flight unit designated for spacecraft No. 3 achieved Predelivery Acceptance (PDA), marking a significant milestone in the Gemini program. This rigorous testing milestone confirmed that all components and systems of spacecraft No. 3 had successfully passed comprehensive testing protocols.

The Predelivery Acceptance process ensures that spacecraft components meet stringent requirements and are ready for integration into the final assembly phase. With the completion of PDA, spacecraft No. 3 was cleared for subsequent phases of integration, further testing, and preparation for its upcoming missions.

On June 8, all astronauts assigned to the Gemini program commenced launch abort training using the Ling-Temco-Vought simulator. This marked a pivotal phase in preparing the astronauts for potential emergency scenarios during Gemini missions.

The training regimen was comprehensive. Astronauts in groups 1 (selected April 1959) and 2 (selected September 1962) completed approximately 100 simulated launch abort runs each, and astronauts in group 3 (selected October 1963) completed 32 simulated launch abort runs each.

The Gemini-Titan 3 launch profile was meticulously simulated, incorporating realistic cues such as noise, vibration, pitch, roll programming, and other motion anomalies that astronauts could encounter during various launch phases. The training sessions, designed to simulate and prepare astronauts for critical abort scenarios, concluded on July 30.

On June 10, the Air Force Space Systems Division's procurement strategy for Gemini launch vehicles (GLVs) and associated aerospace ground equipment significantly changed. The existing cost-plus-fixed-fee contract with Martin was replaced with a more dynamic cost-plus-incentive-fee contract, reflecting evolving practices in aerospace contracting.

Key details of the new contract included negotiations conducted between March 15 and April 30, 1964. This cost-plus-incentive-fee contract incorporated cost, performance, and schedule incentives with a Target cost of $111 million, a Target fee of $8.88 million, a Maximum fee of $16.65 million, and a Minimum fee of $3.33 million from July 1, 1963, through December 31, 1967.

Deliverables: Included 14 GLVs (one already delivered) and associated equipment and services for checkout and launch operations.

On June 12, representatives from NASA, McDonnell, Weber Aircraft, and Air Force 6511th Test Group convened to outline the program's objectives to validate the functional reliability of the Gemini personnel recovery system under simulated operational conditions. This initiative stemmed from discussions dating back to October 30, 1963, during a coordination meeting focused on the ejection seat system.

Program Outline: The planned program aimed to demonstrate the compatibility and reliability of the recovery system through a series of tests:

Static Ground Test: Initially scheduled for September but delayed by pyrotechnics availability issues. It eventually occurred on October 15, utilizing pyrotechnics from the paraglider tow-tail vehicle (TTV) seat. This test successfully demonstrated the system's compatibility with the aircraft but lacked certain components required for a comprehensive evaluation.

Full System Tests: These were originally intended to commence shortly after that but faced further delays until January 28, 1965, due to ongoing pyrotechnics supply challenges. The full system tests involved using a production configuration of the recovery system to assess its operational effectiveness under simulated recovery scenarios.

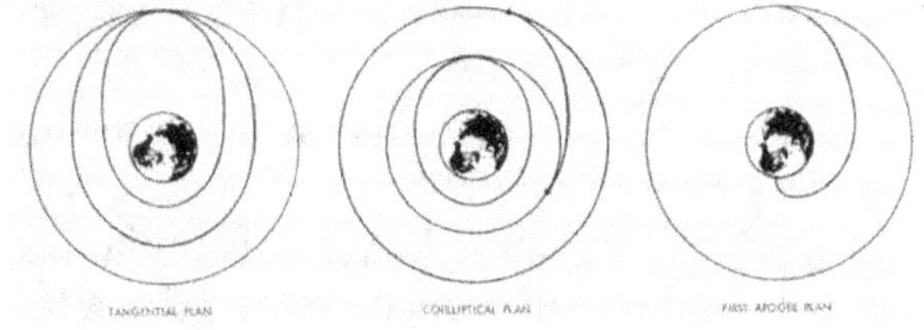

The three basic rendezvous plans being considered for the first Gemini rendezvous mission.

Despite initial setbacks, the program underscored NASA's commitment to ensuring the safety and reliability of the Gemini personnel recovery system, crucial for the success of manned missions during this era of pioneering space exploration.

On June 12, Christopher C. Kraft, Jr., Assistant Director for Flight Operations at the Manned Spacecraft Center, outlined three fundamental plans under consideration for Gemini rendezvous missions. This pivotal decision-making process aimed to optimize mission success and operational flexibility.

Rendezvous at First Apogee: Initially considered but likely to be rejected due to potential orbital dispersions that could necessitate plane changes, complicating mission logistics.

Rendezvous from Concentric Orbits: It is identified as advantageous due to its flexibility in selecting the geographic position for rendezvous points.

Tangential Rendezvous: Received significant attention and effort in preliminary studies.

Decision and Implementation: Following a thorough evaluation, the concentric orbit plan was selected for Gemini-Titan 6, slated as the first rendezvous mission. This strategic decision reflected NASA's commitment to leveraging optimal orbital mechanics and operational planning to achieve successful rendezvous in future Gemini missions.

On June 16, Lockheed initiated test-firing of the propulsion test vehicle assembly at its Santa Cruz Test Base, marking a crucial phase in adapting the Agena D for Gemini missions. This comprehensive testing program was essential due to the extensive modifications required in the propulsion system.

Testing Phases:

First Series (June 16 - July 16): This series established baseline performance for both the Primary Propulsion System (PPS) and Secondary Propulsion System (SPS). It included subjecting one SPS module to dynamic and acoustic environments created by 55 seconds of PPS firing.

Second Series (July 16): Successfully simulated a full Gemini mission profile, incorporating multiple firings and various coast and burn times for both PPS and SPS units.

Third Series (August 7): Concluded the test program with maximum start and minimum impulse firings on both PPS and SPS units. All firings were successful, and post-test analysis revealed only minor anomalies.

Test Results: The entire program encompassed:

27 PPS firings totaling 545 seconds.

30 SPS Unit I firings totaling 286 seconds.

11 SPS Unit II firings totaling 268 seconds.

Post-test inspections confirmed no physical damage to any equipment, validating the robustness and reliability of the adapted propulsion systems for Gemini missions.

On June 17, 1964, a pivotal development unfolded within the Air Force Space Systems Division's procurement efforts for the Gemini launch vehicle. The existing cost-plus-fixed-fee contract with Aerojet-General, responsible for engines and related aerospace ground equipment, transformed. This marked the transition to a more incentivized structure under a newly minted cost-plus-incentive-fee contract.

The negotiations lasted from May 25 to June 17, 1964, and culminated in a detailed agreement to enhance cost efficiency and performance reliability. The revised contract encompassed acquiring 14 complete sets of engines, including one already delivered, alongside associated equipment. This procurement phase was scheduled to unfold between July 1, 1963, and December 31, 1967.

The agreement has a dynamic range of incentives covering cost, performance benchmarked, and stringent schedule adherence. These measures were designed to optimize operational outcomes while ensuring fiscal prudence. Notably, the contract outlined a potential maximum fee reaching $5,885,250, contingent upon superior performance metrics, juxtaposed against a minimum fee baseline of $1,177,050.

The contract set ambitious yet structured goals, initially targeting a cost estimate of $39,235,000 with an accompanying fee aspiration of $3,138,800. This framework underscored a strategic commitment to advancing propulsion capabilities crucial for the Gemini missions, epitomizing the era's pioneering spirit in aerospace engineering and exploration.

In mid-June 1964, a pivotal event unfolded as representatives from the Air Force Space Systems Division (SSD), Aerospace Corporation, and NASA convened for the official roll-out inspection of GLV-2. This meticulous inspection marked a crucial milestone in the preparation and validation process for the Gemini launch vehicle, reinforcing rigorous quality standards and readiness protocols.

On June 19, 1964, the first stage of the Gemini launch vehicle 3 was carefully erected within the vertical test facility at Martin-Baltimore. This significant step underscored the meticulous planning and engineering precision essential for the subsequent stages of the Gemini program. Each phase of assembly and testing was conducted with meticulous attention to detail, reflecting the collective expertise of aerospace engineers and technicians committed to mission success.

By June 22, 1964, Stage II of the Gemini launch vehicle three was successfully erected, further advancing the comprehensive testing and validation process at Martin-Baltimore. The sequential assembly of each stage underscored the systematic approach taken to ensure structural integrity and operational readiness for future missions.

On June 29, 1964, a significant milestone was achieved as power was first applied to the Gemini launch vehicle 3, initiating a series of comprehensive functional verification tests. These tests rigorously evaluated the performance and integration of subsystems critical to mission success. Over the ensuing weeks, thorough testing continued, culminating in the conclusive verification of subsystem functionality by July 31, 1964.

These pivotal dates and events illustrate the meticulous preparations and rigorous testing protocols integral to the Gemini program, highlighting a concerted effort to achieve new frontiers in manned space exploration.

Gemini Recovery School and GLV-2 Acceptance

On June 22, 1964, significant developments unfolded within the Gemini program, marking key readiness and operational preparation milestones.

At Kindley Air Force Base, Bermuda, operations commenced for the inaugural Gemini Recovery School. Organized by the Landing and Recovery Division of the Flight Operations Directorate, this specialized training course represented a crucial initiative in preparing recovery personnel for upcoming Gemini missions. The curriculum catered to diverse roles, including pararescue crews, Air Force navigators, and maintenance personnel, emphasizing the critical skills and procedures essential for safe and effective spacecraft recovery operations.

Simultaneously, on June 22, 1964, the Air Force Space Systems Division (SSD) formally accepted GLV-2 following rigorous evaluations to ensure compliance with exacting specifications and operational requirements. This milestone underscored the culmination of extensive testing and preparation efforts to validate the readiness of the Gemini launch vehicle for future missions.

GLV-2 was originally slated for delivery to the Eastern Test Range (ETR), formerly the Atlantic Missile Range, on June 22, 1964. However, the transport was rescheduled to July 10. This adjustment in the timeline was necessary to accommodate additional modifications and final preparations essential for optimizing the vehicle's readiness and performance capabilities. The rescheduling underscored a proactive approach to ensuring comprehensive readiness and adherence to stringent operational standards as the Gemini program advanced toward critical testing phases.

On June 24, 1964, a pivotal milestone was achieved with the completion of construction for the Gemini-Agena facilities at Complex 14. This marked a significant advancement in infrastructure development crucial for supporting upcoming missions within the Gemini program.

Following the construction milestone, General Dynamics installed and comprehensively checked equipment within the Launch Operations Building. This intricate process, completed by July 20, 1964, ensured the operational readiness of critical systems essential for integrating and launching Gemini-Agena missions.

Concurrently, Lockheed installed and meticulously checked their equipment within the Launch Operations Building. This phase was successfully concluded by July 31, 1964, underscoring the systematic approach and collaborative efforts undertaken to establish robust operational capabilities at Complex 14.

On June 25, 1964, Martin-Baltimore reached a critical milestone in preparing Gemini launch vehicle (GLV) 5, marking significant progress towards readiness for upcoming missions.

Martin-Baltimore received the propellant tanks for GLV-5 from Martin-Denver, which initiated fabrication in October 1963. This meticulous fabrication ensured the tanks met stringent specifications and readiness criteria for spaceflight operations.

Aerojet-General delivered the flight engines designated for GLV-5 on November 5, further advancing the assembly and integration phases of the launch vehicle. The timely delivery underscored collaborative efforts and adherence to scheduled milestones critical for mission planning and execution.

Following the delivery of components, Martin-Baltimore proceeded with crucial assembly tasks. Tank splicing, a complex operation to integrate and prepare propellant tanks for launch, was completed by December 5. Engine installation swiftly followed on December 9, marking decisive steps towards the final assembly of GLV-5.

The culmination of preparation efforts was marked by completing final horizontal tests on January 7, 1965. These comprehensive tests rigorously evaluated the integrated functionality and operational performance of GLV-5, ensuring readiness for the rigorous demands of upcoming missions in the Gemini program.

June 30, 1964, marked a pivotal moment in the testing phase of the Gemini spacecraft. This phase focused specifically on its readiness for water impact landing—an essential aspect of ensuring astronaut safety and mission success.

McDonnell initiated the first of two critical tests to qualify the spacecraft for a water impact landing on this date. Using Static Article No. 4, the test involved dropping the spacecraft from the landing system test rig and heatshield forward. Remarkably, the article sustained no damage during this initial test, highlighting the robust design and structural integrity crucial for withstanding impact forces.

Subsequently, on July 13, McDonnell conducted the second testing phase by dropping the unit conical section forward. Following this drop, the cabin's meticulous pressure decay test revealed a minor leak. However, the spacecraft unit was deliberately left submerged in water for two weeks as part of the qualification process.

During its immersion, the spacecraft unit consumed approximately a pint of water. Importantly, this outcome met stringent qualification requirements, affirming the spacecraft's capability to withstand prolonged exposure to water post-landing and validating its readiness for future Gemini missions.

On July 3, 1964, a significant phase commenced in preparing Gemini spacecraft No. 2, marking a series of critical tests and milestones leading up to its eventual deployment for space missions.

Following the successful mating of its modules, Gemini Spacecraft No. 2 initiated Phase II of Spacecraft Systems Tests (SST) at McDonnell. This comprehensive testing phase, spanning from July through September, aimed to validate all spacecraft systems' integrated functionality and readiness crucial for manned space missions. Throughout August and September, operations alternated between rigorous testing protocols and installing flight items essential for operational deployment.

From August 20 to 24, vibration testing of the spacecraft and its systems was successfully conducted, ensuring structural integrity and operational reliability under simulated launch and mission conditions. These tests were pivotal in verifying the spacecraft's ability to withstand the rigors of spaceflight environments.

Attitude chamber tests were omitted for spacecraft No. 2, as the upcoming Gemini-Titan 2 mission was planned to be unmanned. This strategic decision underscored the mission-specific testing approach tailored to operational requirements and mission objectives.

Phase II mated SST culmination was marked by the Simulated Flight Test conducted from September 3 to 15, 1964, validating mission-critical systems and operational scenarios. Following a thorough spacecraft acceptance review on September 17-18, Gemini spacecraft No. 2 was cleared for deployment and flown to Cape Kennedy on September 21, 1964, signaling readiness for upcoming Gemini missions.

During the week of July 6-12, 1964, a significant milestone was reached in developing the extravehicular life support system (ELSS) chest pack for the Gemini program. This period marked the inaugural design review conducted to evaluate the feasibility and functionality of this crucial component essential for spacewalking missions.

The Manned Spacecraft Center (MSC) thoroughly evaluated AiResearch's basic design for the ELSS chest pack. While the basic design was conditionally approved, MSC recommended specific modifications to enhance performance, reliability, and astronaut safety during extravehicular activities (EVAs).

The design review process identified areas where enhancements could optimize the chest pack's operational capabilities. Recommendations focused on refining technical specifications, improving ergonomics, and ensuring compatibility with the spacecraft's overall life support systems.

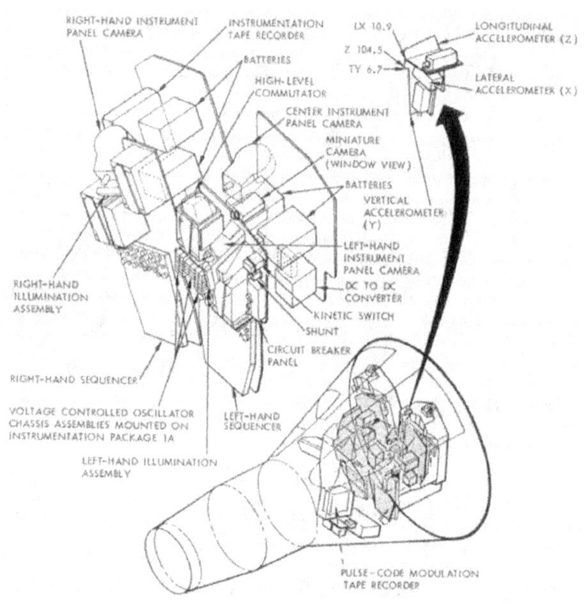

Special instrumentation pallets to be installed in Gemini spacecraft No. 2 in the same positions that astronauts would occupy in later flights. (NASA Photo S-652263, undated.)

The conditional approval with recommendations set the stage for AiResearch to implement necessary changes and refinements to the ELSS chest pack design. This iterative process was crucial in aligning the technology with stringent operational requirements and safety standards for manned space missions.

On July 7, 1964, McDonnell submitted a significant proposal to convert the existing Gemini spacecraft contract to a cost-plus-incentive-fee (CPIF) structure. This pivotal initiative marked a strategic shift in contractual terms and financial incentives to optimize project management and operational outcomes within the Gemini program.

McDonnell, the primary contractor responsible for the Gemini spacecraft, presented its proposal to transition from the current contractual framework to a CPIF contract. This proposed adjustment aligned financial incentives with performance metrics, cost efficiency, and timely delivery of spacecraft components and systems.

Upon receipt of McDonnell's proposal, the Manned Spacecraft Center (MSC) embarked on a rigorous analysis and evaluation process. This comprehensive review aimed to assess the feasibility, benefits, and potential impacts of adopting a CPIF contract structure for the Gemini spacecraft program.

The proposed conversion aimed to enhance project oversight, incentivize cost-effective practices, and foster innovation while ensuring stringent quality and safety standards. By linking contract incentives to performance goals, McDonnell and MSC aimed to optimize operational efficiency and mitigate risks associated with complex aerospace projects.

On July 10, 1964, significant developments unfolded within the Gemini Program Office as plans for future Gemini-Titan (GT) missions 4 through 7 were reviewed and evaluated, outlining ambitious objectives and mission profiles crucial for advancing manned space exploration.

Manager Charles W. Mathews provided insights into the proposed missions:

GT-4: Planned as a four-day mission utilizing battery power, testing endurance and operational capabilities under extended mission durations.

GT-5: Envisaged to incorporate radar systems and a rendezvous evaluation pod, facilitating critical exercises in rendezvous

techniques early in the flight. The mission's duration was expected to be open-ended, spanning up to seven days, contingent upon the availability and performance of fuel cells.

GT-6: Designed as a standard rendezvous mission, likely lasting approximately two days, focusing on refining and validating rendezvous procedures critical for future space operations.

GT-7: Conceptualized as a long-duration mission with an open-ended potential lasting up to 14 days, testing spacecraft endurance, crew performance, and life support systems over extended periods in space.

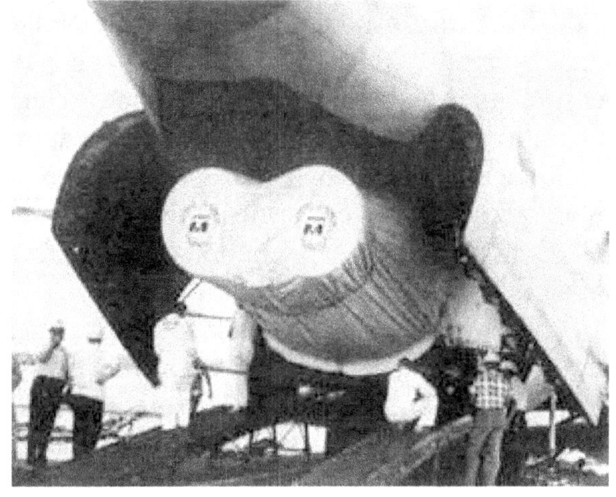

The first stage of Gemini launch vehicle 2 being unloaded from an Air Force C-133 at Cape Kennedy. (KSC 64-14608, July 11, 1964.)

George E. Mueller, NASA's Associate Administrator for the Office of Manned Space Flight, was actively reviewing these ambitious plans. His oversight and strategic direction played a pivotal role in shaping the trajectory of the Gemini program, ensuring alignment with broader space exploration objectives and technological advancements.

The outlined missions underscored NASA's commitment to pushing the boundaries of manned spaceflight capabilities. They encompass diverse objectives, from endurance testing to advanced rendezvous maneuvers and extended-duration missions. These missions were pivotal in preparing astronauts and spacecraft for the challenges of future lunar missions and beyond.

July 11, 1964, stands as a pivotal date in the Gemini program, symbolizing a crucial advancement in the readiness and operational progression of the Gemini Launch Vehicle (GLV) 2.

On this day, GLV-2 was airlifted to the Eastern Test Range (ETR), formerly known as the Atlantic Missile Range. This relocation marked a significant milestone in the Gemini program's timeline, signifying GLV-2's readiness for rigorous testing and validation for upcoming manned space missions.

The airlifting of GLV-2 to the ETR underscored meticulous preparations to ensure the launch vehicle's readiness and operational capability. This relocation facilitated comprehensive testing and integration activities critical for verifying GLV-2's performance, reliability, and safety in simulated mission environments.

The transfer to ETR enabled NASA and its partners to conduct rigorous tests, evaluations, and launch rehearsals essential for achieving mission success and astronaut safety. It marked a pivotal step towards advancing manned space exploration goals during the dynamic Gemini era.

In mid-July, the Eastern Test Range welcomed the arrival of Gemini launch vehicle 2, marking a pivotal moment in America's ambitious space exploration program. The spacecraft's journey began with the erection of Stage I at Complex 19 on July 13, swiftly followed by Stage II on July 14. These critical components of the mission's propulsion and systems architecture set the stage for an intensive series of tests and preparations.

By July 20, the meticulous process of powering up the vehicle commenced a crucial step preceding the Subsystems Functional Verification Tests scheduled to begin on July 21. These tests were designed to rigorously assess and validate every spacecraft subsystem, ensuring its readiness for the challenging mission ahead.

During July 16-17, the Flight Crew Support Division raised significant concerns regarding McDonnell's procedures for conducting ejection seat sled tests. These tests were pivotal in ensuring the safety and reliability of the ejection seats intended for use in the Gemini spacecraft. However, the division found the procedures inadequate to instill confidence in the seats' suitability for manned missions.

Of particular concern were the methods used to rig dummies for these tests. The dummies were subjected to extreme restraint-harness tensions and highly torqued joints, conditions that could not accurately simulate human responses. This discrepancy prompted the Flight Crew Support Division to request McDonnell reassess their approach and provide a detailed report to the Gemini Program Office.

Between July 19 and July 25, the Gemini Program Office encountered a significant setback during the testing phase of section I of the fuel cells intended for the upcoming Gemini-Titan 5 mission. Initial tests revealed a troubling issue: output decay, indicating a failure in the expected performance standards.

The discovery prompted an immediate and thorough investigation into the root cause of the malfunction. Engineers and technicians conducted a comprehensive analysis to pinpoint factors contributing to the output decay and devise corrective measures. This rigorous investigation was crucial for ensuring the success of the Gemini-Titan 5 mission and maintaining the high standards of reliability and safety that defined NASA's pioneering efforts in manned space exploration.

Gemini-Titan 4 Mission

On July 27, NASA announced the crew assignments for the upcoming Gemini-Titan (GT) 4 mission, scheduled for the first quarter of 1965. Astronauts James A. McDivitt and

Edward H. White II were named as the command pilot and pilot, respectively, with Frank Borman designated as the backup command pilot and James A. Lovell, Jr. as the backup pilot.

The GT-4 mission was slated to last up to four days and would involve the execution of 10 to 11 experiments aimed at advancing scientific understanding in space. These experiments included medical tests, radiation measurements, and the study of Earth's magnetic field. These endeavors underscored NASA's commitment to leveraging manned space missions as scientific research and exploration platforms.

At a subsequent press conference on July 29 at the Manned Spacecraft Center, Deputy Gemini Program Manager Kenneth S. Kleinknecht outlined further details about the mission. He highlighted a significant experiment wherein an astronaut would be exposed to the hazards of outer space without the full protection of the spacecraft. Initially described as potentially involving "stepping into space," Kleinknecht later clarified that it might entail opening a hatch and standing up.

On July 27, the inaugural meeting of the Gemini Configuration Control Board marked a pivotal milestone in the Gemini program. This board convened for the first time and set the stage for rigorous oversight and management of the spacecraft's configuration—a critical aspect in ensuring mission success and safety.

At this meeting, McDonnell presented their proposal for implementing the spacecraft configuration management system. The program office carefully reviewed this system, Which was essential for maintaining consistency and reliability across all components of the Gemini spacecraft. Initial steps towards implementation were already underway, marking a proactive approach to streamlining processes and enhancing operational efficiency.

Scheduled to convene weekly after that, these meetings reflected NASA's commitment to meticulous planning and coordination. They provided a platform for stakeholders to discuss and resolve issues pertaining to spacecraft design, modifications, and system integration—a framework vital for navigating the complexities of manned space missions.

On July 29, Flight Crew Support Division personnel embarked on a crucial simulation at Langley Research Center, focused on refining the optical rendezvous maneuver for the Gemini spacecraft. This simulated scenario played out inside a 40-foot diameter radome, meticulously recreating the commanding view from the spacecraft's pilot station and window port.

Central to the simulation was the projection of a flashing target against a backdrop of stars. This setup aimed to replicate the intricate conditions of space, where precise alignment and control are paramount during rendezvous operations. A notable discovery during the demonstration was the utility of a lighted window reticle, enhancing the accuracy of line-of-sight control tasks critical to the maneuver's success.

On July 29, North American Aviation achieved a significant milestone in the Paraglider Landing System Program with the first tow test vehicle (TTV) captive-flight test. Conducted under controlled conditions, a helicopter towed the TTV to an altitude of 2600 feet. The test pilot successfully maneuvered the vehicle for about 20 minutes before executing a smooth three-point landing. Immediately after touchdown, the tow cable was released, followed by the wing deployment approximately four seconds later. This initial flight was hailed as highly successful, demonstrating promising progress for the program.

However, subsequent developments on August 7 revealed challenges during a free-flight test. Despite being towed to a higher altitude of 15,500 feet and then released, the TTV entered into a series of uncontrolled turns, necessitating the pilot to bail out for safety. In response, North American Aviation initiated an intensive investigation and improvement effort to isolate and rectify the malfunction.

This effort included a rigorous test program involving 14 radio-controlled, half-scale TTV flights between August 24 and December 13. These tests aimed to refine the vehicle's control systems and operational parameters. By December, successful radio-controlled, full-scale TTV free flights on December 15 and 17 provided renewed confidence. This progress paved the way for another pilot-controlled flight attempt on December 19, which concluded with excellent results, marking a significant turnaround in the program's trajectory.

Chapter 8 - Exploring Future Frontiers

On July 30, prompted by a request from NASA Headquarters, the Gemini Program Office (GPO) presented a visionary study outlining potential missions beyond 12 Gemini flights. Titled "The Advanced Gemini Missions Conceptual Study," this document proposed 16 additional missions. These envisioned missions spanned diverse objectives, reflecting NASA's evolving goals in manned space exploration.

Among the proposed missions were groundbreaking concepts such as experiments aboard a space station, satellite rendezvous, capture missions, hypothetical lifeboat rescue missions, and missions to orbit the Moon and even circumnavigate it. Each mission concept aimed to push the boundaries of human spaceflight capabilities and expand scientific knowledge in uncharted realms.

However, by February 28, 1965, the Gemini Program Office reported that a preliminary proposal for follow-on missions, specifically testing a land landing system, had not received approval. Consequently, spare Gemini launch vehicles designated for flights 13, 14, and 15 were canceled, marking a pivot in the program's trajectory. At that time, there were no immediate plans for missions beyond the originally approved 12-flight program.

From August 4 to 6, the Manned Spacecraft Center's Propulsion and Power Division undertook a crucial Gemini fuel cell system test. Designed to power the spacecraft during missions, this test was intended to validate the system's reliability and performance under operational conditions.

However, the test encountered an unexpected setback when the system was inadvertently operated for 15 minutes under a short circuit condition before the scheduled test began. This incident compromised the system's performance, revealing that two of the cells could not sustain loads of six amperes as required. Recognizing the implications of these findings, the test was promptly terminated to prevent further potential damage.

Subsequent analysis of the test yielded a product water sample that exhibited highly acidic properties, suggesting a possible membrane failure within the fuel cell. This discovery underscored the importance of rigorous testing and quality control measures in aerospace engineering, particularly in ensuring the robustness of vital spacecraft components.

On August 7, the formal Combined Systems Acceptance Test (CSAT) of Gemini launch vehicle (GLV) 3 was successfully conducted, marking a critical milestone in its preparation for future missions. This comprehensive test, aimed at validating the launch vehicle's integrated performance, confirmed its operational readiness.

Following the CSAT, the Vehicle Acceptance Team (VAT) convened on August 17 to meticulously review the test results alongside other pertinent data from the testing and manufacturing phases. Despite GLV-3 not being immediately required at Cape Canaveral, the Manned Spacecraft Center, in alignment with Aerospace recommendations, opted to implement all engineering changes at Baltimore rather than on-site at the Cape.

After thoroughly assessing these modifications, the VAT instructed Martin to proceed with a second CSAT upon their completion. These necessary adjustments were finalized by September 15, followed by a subsystems retest concluding on September 28. The subsequent second CSAT, validating the effectiveness of these modifications, was successfully executed on September 30.

During a NASA-McDonnell Management Panel meeting on August 14, a critical issue arose regarding the extravehicular activity (EVA) chest pack size. It was noted that if stowed aboard Spacecraft No. 6, the chest pack would occupy space otherwise allocated for conducting experiments during that mission. This concern also extended to subsequent missions, prompting a thorough discussion on potential solutions.

In response to this challenge, McDonnell was tasked with conducting a study to assess the impact of the EVA chest pack on spacecraft design and operation. Additionally, the panel requested alternative proposals to mitigate the space constraint issue. One proposed alternative involved storing some experiments in the adapter section of the spacecraft. However, this solution necessitated that EVA would become a prerequisite for conducting those specific experiments.

This deliberation highlighted the delicate balance between accommodating essential equipment for astronaut activities and maximizing the scientific utility of spacecraft missions. It underscored the collaborative efforts between NASA and its contractors to innovate and adapt spacecraft configurations to optimize mission objectives while ensuring operational efficiency and safety in the pioneering era of manned space exploration.

On August 16, Martin-Baltimore received the propellant tanks destined for Gemini launch vehicle 6, marking a significant step in the vehicle's preparation for an upcoming mission. These tanks, fabricated by Martin-Denver starting in April, underwent meticulous inspection upon arrival at Martin-Baltimore.

Following inspection, the propellant tanks were carefully stored at the facility. They remained in storage until December 18, awaiting their integration into the Gemini launch vehicle. This storage period allowed for continued preparations and integration activities to ensure the vehicle's readiness and reliability for its mission.

On August 17, a severe electrical storm near Complex 19 disrupted the testing of Gemini launch vehicle (GLV) 2. Reports indicated a lightning strike in the vicinity, prompting an immediate halt to all testing activities. This incident triggered a comprehensive investigation into the electromagnetic effects (EME) potentially caused by the lightning strike.

The investigation concluded on September 2, revealed no physical signs of direct impact but identified multiple failed components. Most failures occurred in aerospace ground equipment (AGE), with some affecting GLV-2 itself. While Complex 19 was not directly hit, the damage resulted from electromagnetic interference or static charges induced by the nearby lightning strike.

In response, a meticulous recovery plan was devised to restore confidence in all systems associated with the launch vehicle, including AGE, ground instrumentation equipment, and facility systems. The plan included replacing all semiconductor-containing components and mandating a comprehensive retesting regimen equivalent to GLV-2's initial arrival at the Eastern Test Range.

On August 22, the Manned Spacecraft Center's (MSC) Procurement and Contracts Division reported a crucial update regarding the Gemini flight suit contract. An amendment covering G3C flight suits and related equipment for the upcoming Gemini-Titan (GT) 3 mission was sent to the contractor, David Clark Company.

The first four Gemini flight suits earmarked for GT-3 were delivered by late August. This milestone marked a significant step in outfitting astronauts for the mission. To mitigate earlier issues with fitting training suits, astronauts underwent preliminary fittings of the flight suits before final delivery, ensuring optimal comfort and functionality during training and mission activities.

On August 22, the Crew Systems Division made a significant announcement regarding integrating displays and associated circuitry into the extravehicular life support system (ELSS) for the astronaut Modular Maneuvering Unit (MMU). AiResearch was formally instructed to commence immediate integration efforts, marking a crucial step in advancing astronaut mobility and operational capabilities during extravehicular activities (EVAs).

The MMU, slated for deployment in Gemini-Titan 9 as Department of Defense experiment D-12, represented a pioneering initiative to enhance astronaut maneuverability and efficiency in space. Scheduled for delivery in January 1965, the prototype ELSS was integral to supporting the MMU's functionality and ensuring astronaut safety during its operational deployment.

On August 22, the Flight Crew Support Division announced the comprehensive training plan for egress and recovery operations tailored for the first manned Gemini flight crew. This structured program was designed in three distinct phases to ensure thorough preparedness and safety for astronauts during critical mission scenarios.

Phase I: Initially, astronauts underwent a detailed review of egress procedures within the McDonnell Gemini mock-up. This phase focused on familiarizing crew members with the spacecraft's layout and practicing the initial steps of exiting the spacecraft under simulated conditions.

Phase II: The training progressed to a comprehensive review of egress development results, integrating hands-on training with the trainer apparatus and the Ellington flotation tank. This phase aimed to refine and validate egress techniques in a controlled environment, emphasizing safety protocols and effective maneuvering.

Phase III: The final phase involved practical egress exercises in open water, collaborating closely with essential recovery forces. This phase simulated real-world scenarios, ensuring astronauts could safely exit the spacecraft and initiate recovery procedures under varying conditions, including emergency water landings.

Weathering the Storms

Cape Kennedy faced significant challenges in late August and early September as hurricanes Cleo and Dora swept through the area, impacting preparations for the Gemini launch vehicle (GLV) 2.

Hurricane Cleo (August 27): Hurricane Cleo struck Cape Kennedy, prompting immediate precautions for GLV-2. Stage II of the launch vehicle was safely lowered and stored while the erector was positioned horizontally. Stage I remained lashed in its vertical position. Operations resumed on September 1 when Stage II was reerected. Power was restored to the launch vehicle on September 2, and Subsystem Functional Verification Tests (SSFVT) resumed on September 3.

Hurricane Dora (September 8-14): As Hurricane Dora approached Cape Kennedy, forecasts necessitated further protective measures. On September 8, both stages of GLV-2 were once again lowered (deerected) and securely stored in the Missile Assembly Building to safeguard against potential damage. Hurricane Ethel subsequently posed additional threats, prompting the stages to remain in protective storage until September 14. They were then returned to complex 19 and reerected.

Resumption of Operations: SSFVT resumed operations on September 18 and concluded successfully on October 5, marking successful navigation through the tumultuous hurricane season.

On August 31, the Manned Spacecraft Center (MSC) reported significant ongoing efforts to address production challenges at Ordnance Associates in Pasadena, California, a key pyrotechnics contractor for the Gemini program. These challenges were proving more extensive than initially anticipated, causing disruptions and delays.

The issues primarily stemmed from deficiencies in production planning, fabrication, testing, and quality control. Reports indicated a lack of systematic record-keeping, complicating the production process's tracking and routing of parts. Additionally, there were concerns that development engineers were bypassing critical checkpoints to expedite parts release for testing programs, further exacerbating the situation.

These production inefficiencies necessitated a temporary halt in production efforts, significantly delaying the overall Gemini program timeline. The disruptions challenged maintaining the rigorous standards required for safe and reliable spacecraft operations.

Efforts were underway to address these complexities and restore production efficiency. Steps included tightening quality control measures, enhancing record-keeping practices, and reinforcing adherence to established testing protocols. These measures aimed to mitigate future setbacks and ensure that production resumed smoothly to support upcoming Gemini missions.

On August 31, the Gemini Program Office (GPO) reported significant developments and challenges in the research and development testing of crucial spacecraft systems, affecting the overall program timeline and testing strategies.

Completion of R&D Testing: GPO announced the substantial completion of all research and development testing for components, including thrust chamber assemblies, of the reentry control system (RCS) and orbit attitude and maneuver system (OAMS) configured for spacecraft Nos. 2 through 5. System testing of two RCS units was actively underway, with expectations to conclude the test program by the end of 1964.

Ongoing Challenges and Adjustments: Despite progress, delays persisted, particularly in completing research and development testing for the OAMS configuration intended for spacecraft Nos. 2 through 5 within the next three months. Compounding these delays was the lack of approved plans for testing the spacecraft No. 6 configuration.

Qualification Test Program Adjustments: Recognizing the impact of delays, GPO reviewed the qualification test program to streamline schedules without compromising data integrity. Adjustments included eliminating some test requirements and focusing on conducting more tests on individual units to reduce hardware demands. This strategic shift aimed to expedite testing timelines, crucially addressing previous delays attributed to hardware shortages.

Future Reliability Testing: Anticipating the completion of qualification testing, GPO planned to initiate reliability testing on qualifying hardware. This phased approach ensured that reliability assessments aligned with final testing outcomes, optimizing the efficiency and effectiveness of the overall testing regimen.

On September 4, the Air Force Space Systems Division (SSD) and launch vehicle contractors recommended proceeding with the scheduled flight of Gemini Launch Vehicle (GLV) 2. This recommendation came despite concerns raised by the Manned Spacecraft Center (MSC) regarding potential impacts from recent incidents.

MSC had proposed removing GLV-2 from the Gemini program due to concerns arising from two significant events: an electromagnetic incident on August 17 and the effects of Hurricane Cleo. These events prompted uncertainties about the launch vehicle's readiness and reliability.

Following thorough evaluations of the incidents, their potential impacts, corrective measures taken, and subsequent retesting efforts, SSD, Martin, Aerospace, and Aerojet-General collectively recommended that GLV-2 proceed as planned. Their assessment concluded that the corrective actions implemented were sufficient to mitigate any identified risks, ensuring the launch vehicle's readiness for flight.

NASA accepted the recommendation put forth by SSD and the contractors, endorsing the decision to maintain GLV-2 in the Gemini program lineup. This decision preserved the planned flight sequence, with GLV-3 identified as a suitable substitute for subsequent missions if necessary, ensuring continuity and progress within the Gemini program.

In early September, McDonnell initiated the crucial phase of final checkout and calibration tests for the Gemini translation and docking simulator. This pivotal simulator, designed to replicate the intricate maneuvers required for space docking, underwent rigorous engineering data runs starting September 12. These tests, integral to evaluating the simulator's control system, were slated to span two weeks, concluding around late October.

The simulator's calibration tests were meticulously planned to ensure every aspect of its functionality aligned with the exacting standards set by Project Gemini. This included simulations of translational movements and docking procedures essential for astronaut training. Upon completion of these tests, scheduled by late October, the simulator would transition seamlessly into crew training sessions, marking a critical milestone in the preparation for manned Gemini missions.

In mid-September, McDonnell embarked on the final assembly of Gemini spacecraft No. 3, marking a pivotal phase in NASA's Gemini program. Over a meticulous two-week period, from September 13 to 27, engineers meticulously integrated and tested

the various modules comprising the spacecraft.

Simultaneously, the second phase of Spacecraft Systems Tests (SST) commenced, aiming to evaluate the spacecraft's readiness for manned missions rigorously. Vibration testing, conducted on November 7-8, subjected the spacecraft to simulated launch conditions to validate structural integrity. Following this, altitude chamber tests commenced on November 12, simulating the varying atmospheric pressures encountered during spaceflight.

A significant milestone arrived during the manned altitude tests conducted from November 15 to 19. The Gemini-Titan 3 prime and backup crew's space suits underwent thorough checks, confirming their operational readiness without notable issues.

The culmination of these tests was the Simulated Flight Test, conducted from December 6 to 21. This test rigorously simulated the entire mission profile to ensure all systems functioned flawlessly in concert.

On September 21, Gemini Spacecraft No. 2's journey reached a crucial phase upon arrival at Cape Kennedy. Immediately installed within the Cryogenic Building of the Merritt Island Launch Area Fluid Test Complex, meticulous inspections commenced alongside its integration with aerospace ground equipment (AGE). Hypergolic and cryogenic servicing followed swiftly, essential for ensuring the spacecraft's operational readiness.

Dynamic testing of the reentry control system and orbit attitude maneuver engines was conducted with static firings on October 4-5, validating crucial propulsion systems under simulated operational conditions.

Subsequently, on October 10, Gemini Spacecraft No. 2 was relocated to the Weight and Balance Building for the meticulous assembly of pyrotechnics and installation of seats and pallets, a process completed by October 17. The following day, the spacecraft was transferred to Complex 19, where it underwent meticulous preparations for mating with Gemini launch vehicle 2.

From October 21 to 27, premate systems testing was rigorously executed, ensuring seamless integration and functionality between the spacecraft and launch vehicle systems. This phase culminated with the Premate Simulated Flight Test, meticulously conducted and completed by November 4.

On September 23, a pivotal announcement from the Manned Spacecraft Center during a Trajectories and Orbit Panel meeting marked significant adjustments to the mission plan for Gemini-Titan 6. These changes addressed critical aspects of the mission's strategy, particularly in response to challenges concerning the Agena spacecraft's operational lifetime.

Previously assuming a 20-day operational window for the Agena, it was now determined that this duration would not be feasible without modification. Consequently, the focus shifted towards optimizing the timing of the spacecraft launch to coincide closely with the Agena's operational period. This adjustment aimed to increase the likelihood of achieving rendezvous within the limited timeframe during which the Agena remained a viable target.

A key modification included removing restrictions on Agena maneuvers, a decision intended to enhance flexibility and maximize the probability of successful rendezvous. By relieving these constraints, the mission planners sought to improve the operational dynamics between the Gemini spacecraft and the Agena, thereby optimizing the mission's chances of achieving its rendezvous objectives within the designated timeframe.

On September 24, Lockheed achieved a critical milestone in the Gemini program by completing the modification and final assembly of Gemini Agena target vehicle 5001. This pivotal spacecraft, integral to upcoming Gemini missions, underwent meticulous assembly processes at Lockheed's facilities.

After completion, Gemini Agena target vehicle 5001 was promptly transferred to systems test complex C-10 at the Lockheed plant. The next day, on September 25, Lockheed engineers began integrating and hooking up the vehicle for comprehensive systems testing.

From September 25-26, pivotal preliminary flight tests of the rendezvous radar were conducted at White Sands Missile Range by Instrumentation and Electronics Division representatives. These tests marked a critical phase in validating the radar system's functionality and performance under operational conditions.

During the initial testing phase, operations were temporarily interrupted due to major maintenance on the T-33 aircraft used for the tests. Testing activities resumed on October 19, demonstrating resilience and meticulous planning in the testing schedule.

The flight testing campaign continued rigorously, culminating in conclusive tests completed on December 8. These tests comprehensively evaluated the rendezvous radar's capabilities to accurately track and guide spacecraft during critical rendezvous maneuvers, essential for the success of future Gemini missions.

On September 29, Gemini Program Manager Charles W. Mathews presented a revised flight schedule to the Gemini Management Panel, necessitated by recent weather-related challenges, including a lightning strike and hurricane conditions. The updated schedule outlined key launch dates for upcoming missions, reflecting NASA's adaptive approach to managing project timelines effectively.

The revised flight schedule was detailed: Gemini-Titan (GT) 2 was scheduled for November 17; GT-3 was set for January 30, 1965; and GT-4 was targeted for April 12. Notably, for missions GT-4 through GT-7, launch intervals were planned at three-month intervals. As the program progressed beyond GT-7, these intervals were slated to shorten to two and a half months, demonstrating NASA's commitment to accelerating mission cadences as operational readiness improved.

During a pivotal meeting of the Gemini Management Panel on September 29, discussions centered on the power sources for the upcoming Gemini-Titan (GT) 5 mission, which is explicitly geared for long-duration spaceflight. The panel reviewed a comprehensive study proposing a hybrid approach utilizing fuel cells and batteries to optimize mission power requirements.

The proposed strategy outlined that batteries would be employed during peak load demands, while the fuel cell would serve as the primary power source for the remainder of the mission's electrical needs. The panel enthusiastically accepted this approach, directing McDonnell to implement the plan.

In addition to planning for GT-5, decisions were made regarding other missions within the Gemini program. The panel decided to remove the fuel cell from GT-4 and substitute it with batteries, pending approval from NASA Headquarters. Furthermore, it was decided to utilize older fuel cell versions for GT-2, reserving the redesigned version for subsequent manned flights. This strategic decision aimed to accumulate valuable flight experience with the fuel cell component before transitioning to its enhanced iteration.

On September 30, manned at-sea tests of the Gemini spacecraft commenced, utilizing static article No. 5 as part of rigorous testing procedures. Over two days, spacecraft postlanding systems operated satisfactorily, though both crew members experienced discomfort while wearing their pressure suits. In response, the comfort level was improved by removing the suits despite high cabin heat and humidity levels.

Regrettably, the test was prematurely halted after 17 hours due to Hurricane Hilda's approach. However, efforts continued to refine operational protocols. On November 13, a subsequent test explored the feasibility of alleviating heat and humidity issues by opening the spacecraft hatch. This adjustment effectively lowered internal temperatures, enhancing comfort for test subjects.

Another critical test was conducted three days later to validate water egress procedures. During this test, astronauts successfully exited the spacecraft and demonstrated the capability to close and latch the hatch securely behind them. This successful demonstration underscored the spacecraft's robust design, ensuring that the reentry vehicle could be safely recovered even if astronauts needed to evacuate.

At-sea egress training in Galveston Bay. (NASA Photo No. 65-H-641, released Apr. 14, 1965.)

In early September, Bell Aerosystems initiated a crucial test program to identify the root cause of a failure in the Secondary Propulsion System (SPS) Unit II thrust chamber. This failure occurred during Preliminary Flight Rating Tests when the chamber wall burned through near the injector face before reaching the specified accumulated firing time of 400 seconds.

The test program was designed to consist of six series, each comprising three 50-second firings separated by 30-minute coast periods. Variations in the fuel and oxidizer temperature ranges were planned for each series to pinpoint the conditions leading to the chamber failure. Initially scheduled for completion within two weeks, the test program faced delays due to test cell issues and was extended until mid-November.

Despite encountering setbacks, four test series were conducted, yielding crucial findings. It was established that the chamber wall burned through specifically when fuel and oxidizer temperatures were elevated (above 100 degrees F) and during burn times approaching 50 seconds.

The Gemini Project Office concluded that this issue posed no mission threat, affirming Lockheed's analysis of SPS operation, which indicated that the maximum propellant temperature range in orbit was 0 degrees to 85 degrees F, with a 30-degree F safety margin. The nominal operational temperature range was identified as 30 degrees to 55 degrees F.

Hurricane Isbell

On October 6, the Prespacecraft Mate Combined Systems Test (CST) for Gemini launch vehicle two was successfully conducted at Complex 19. As the CST performed at the Martin plant, this critical test involved an abbreviated countdown and simulated flight events. A spacecraft simulator replicated the electrical characteristics of the actual spacecraft, aiming to validate and instill confidence in the launch vehicle's operational readiness.

Following this milestone, Electrical Electronic Interference Tests were completed by October 12, further ensuring the integrity of electronic systems crucial for mission success.

The operational schedule was briefly disrupted when Hurricane Isbell threatened the area on October 14-15. Although the storm's trajectory remained sufficiently south of Cape Kennedy to avoid direct impact, precautionary measures were taken, resulting in temporary curtailment of testing activities.

On October 7, the vehicle acceptance team convened for the second time at Martin-Baltimore to meticulously review test and manufacturing data about the Gemini launch vehicle (GLV) 3. This comprehensive review aimed to ensure the vehicle's readiness for upcoming missions within the Gemini program.

The meeting concluded on October 9, with GLV-3 deemed acceptable after thorough scrutiny and deliberation. Martin was subsequently authorized to remove the vehicle from the vertical test cell. Final checks, including weighing and balancing procedures, were conducted meticulously in preparation for the next phase.

On October 27, GLV-3 passed the roll-out inspection, marking a significant milestone in its readiness assessment. Subsequently, the vehicle was formally handed over to the Air Force after meeting all stipulated criteria and addressing any identified discrepancies.

The Air Force Space Systems Division formally accepted GLV-3, underscoring the culmination of rigorous testing, meticulous quality assurance processes, and collaborative efforts between NASA and its aerospace partners. This acceptance paved the way for GLV-3's pivotal role in advancing manned space exploration during the Gemini missions.

Backup and prime crews for Gemini-Titan 3 mission at Gemini launch vehicle 3 roll-out inspection. Left to right: Thomas P. Stafford, Walter M. Schirra, Jr., John W. Young, and Virgil I. Grissom. (NASA Photo No. 64-H-2598 [Gemini], Oct. 28, 1964.)

From October 9 to 17, NASA conducted the first major tests of its worldwide tracking network as part of the Gemini program to prepare for manned orbital flights. These tests represented a crucial milestone in the readiness assessment for upcoming manned missions.

Over nine days, simulated flight missions were executed, involving coordination between the Goddard Space Flight Center, the Mission Control Center at Cape Kennedy, and eight remote sites within the global tracking network. The primary objectives were to evaluate the effectiveness of tracking and communications equipment, validate flight control procedures, and ensure seamless operational integration across the network.

These tests culminated in extensive efforts to update and adapt the Manned Space Flight Tracking Network to support the complex requirements of Gemini missions. The conversion process from the Mercury network to accommodate Gemini had spanned two years and incurred $50 million, underscoring the scale and investment required to advance NASA's capabilities in manned spaceflight operations.

On October 10, a significant milestone was achieved in the Gemini Program. The Gemini Program Office reported completing the predelivery acceptance test for the first production rendezvous radar. This radar system was designated for installation in Gemini spacecraft No. 5, marking a crucial advancement in the program's capability for orbital rendezvous operations.

Completing this test underscored the radar's readiness to fulfill its critical role in facilitating rendezvous and docking maneuvers in space. This technology was pivotal for enabling astronauts to maneuver and dock with other spacecraft, a capability essential for achieving complex mission objectives during the Gemini program.

On October 15, McDonnell achieved a significant milestone by completing the final assembly and systems tests of Gemini spacecraft No. 3A. Following this accomplishment, the spacecraft was delivered

to the laboratory for thermal balance testing, marking a critical phase in its preparation for future missions.

Spacecraft No. 3A was designated as a thermal qualification test unit to validate the spacecraft's readiness to withstand the thermal challenges of spaceflight. All systems and subsystems were confirmed to be flightworthy, except for certain easily replaceable components like the heatshield and ejection seats, which were temporarily substituted with non-flight articles approved by NASA.

The thermal qualification testing process included rigorous mission simulations within altitude chambers, ensuring that all systems could operate reliably throughout their designated duty cycles. Over the next two months, the spacecraft underwent installation in the altitude chamber, conducted dry run tests, and underwent thorough reviews during readiness meetings.

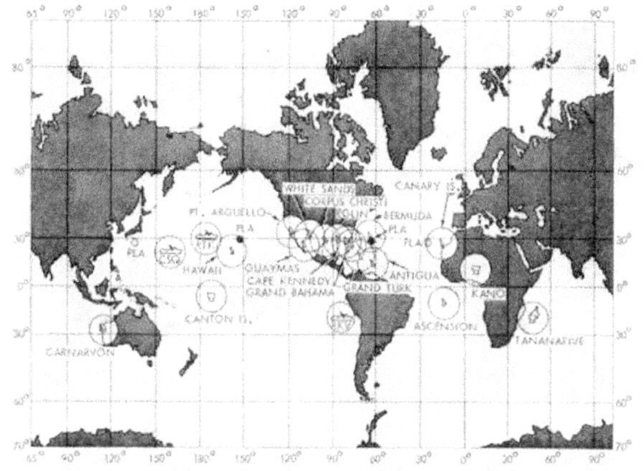

The Gemini Network. See Appendix 4 tabulation of equipment at each site. (NASA Photo S-65-4007, undated.)

Formal thermal qualification testing commenced on December 19, emphasizing McDonnell's commitment to rigorously vetting the spacecraft's performance under simulated space conditions. This meticulous testing regimen was essential to certify the spacecraft's readiness for upcoming Gemini missions, highlighting NASA's dedication to ensuring mission success through comprehensive testing and validation processes.

On October 17, the Flight Crew Support Division announced a significant milestone as the primary crew for Gemini-Titan (GT) 3 completed egress practice. This training took place using boilerplate No. 201 within the Ellington Air Force Base flotation tank, ensuring emergency readiness.

Scheduled next was similar training for the backup crew of GT-4, set for October 23. Looking ahead, full-scale egress and recovery training for GT-3 and GT-4 crews was slated to commence around January 15. This comprehensive training regimen would include parachute refresher courses, crucial for preparing astronauts to safely exit and recover from the spacecraft upon return to Earth.

Water egress training in the flotation tank at Ellington Air Force Base, Texas. (NASA Photo S-65-2503, Feb. 5,1965.)

On October 17, the Crew Systems Division reported significant progress in developing extravehicular activity (EVA) suits for the Gemini program. The first Gemini extravehicular prototype suit had been received from the contractor and assigned to Astronaut James A. McDivitt for evaluation in the Gemini mission simulator.

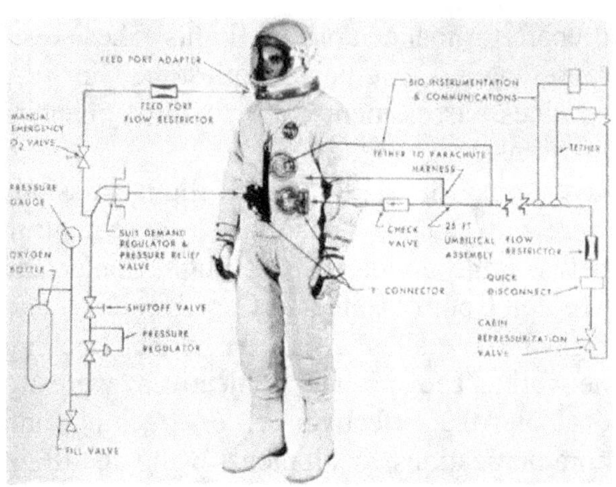

Diagram of the Gemini G4C extravehicular suit. (NASA Photo S-65-4858, May 1965.)

During testing, McDivitt noted concerns regarding the suit's bulkiness and limited mobility in its unpressurized state. However, these issues did not significantly impair mobility when the suit was pressurized, a critical factor for astronauts conducting spacewalks.

Meanwhile, efforts continued to enhance the suit's capabilities. A thermal/micrometeoroid cover layer was installed on a test suit, which was subsequently sent to Ling-Temco-Vought for thermal testing in the space simulator chamber. This testing aimed to validate the suit's ability to withstand the harsh thermal and micrometeoroid conditions of space, ensuring astronaut safety and functionality during extravehicular activities.

On October 17, the Crew Systems Division reported significant progress in evaluating extravehicular life support system (ELSS) ingress techniques for the Gemini program. Zero-gravity (zero-g) tests were conducted at Wright-Patterson Air Force Base to assess the effectiveness of these techniques in realistic space conditions.

During the tests, subjects wearing the ELSS chest pack practiced entering the spacecraft and securing the hatch in a simulated zero-g environment. The results were promising: Subjects could complete these tasks within approximately 50 seconds after practice.

These findings highlighted advancements in astronaut training and readiness for extravehicular activities, crucial for ensuring operational success during Gemini missions. The ability to efficiently ingress and secure the spacecraft hatch in zero-g conditions was essential for astronaut safety and mission effectiveness in Earth's orbit.

On October 26, Russell L. Schweickart embarked on a crucial mission to evaluate Gemini biomedical recording instruments. Over eight days, he conducted extensive tests while wearing a Gemini space suit, aiming to assess the functionality and reliability of these critical instruments in simulated mission conditions.

During this period, Schweickart performed multiple zero-gravity (zero-g) flight profiles, simulating the weightless environment of space. Additionally, he underwent a simulated four-day Gemini mission, replicating the duration and challenges of an actual mission scenario. Furthermore, several centrifuge runs were conducted to study the physiological responses and operational effectiveness of the biomedical recording instruments under varying gravitational forces.

These tests were pivotal in validating the performance and accuracy of the Gemini biomedical recording instruments, essential for monitoring astronaut health and well-being during missions. Schweickart's dedicated evaluation provided valuable insights into the instruments' capabilities and operational suitability, contributing significantly to the preparation and safety of future manned missions in the Gemini program.

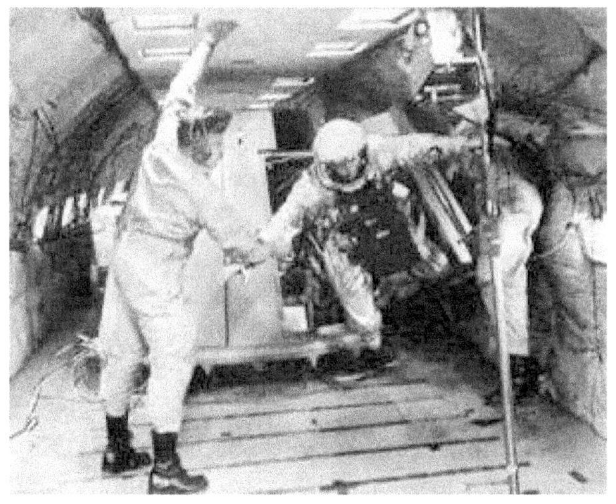

Norman Shyken, McDonnell engineer-pilot, in zero-g tests in an Air Force KC-135 jet transport. (NASA Photo S-64-23051, May 25, 1964.)

On October 28, Gemini launch vehicle 4 reached a significant milestone as it was erected in the vertical test facility at Martin-Baltimore. This marked a crucial step in the preparation process for the upcoming mission.

Following its erection, power was applied to the vehicle for the first time on November 4, initiating essential electrical and subsystem checks. These initial tests were critical for verifying the functionality and readiness of various systems essential for launch and mission success.

Subsequently, Subsystems Functional Verification Tests were completed by November 19, further validating the operational integrity of the launch vehicle's subsystems. These tests ensured that all critical components, from propulsion systems to guidance and control mechanisms, were operating as expected and ready for the next phase of preparations.

On October 28, Bell Aerosystems achieved a significant milestone by successfully testing the Agena secondary propulsion system (SPS) to assess its ability to withstand extended launch hold conditions. This test was crucial for ensuring the reliability and readiness of the SPS for upcoming missions in the Gemini program.

The test involved subjecting the SPS to a rigorous evaluation process. Initially, the system underwent a 20-day dry (unloaded) period to simulate conditions before launch. Following this, it endured a 20-day wet (loaded) period to simulate the pressures and stresses it would encounter during actual mission scenarios.

During the test, the SPS was reverted to a hold condition and refired on November 2, demonstrating its capability to sustain functionality after extended periods of inactivity and under various operational conditions.

On November 5, a pivotal moment in the Gemini program occurred as Gemini launch vehicle 2 and spacecraft No. 2 were mechanically mated at complex 19. This milestone marked a critical step towards integrating the launch vehicle and spacecraft for an upcoming mission.

Following the mechanical mating, rigorous validation processes were undertaken to ensure seamless compatibility and operational readiness. The Electrical Interface Integrated Validation was completed by November 9, confirming the compatibility between the launch vehicle and spacecraft. This validation also included checking redundant circuits connecting the interface, crucial for ensuring reliable communication and control during the mission.

Subsequently, the Joint Guidance and Control Test was conducted and completed by November 12. This test aimed to establish the proper functioning of the secondary guidance system, which comprised the spacecraft's inertial guidance system and the launch vehicle's secondary flight control system. The successful completion of this test verified that both systems were integrated effectively and capable of maintaining precise guidance and control throughout the mission.

Astronauts Grissom and Young in the Gemini mission simulator at Cape Kennedy prior to the Gemini-Titan 3 mission. (NASA Photo No. 65-H-415, released Mar. 19, 1965.)

On November 9, the operational readiness of the Gemini mission simulator at Cape Kennedy achieved a pivotal advancement in the Gemini program. Configured to replicate spacecraft No. 3, this simulator was crucial in preparing astronauts for upcoming missions.

Over the subsequent three weeks, the simulator saw extensive utilization, with approximately 40 hours dedicated to flight crew training and an additional three hours for other personnel from the Manned Spacecraft Center. This intensive usage underscored its importance in familiarizing astronauts with spacecraft operations, emergency procedures, and mission scenarios in a realistic, simulated environment.

The deployment of the Gemini mission simulator marked a significant milestone in NASA's preparations for manned orbital flights. It provided astronauts and ground personnel with essential training opportunities to refine their skills and responses, ensuring readiness for the challenges of space missions.

On November 10, a crucial milestone was reached in the Gemini program as Gemini Agena target vehicle (GATV) 5001 completed a simulated flight, encompassing ascent and orbit, at Lockheed's test complex C-10. This comprehensive test aimed to validate the vehicle's performance under conditions simulating actual mission scenarios.

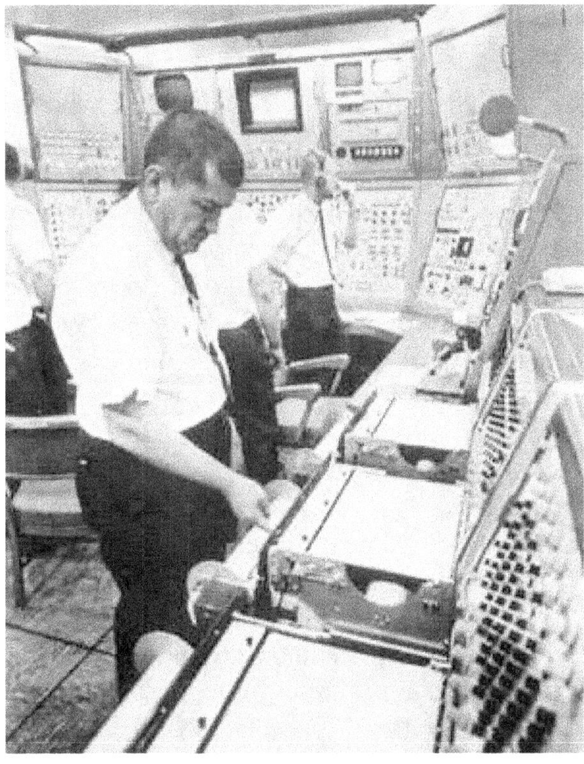

Technicians at the mission simulator console. (NASA Photo No. 65-H-416, released Mar. 19, 1965.)

During the test, minor anomalies were identified, necessitating some portions of the test to be rerun to ensure all systems functioned as intended. Despite these challenges, the completion of GATV 5001 systems tests marked an important step forward in its preparation for subsequent captive-firing tests scheduled at Lockheed's Santa Cruz Test Base.

Following the conclusion of systems tests, GATV 5001 was prepared for shipment, which took place on November 30. This shipment signaled the vehicle's readiness for

further testing and integration with the Gemini spacecraft, advancing NASA's preparations for upcoming missions in the Gemini program.

On November 17, a significant milestone was achieved in the Gemini program: Gemini launch vehicle 2 and spacecraft No. 2 were electrically mated at complex 19. This critical step integrated the launch vehicle and spacecraft systems in preparation for upcoming missions.

The following day, the Joint Combined Systems Test was conducted, representing the first comprehensive test of the combined systems of the launch vehicle and spacecraft. This test included an abbreviated countdown sequence and two flight simulations. The primary objective was to validate the functionality of both the primary and secondary guidance systems during simulated flight scenarios.

Building on this success, a second test known as the Flight Configuration Mode Test (FCMT) was completed on November 21. Similar to previous combined systems tests, FCMT differed in that all umbilicals were dropped, simulating conditions closer to an actual launch environment. This test was a crucial step in preparing for the Wet Mock Simulated Launch, further refining procedures, and ensuring readiness for upcoming missions.

On November 24, Gemini-Titan (GT) 2 achieved a significant milestone with successfully completing the Wet Mock Simulated Launch. This full-scale countdown exercise included propellant loading, marking a pivotal step in preparing for the upcoming mission.

Procedures for flight crew suiting and spacecraft ingress were meticulously practiced during the exercise. The primary crew for Gemini-Titan 3 participated in this simulation, wearing training suits and full biomedical instrumentation. They were supported by bio instrumentation and aeromedical personnel who would be integral to the GT-3 launch operation.

Key operational procedures were refined based on the outcomes of this practice run. It was decided that all physical examinations, attachment of bio instrumentation sensors, and suit donning would subsequently take place in the pilot-ready room at complex 16, streamlining pre-launch preparations.

Following the Wet Mock Simulated Launch, the final readiness of the vehicle for flight was confirmed through the Simulated Flight Test conducted on December 3. This test involved a repeat of the Joint Combined Systems Test for the launch vehicle, focusing on system functionality. At the same time, the spacecraft underwent a detailed mission simulation to ensure all systems operated flawlessly under simulated mission conditions.

On November 24, preparations were underway for the shipment of Gemini launch vehicle (GLV) 3 from Martin-Baltimore to Cape Kennedy. However, the shipment was postponed due to the delay in the launch of GLV-2 and necessary modifications. These modifications, initially planned for Cape Kennedy, were completed at the Martin-Baltimore facility by January 14, 1965.

Following the completion of the modifications, GLV-3 underwent thorough reinspection to ensure readiness for delivery. By January 20, all preparations for shipment were finalized. Stage II of the launch vehicle was airlifted to Cape Kennedy on January 21, with Stage I following closely on January 23.

On November 25, the Gemini launch vehicle (GLV) 4 underwent the Combined Systems Acceptance Test, marking a crucial step in its journey toward deployment. This comprehensive test involved meticulous inspections and the vehicle acceptance team reviewed extensive test and manufacturing data. From December 11 to 13, these evaluations ensured that GLV-4 met stringent standards before authorization was granted to remove it from the vertical test cell at Martin's facilities.

Over the following three months, GLV-4 underwent significant enhancements, with 27 engineering changes installed to optimize its performance and reliability. These modifications were essential for ensuring the vehicle's readiness for its upcoming mission. Final integrity checks and weighing and balancing procedures were completed by March 8, 1965, confirming the vehicle's preparedness for deployment to Cape Kennedy.

On November 30, Lockheed initiated a critical phase in testing the Gemini Agena target vehicle (GATV) 5001 by shipping it to the Santa Cruz Test Base. The primary objective of this phase was to conduct captive-firing tests to verify the operational capabilities of the GATV's primary and secondary propulsion systems under actual firing conditions. These tests were crucial for validating operational procedures and techniques related to vehicle handling, launch preparation, servicing, countdown procedures, and postfire servicing.

Additionally, ground equipment specific to the Gemini program, including the pulse-code-modulated telemetry ground station, was tested to ensure seamless integration with the GATV operations. The testing also involved installing and testing the target docking adapter (TDA), manufactured by McDonnell, designed to facilitate docking maneuvers in space.

However, challenges emerged during the testing process. On December 17, when attempting to physically mate the TDA with the GATV on the test stand, issues with wiring continuity were discovered, complicating the interface between the two vehicles. Despite these initial difficulties, the teams persisted, successfully achieving the physical mate after resolving the discrepancies.

Due to these challenges, the scheduled captive flight test was postponed until January 20, 1965. This allowed sufficient time to address and rectify the wiring issues and ensure the integrity of the GATV for future missions in the Gemini program.

On November 30, astronauts James McDivitt and Edward White, designated as the command pilot and pilot respectively for the Gemini-Titan 4 mission, commenced their crew training at Gemini mission simulator No. 2 in Houston. This training marked the beginning of their intensive preparation for the upcoming mission.

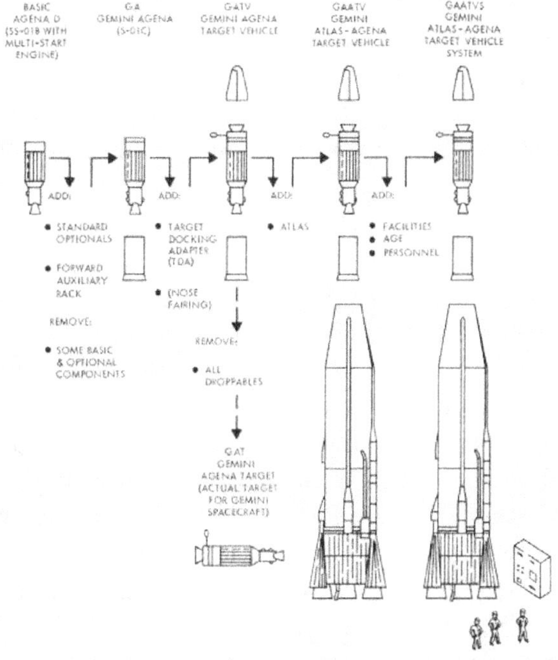

The first week of training focused on familiarizing the crew with the interior layout and operational procedures of the Gemini spacecraft. This training phase was crucial as it allowed McDivitt and White to become accustomed to the controls, instrumentation, and overall environment within the spacecraft simulator.

As part of their training regimen, the astronauts would progressively advance to more complex simulations and scenarios over the coming weeks. This training aimed to enhance their familiarity with the spacecraft and prepare them for various mission contingencies and emergency procedures they might encounter during the Gemini-Titan 4 mission.

On December 1, the roll-out inspection and delivery of the first Atlas standard launch vehicle (SLV-3) for the Gemini program were completed at the San Diego General Dynamics/Convair plant. This milestone marked a significant step forward in preparing for upcoming Gemini missions.

Originally scheduled for November 23, the inspection process was delayed due to the discovery of scored fuel and oxidizer lines, necessitating additional checks and maintenance before final acceptance. Once accepted by the Air Force, the Atlas SLV-3 was transported by truck to the Eastern Test Range, where it arrived on December 7.

Terminology for the Gemini Agena target vehicle program. (Lockheed, Gemini Agena Target Press Handbook, LMSC A766871, Feb. 15, 1966, p. 1-1.)

On December 3, NASA informed North American that no further funds were available for flight testing in the Paraglider Landing System Program following the completion of full-scale test vehicle flight test No. 25. Despite the funding constraints, NASA authorized North American to continue using the test vehicles and equipment for a contractor-supported flight test program.

In response, North American conducted a focused two-week test program, culminating in a significant milestone with a highly successful manned tow-test vehicle flight on

December 19. This flight demonstrated the viability and functionality of the Paraglider Landing System under real-world conditions, validating years of development and testing efforts.

Although NASA's decision marked the conclusion of direct funding for flight testing, allowing North American to proceed with contractor-supported tests ensured that valuable data continued to be gathered. The successful outcome of the manned tow-test vehicle flight underscored the potential of the Paraglider Landing System for future NASA missions despite the program's shift in financial support dynamics.

From December 7 to December 11, a critical four-day Gemini space suit comfort test was conducted as part of its qualification test program. The test involved a human volunteer who wore the suit continuously during the entire period. The objective was to evaluate the suit's comfort over an extended duration, ensuring it could support astronauts effectively during missions.

The suited subject interacted with Gemini food and bio instrumentation systems and the Gemini waste management hardware throughout the test. This comprehensive evaluation aimed to simulate real mission conditions, assessing how well the suit accommodated various operational tasks and maintained comfort levels for prolonged wear.

On December 9, the Gemini-Titan (GT) 2 launch countdown commenced at 4:00 a.m. EST. The countdown progressed smoothly with minor holds until just after engine ignition, when a critical issue arose. A shutdown signal from the master operations control set (MOCS) abruptly halted the launch attempt.

The root cause was the loss of hydraulic pressure in the primary guidance and control system of stage I of the launch vehicle. This triggered an automatic switchover to the secondary guidance and control system.

During a 3.2-second holddown period following the ignition command, this switchover was misinterpreted as a shutdown command, leading the MOCS to terminate the launch.

Subsequent investigation revealed that the primary servo-valve in one of the four tandem actuators responsible for controlling the movement of the stage I thrust chambers had failed. As a corrective measure, all four stage I tandem actuators were replaced with redesigned versions to prevent the recurrence of the issue.

On December 9, during the launch attempt of Gemini-Titan 2 (GT-2), the Mission Control Center at Houston played a crucial role alongside its counterpart at Cape Canaveral. This dual-operation approach primarily validated the computer launch programs and ensured redundancy in monitoring and control.

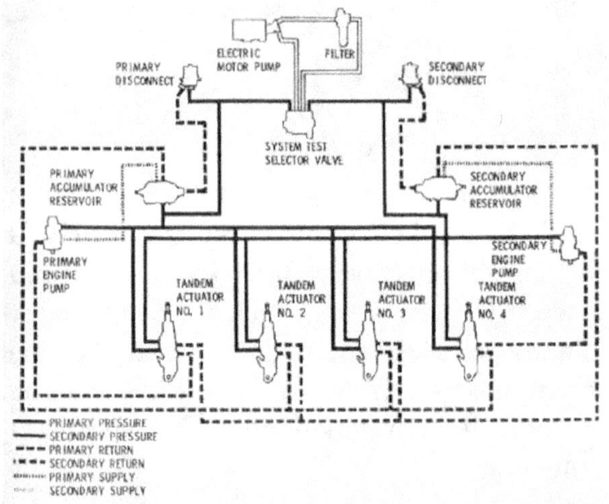

Gemini launch vehicle stage I hydraulic system. (Martin Photo 8B65778, undated.)

The Houston center operated passively, processing and displaying live and simulated data from the Gemini launch vehicle and spacecraft. This included telemetry data and television display formats, providing real-time insights into the status of the mission components.

In December 1964, the Gemini Program Office (GPO) streamlined its Agena target vehicle program by initiating contractual actions to eliminate the eighth Agena. This decision was part of ongoing efforts to optimize resources and focus on key mission priorities within the Gemini program.

By March 6, 1965, the GPO removed the seventh Agena from the program. These adjustments aimed to rationalize the number of Agena vehicles planned for use, possibly reflecting changes in mission requirements, budget considerations, or operational efficiencies identified during program development.

On December 15, the Gemini Phase II centrifuge training program concluded, marking a significant milestone in astronaut preparation for upcoming missions. This training phase focused on refresher sessions for the Gemini-Titan 3 and 4 flight crews, who conducted their centrifuge runs while wearing pressure suits, and familiarization training for astronauts not yet assigned to specific missions under shirt-sleeve conditions.

The program commenced in early November, providing crucial training in simulated high-gravity conditions to ensure that astronauts were well-prepared for the physiological challenges of spaceflight. This rigorous training regimen enhanced crew readiness and safety during the upcoming Gemini missions, emphasizing operational proficiency and crew comfort in various scenarios.

Atlas SLV-3's Historic Maiden Flight

On December 16,, a pivotal moment unfolded at Complex 14 of the Eastern Test Range. Standing proudly was the Atlas standard launch vehicle (SLV-3) 5301, poised for a mission to mark a significant milestone for the Gemini program. Its impending launch represented the inaugural flight of the Gemini-specific Atlas variant and the debut of SLV-3 from a brand-new launch complex.

The preparations were meticulous and extensive. Tests commenced immediately to validate every aspect of the pad and its crucial Aerospace Ground Equipment (AGE). These systems were not just components; they were the backbone ensuring the success of the upcoming mission.

On December 17, the crucial Phase III tests commenced to validate the reliability of the Gemini parachute recovery system. Static article No. 7 was the first to undergo evaluation, followed by static article No. 4A in a series of 10 meticulously planned drops. These tests marked a pivotal phase in the Gemini program, aiming to ensure astronauts' safe return from space.

Each drop was executed flawlessly Throughout these tests until February 11, 1965, devoid of any parachute or sequencing anomalies. The meticulous planning and flawless execution culminated in the successful completion of Phase III, confirming the readiness and effectiveness of the Gemini parachute system.

On December 17, a pivotal moment unfolded as the Air Force Space Systems Division formally approved Agena D (AD-82) for integration into the Gemini program. Following acceptance, Lockheed promptly transferred AD-82 to the vehicle final assembly area for critical modifications to transform it into Gemini Agena target vehicle 5002.

Scheduled to commence in mid-January 1965, the meticulous modification work aimed to adapt AD-82 to meet the stringent requirements of the Gemini missions.

On December 18, Martin-Baltimore initiated a critical phase by retrieving the propellant tanks for Gemini Launch Vehicle (GLV) 6 from storage. These tanks were meticulously cleaned and purged with nitrogen, marking the initial preparation steps for upcoming missions. This thorough maintenance was completed by February 5, 1965.

Following this preparatory phase, Aerojet-General delivered the flight engines designated for GLV-6 on February 1, contributing essential components to the assembly process. Subsequently, tank splicing operations were finalized by February 23, ensuring structural integrity and readiness for flight. The installation of the flight engines followed closely on February 25, marking another milestone in the assembly process.

Horizontal testing of GLV-6, a crucial step to validate its operational readiness, concluded successfully on April 3.

On December 19, Gemini spacecraft No. 3A embarked on thermal qualification tests within McDonnell's altitude chamber, marking a critical phase in its preparation for space missions. However, an unexpected setback occurred during the initial test from December 19 to 21 when the spacecraft's coolant system froze. This incident prompted a rigorous three-week retesting and redesigning of the coolant system to ensure reliability and functionality under extreme conditions.

The redesigned coolant system, incorporating lessons learned from the initial test, was subsequently implemented across other spacecraft in the program. Test No. 2 commenced from January 6 to 13, followed by the conclusion of the test program on February 19 with the third test run. These tests simulated over 220 orbits, rigorously evaluating the spacecraft's ability to regulate internal temperatures across varying altitudes and conditions.

Evolution of the Gemini Extravehicular Space Suit

On December 28, the Crew Systems Division received a prototype G4C extravehicular Gemini space suit, marking a significant step in astronaut equipment development. This advanced suit featured a thermal/micrometeoroid cover layer, a redundant closure system, and an open visor assembly designed to provide crucial visual, thermal, and structural protection during spacewalks.

Initial zero-gravity tests conducted in January 1965 revealed the suit's overall satisfactory performance. However, astronaut feedback highlighted challenges related to maneuverability due to the heavy bulk of the cover layer. In response, engineers undertook a redesign effort to streamline the cover layer, aiming to reduce unnecessary bulk without compromising protective capabilities.

By February, the redesigned cover layer underwent rigorous testing and proved highly satisfactory, effectively addressing previous mobility concerns.

The Gemini G4C extravehicular suit with chestpack ventilation control module and gold-coated umbilical line. (NASA Photo S-65-27424, May 28, 1965.)

Chapter 9 - Flight Tests

January 1965 through December 1965

On January 4, 1965, McDonnell, NASA's Gemini program contractor, delivered the third Gemini spacecraft to Cape Kennedy. This pivotal moment marked a significant step forward in America's quest to conquer space. Following a meticulous receiving inspection on January 6, the spacecraft was swiftly relocated to the Merritt Island Launch Area Radar Range.

Gemini spacecraft No. 3 being unloaded at Cape Kennedy. (NASA Photo 104-KSC-65-00003, Jan. 4, 1965.)

A crucial phase awaited Gemini No. 3: the communications radiation test, a specialized assessment reserved for spacecraft slated for imminent manned missions. Conducted on January 7, this rigorous test replicated the radio-frequency conditions the spacecraft would encounter during its mission. Designed to ensure robust communication capabilities in the harsh realities of space, this test was a critical milestone before proceeding to the next phase: preparations for static firing.

On January 5, 1965, NASA Headquarters took decisive action in refining the flight plan for the upcoming Gemini-Titan (GT) 3 mission, anticipating potential challenges with the retrorocket system. The primary concern was ensuring astronauts' safe return even in retrorocket failure, which demanded meticulous contingency planning.

The proposed solution from Headquarters entailed integrating three orbit attitude and maneuver system maneuvers into the flight plan. These maneuvers were strategically designed to establish a fail-safe orbit. This orbit would enable the spacecraft to reenter the Earth's atmosphere safely, regardless of whether the retrorockets fired as planned or encountered difficulties.

Under the guidance of the Mission Planning and Analysis Division, the proposal underwent thorough refinement to ensure efficacy and reliability. Ultimately, these revisions were adopted into the flight plans not only for GT-3 but also for the subsequent GT-4 mission.

This proactive approach underscored NASA's commitment to astronaut safety and mission success during the pioneering days of the Gemini program, showcasing the agency's ability to adapt and innovate in the face of technical challenges.

January 6, 1965, was pivotal for NASA's ambitious Gemini program as the Manned Spacecraft Center released the comprehensive Gemini Program Mission Planning Report. Prepared by the Gemini Program Office, this seminal document formalized the overarching goals and detailed guidelines for each mission within the program.

The report meticulously outlined the configuration of the spacecraft to be utilized, specifying primary mission objectives and detailing the planned sequence of missions. It provided a structured framework encompassing all aspects of Gemini missions, from technical specifications to operational protocols.

A significant highlight of the report was its emphasis on integrating extravehicular operations (EVA) into Gemini missions. This strategic initiative underscored NASA's commitment to advancing capabilities beyond the confines of spacecraft, outlining the necessary equipment, operational objectives, and proposed procedures starting from the fifth mission onwards.

In addition to operational guidelines, the report comprehensively detailed the experiments slated for each Gemini mission, aligning specific objectives with assigned missions. It emphasized flexibility and adaptability, intending periodic revisions to accommodate evolving mission requirements.

Anticipating the dynamic nature of space exploration, the report laid the groundwork for future mission directives. Detailed planning was expected approximately six months before each mission's scheduled launch. This approach ensured that each mission was meticulously prepared to achieve its scientific and operational objectives.

The Atlas SLV-3 continued to mark a milestone in space exploration. At the new launch complex, The preparations were meticulous and extensive. Tests commenced immediately to validate every aspect of the pad and its crucial Aerospace Ground Equipment (AGE). These systems were not just components; they were the backbone ensuring the success of the upcoming mission. Over the ensuing weeks, engineers diligently scrutinized every detail, ensuring that the complex and its AGE were primed for the demanding task ahead.

By mid-January of 1965, the stage was set for propellant loading tests, a critical phase to ensure the vehicle's readiness for flight. These tests were not merely procedural; they were the final checks to confirm that every system, from fuel lines to ignition sequences, operated flawlessly under the pressures of actual launch conditions.

January 6, 1965, marked a pivotal moment in the preparations for the Gemini-Titan 2 mission as redesigned stage I tandem actuators arrived and were promptly installed in the Gemini launch vehicle (GLV) 2. This critical upgrade followed a period of retesting that commenced after the initial scrubbing of the Gemini-Titan 2 mission on December 9, 1964.

Upon receiving the redesigned actuators, activities to prepare GLV-2 for another launch attempt resumed in earnest. Subsystems underwent meticulous retesting to ensure optimal performance and reliability.

A significant milestone in the preparations was completing the final combined systems test, the Simulated Flight Test, on January 14. This comprehensive test simulated the conditions and operations expected during an actual mission, validating the readiness of the spacecraft and launch vehicle systems.

Preparations progressed smoothly, and the launch of GLV-2 was scheduled for January 19. This date marked the culmination of intensive efforts to address technical challenges and ensure that all systems functioned flawlessly for a successful mission.

On January 11, 1965, an intensive testing program was begun to qualify the Gemini escape-system personnel parachute. This critical phase was pivotal in ensuring the safety and reliability of emergency procedures for astronauts aboard the Gemini spacecraft.

The testing began with two low-altitude dummy drops, revealing issues separating the backboard and egress kit. Prompt corrective measures were implemented to resolve the interference, leading to successful tests with two additional drops on January 15. These tests validated improvements and set the stage for further evaluations.

After the initial tests, four high-altitude dummy drops were conducted during the week of January 18. While the system sequencing proved satisfactory, two drops encountered deployment issues, with the ballute deploying too slowly. Engineers promptly addressed this concern, conducting corrective tests on February 12 and 16, which confirmed the effectiveness of the adjustments made.

Concurrently, low-altitude live jump tests commenced on January 28, marking a significant transition to real-world scenarios for the parachute system. Throughout several tests, including the final test on February 10, the parachute system demonstrated reliability amidst the rigorous conditions of actual human trials.

Building on the success of the initial phases, high-altitude live jump tests began on February 17. This progression represented a critical advancement towards validating the parachute system's performance under more extreme conditions and further enhancing astronaut safety.

Throughout these meticulous testing phases, from dummy drops to live jump tests, NASA's rigorous approach underscored its commitment to ensuring the Gemini escape system's readiness for potential emergencies during manned missions. These efforts were crucial in fortifying the safety protocols integral to the Gemini program's operational success.

On January 12, 1965, NASA initiated crucial flight tests using the zero-gravity mock-up of the Gemini spacecraft. This mock-up, housed within a KC-135 aircraft, served as a pivotal training platform for astronauts to simulate extravehicular activities (EVAs) under conditions of weightlessness.

The Gemini-Titan (GT) 3 flight crew actively participated in these inaugural exercises, leveraging the unique environment provided by the KC-135 aircraft to hone their skills in zero-gravity operations. The exercises were meticulously designed to familiarize astronauts with the challenges and dynamics of spacewalks, ensuring they were well-prepared for the demands of upcoming missions.

Following the GT-3 crew's session, the GT-4 flight crew duplicated the exercises the following day. This sequential training approach allowed each crew to benefit from

hands-on experience in a simulated microgravity environment, essential for mastering tasks critical to the success of their missions.

January 14, 1965, marked a significant milestone in NASA's efforts to streamline and accelerate the Gemini program. A dedicated task force within the Office of Manned Space Flight completed a two-month study to determine the feasibility of reducing the interval between Gemini flights from three months to two.

The study's findings, presented to George E. Mueller, NASA's Associate Administrator for Manned Space Flight, on January 19, outlined crucial requirements and proposed recommendations for achieving an accelerated launch schedule starting with Gemini-Titan 6. Central to this plan was the need for flight-ready vehicles to be delivered directly from the factory, minimizing testing requirements at the Cape.

Several significant changes were identified as necessary to implement this accelerated schedule effectively:

Factory Testing Emphasis: Shifted most testing activities to the factory, particularly spacecraft altitude testing conducted exclusively at McDonnell.

Facility Activation: The second cell in the vertical test facility at Martin-Baltimore was activated to enhance testing capabilities.

Simplified Testing at Cape: Streamlined subsystem testing at the Cape, optimizing efficiency without compromising safety.

Electronic interference testing and the Flight Configuration Mode Test were removed from the pre-launch testing regimen to expedite preparations.

By adopting these strategic changes, NASA aimed to significantly enhance the efficiency of its operations while maintaining rigorous safety standards. This proactive approach underscored NASA's commitment to meeting ambitious mission timelines and advancing the capabilities of the Gemini program amidst the pressures of the space race.

The decisions made on January 14 set the stage for a more agile and responsive approach to manned space missions, marking another step forward in America's quest for space exploration supremacy during the Gemini era.

On January 15, 1965, a pivotal phase in the Gemini program unfolded as Gemini spacecraft No. 3 thrusters underwent static firing. This marked a crucial step in a comprehensive propulsion system verification test program on spacecraft Nos. 2 and 3.

The primary objective of these tests was to conduct a thorough end-to-end verification of the propulsion systems. This included validating servicing procedures and equipment essential at the launch complex. By performing these tests early in the process, NASA aimed to identify and rectify any potential issues well ahead of actual launch operations.

Moreover, these tests also completed the development and systems testing of the Gemini spacecraft's hypergolic systems. Hypergolic propellants were critical for maneuvering and attitude control in space, making their reliability paramount for mission success. Completing these tests bolstered confidence in the propulsion systems before committing them to flight.

Following the static firing and verification tests, deservicing the propulsion system continued until January 21. This phase ensured that all systems were safely prepared and maintained in optimal condition for subsequent stages of spacecraft preparation.

NASA's Engineering and Development Directorate reported a significant achievement on January 15, 1965: its Crew Systems Division qualified the Gemini spacecraft bioinstrumentation equipment.

Bioinstrumentation equipment was crucial in monitoring astronauts' health and physiological responses during missions. Qualifying this equipment involved rigorous testing and verification to ensure its reliability and accuracy in demanding spaceflight conditions.

This achievement underscored the meticulous preparation and attention to detail that characterized NASA's approach to crew safety and operational readiness within the Gemini program. With the qualified bioinstrumentation equipment, astronauts could proceed confidently, knowing that their vital signs and physiological data would be accurately monitored throughout their missions.

The successful qualification of the bioinstrumentation equipment paved the way for upcoming manned missions, ensuring that astronauts would be equipped with state-of-the-art technology to support their health and well-being in the challenging space environment.

On January 16, 1965, NASA's qualification testing for simulated off-the-pad ejection (SOPE) resumed after a delay caused by pyrotechnics availability issues. This testing, critical for ensuring astronaut safety in emergencies, faced initial challenges but ultimately succeeded through rigorous troubleshooting and redesign efforts.

Simulated off-the-pad ejection test No. 13 at U.S. Naval Ordnance Test Station, China Lake, California. (NASA Photo No. 65-H-197, released Feb. 12, 1965.)

During SOPE No. 12, the left seat performed satisfactorily, but the right seat rocket catapult fired prematurely due to a malfunction in the right hatch actuator. This resulted in the seat colliding with the hatch and failing to eject as intended. Promptly addressing this issue, NASA modified all hatch actuators to prevent the recurrence of similar failures.

Following modification and testing, redesigned hatch actuators were implemented in SOPE No. 13 on February 12. This proved successful, with all systems functioning correctly. This marked a significant milestone in the qualification test program, validating the effectiveness of the corrective measures taken.

The qualification test program concluded on March 6 with SOPE No. 14, where the complete ejection system operated flawlessly. All equipment involved in the tests was recovered in excellent condition, affirming the ejection system's readiness for deployment in actual missions.

The successful conclusion of the SOPE qualification tests highlighted NASA's commitment to meticulous testing and continuous improvement in astronaut safety protocols. By rigorously testing and refining critical systems like the ejection system, NASA ensured that astronauts aboard the Gemini spacecraft were well-protected in emergencies during launch or ascent.

On January 19, 1965, NASA's Mission Planning and Analysis Division proposed a significant improvement for the Gemini launch vehicle (GLV) countdown procedures based on a detailed report from Space Technology Laboratories.

Following an extensive study that analyzed 325 missile countdowns, 205 missile launches, and all Titan scrubs and holds, Space Technology Laboratories recommended implementing properly located built-in holds within the GLV/Gemini countdown sequence. These holds were designed to optimize the launch process by strategically pausing countdown operations at key points.

The study findings indicated that integrating built-in holds could substantially enhance the efficiency of GLV launches. By preemptively addressing potential issues and allowing for necessary adjustments without necessitating a complete scrub, these holds aimed to reduce the number of mission cancellations and delays.

The proposed inclusion of built-in holds represented a proactive measure to streamline launch operations and mitigate risks associated with technical or environmental factors. This approach aligned with NASA's ongoing efforts to refine operational protocols and maximize the reliability of the Gemini program's launch vehicles.

With the recommendation from the Mission Planning and Analysis Division, NASA moved forward to integrate these improvements into the GLV/Gemini countdown procedures. This initiative reflected NASA's commitment to continuous improvement and readiness to advance human spaceflight capabilities during the Gemini era.

On January 19, 1965, a significant issue arose during the Gemini-Titan (GT) 2 mission countdown when the hydrogen inlet valve of the fuel cell failed to open. This incident prompted extensive efforts to resolve the problem before the scheduled launch.

Despite concerted efforts to rectify the malfunctioning valve, it became evident that further attempts to free it would significantly delay the countdown process. As a result, the decision was made to cease work on the fuel cell system, ultimately leading to the failure to activate it for the flight.

The fuel cell in spacecraft No. 2 was not the current flight design. Following a design change in January 1964, earlier versions of the fuel cells were available, albeit with known defects. Despite these issues, conducting flight testing with the reactant supply system was deemed beneficial, albeit on a non-interference basis with the primary mission objectives.

The incident highlighted the complexities of preparing for manned space missions and underscored the importance of meticulous system testing and readiness. Although the fuel cell could not be activated for the GT-2 mission, the decision ensured that the launch proceeded without compromising safety or mission objectives.

NASA's approach to handling technical setbacks during the GT-2 countdown exemplified its commitment to rigorous safety protocols and operational readiness. This incident provided valuable insights contributing to the ongoing refinement of spacecraft systems and procedures within the ambitious Gemini program.

On January 19, 1965, at 9:04 a.m. E.S.T., the second mission of NASA's Gemini program, Gemini-Titan 2 (GT-2), embarked on a groundbreaking suborbital flight from Cape Kennedy's Complex 19. This mission was a pivotal step in advancing the capabilities of manned spaceflight, focusing on critical evaluations of the Gemini spacecraft's reentry module, structural resilience, and overall system performance.

GT-2's primary objective was to subject the spacecraft's reentry module to the rigors of a maximum-heating-rate reentry, affirming the adequacy of its heat protection systems. From liftoff through reentry, meticulous scrutiny ensured the spacecraft's structural integrity under varying conditions. Concurrently, extensive tests were conducted on communication systems, cryogenics, and the spacecraft's fuel cell and reactant supply systems. However, a setback occurred as the fuel cell system malfunctioned before liftoff, precluding the collection of intended test data.

This suborbital ballistic flight achieved a maximum altitude of 92.4 nautical miles. After 6 minutes and 54 seconds of flight, retrorockets fired to initiate the return trajectory. The spacecraft gracefully landed in the Atlantic Ocean 11 minutes and 22 seconds later, an impressive 1848 nautical miles southeast of its launch site. The primary recovery ship, the aircraft carrier Lake Champlain, executed Swift recovery operations, which successfully retrieved the spacecraft at 10:52 a.m. e.s.t.

GT-2's mission duration of 18 minutes and 16 seconds exemplified NASA's progress in mastering the intricacies of manned spaceflight, setting the stage for subsequent Gemini missions that would further explore the frontiers of space exploration.

On January 20th, a pivotal milestone was achieved in developing the Gemini Agena target vehicle 5001. Located at Lockheed's Santa Cruz Test Base, the vehicle underwent a rigorous hot-firing test to simulate a comprehensive 20,000-second mission. This comprehensive test included multiple firings of both primary and secondary propulsion systems, crucial for validating the vehicle's operational readiness.

The test's real-time operational data was meticulously transmitted to ground stations via PCM (pulse-code-modulated) telemetry. These stations, strategically positioned at the test site and in Sunnyvale, played integral roles in monitoring and assessing the vehicle's performance under simulated mission conditions.

Despite the overall success of the test, a notable anomaly surfaced during the evaluation phase. Specifically, a series of command programmer time-accumulator jumps occurred, totaling seven instances and amounting to 77,899 seconds collectively. This anomaly, while significant, provided valuable insights for further refinement and calibration of the vehicle's systems.

After completing testing activities, Gemini Agena target vehicle 5001 was methodically dismounted from the test stand on February 1st and transported back to Sunnyvale for detailed post-test analysis and adjustments. This meticulous approach ensured that any identified issues were addressed promptly, reaffirming Lockheed's commitment to the flawless execution of the Gemini program's ambitious objectives.

On January 22nd, a pivotal phase commenced in the assembly of Gemini spacecraft No. 3 at Cape Kennedy's industrial area. This meticulous process began with a thorough receiving inspection, ensuring all components met stringent quality standards before integration began.

Throughout the assembly process, culminating in the spacecraft's transfer to complex 19, specific tasks were meticulously executed to prepare it for its upcoming mission. Notable activities included installing pyrotechnic devices essential for various operational functions and integrating flight seats crucial for astronaut safety and comfort during the mission.

Special attention was devoted to tests tailored to spacecraft No. 3's unique requirements. This included a rigorous communications test to verify seamless data transmission capabilities and propulsion verification tests to ensure optimal performance under mission conditions. Concurrently, efforts were made to address manufacturing discrepancies and update the spacecraft's configuration to align with mission parameters.

A significant milestone in this preparation phase was assembling the rendezvous and recovery section, pivotal for the spacecraft's reentry and recovery procedures post-mission. As these tasks were completed, meticulous preparations were made to transition the spacecraft to the launch complex, ensuring all systems were primed for the upcoming mission.

By February 4th, all preparatory activities for spacecraft No. 3 were successfully concluded, marking a crucial step forward in the Gemini program's ambitious goals. Each meticulous task during this period underscored the dedication and precision required to succeed in manned space exploration during the 1960s.

On January 25th, a significant event unfolded at complex 19 of Cape Kennedy as Gemini launch vehicle three was erected. This marked a crucial step in the meticulous preparations for an upcoming mission. This towering structure, emblematic of America's burgeoning space ambitions, stood ready to propel humanity further into the cosmos.

Following its erection, the vehicle received power on January 29th, initiating a series of critical Subsystems Functional Verification Tests (SSFVT). These tests, spanning from January 29th to February 12th, rigorously evaluated each subsystem's operational integrity and performance essential for a successful launch and mission execution. From propulsion systems to guidance mechanisms, every component underwent meticulous scrutiny to ensure flawless functionality under the rigors of space.

econd stage of Gemini launch vehicle 5 being hoisted to the top of the vertical test facility at Martin-Baltimore. (NASA Photo S 65 2867, Feb. 8, 1965.)

The culmination of these tests by February 15th and 16th led to the pivotal Combined Systems Test. This comprehensive evaluation, conducted just before the spacecraft's integration, synchronized the seamless interaction of all systems within the launch vehicle. Such meticulous preparations were essential for ensuring optimal performance during every phase of the upcoming mission.

On January 28th, a pivotal milestone in developing the Gemini spacecraft was reached with the approval of the NASA-McDonnell incentive contract. This significant agreement, endorsed by NASA Headquarters' Procurement Office and the Office of Manned Space Flight, solidified the partnership between the Manned Spacecraft Center (MSC) and McDonnell Aircraft Corporation.

The contract began with preliminary negotiations between MSC and McDonnell, which concluded on December 22, 1964. Following these discussions, the proposed contract was forwarded to NASA Headquarters for approval of MSC's negotiated position, a step formally endorsed on January 5, 1965. Subsequently, final negotiations commenced and were swiftly concluded by January 15th.

With terms finalized, representatives of MSC and McDonnell officially signed the contract, underscoring their commitment to advancing the Gemini program's objectives. The completed contract was promptly submitted to NASA Headquarters on January 21st for final approval, marking a crucial step towards the program's continued progress and success.

On January 28th, the Gemini program resumed its crucial High-Altitude Ejection Test (HAET) with HAET No. 2, marking a significant step towards ensuring the reliability of the Gemini personnel recovery system. This test, conducted under controlled flight conditions, aimed to validate the system's functionality in real-world scenarios.

During HAET No. 2, the recovery system underwent its first ejection from an F-106 aircraft at an altitude of 15,000 feet and a speed of Mach 0.72. Initially planned for 20,000 feet, the altitude was adjusted due to revised mission ground rules for mode 1 abort procedures. Despite this change, the ejection seat and the dummy payload were recovered without incident, demonstrating the system's capability under operational conditions.

The HAET program continued with HAET No. 3, concluding on February 12th. However, an issue arose during this test when the dummy's parachute failed to deploy. Investigation revealed that the aneroid device responsible for initiating chute deployment had malfunctioned, a problem further compounded by a similar failure observed during subsequent qualification tests on February 17th.

These setbacks prompted a redesign of the aneroid device to enhance reliability. Despite these challenges, the Gemini Program Office determined that the test failures were not directly attributable to HAET conditions and thus did not necessitate repeating the test series. Significantly, all other systems performed flawlessly during the tests, underscoring the program's meticulous approach to safety and system reliability.

HAET No. 3, conducted at an altitude of 40,000 feet and a speed of Mach 1.7, represented a critical phase in validating the Gemini personnel recovery system's readiness for manned missions. These tests, conducted with precision and rigorous scrutiny, played a pivotal role in preparing for the challenges of manned space exploration during the Gemini era.

On January 29th, a crucial milestone was achieved in preparation for the Gemini-Titan 3 mission as qualification testing for the food, water, and waste management systems was completed. These essential systems, integral to sustaining astronaut health and mission success in space, underwent rigorous evaluation to ensure their reliability under the demanding conditions of manned spaceflight.

Each food, water, and waste management system component was meticulously scrutinized throughout the qualification testing phase. Testing protocols included assessing performance under simulated mission scenarios to validate functionality, efficiency, and safety protocols.

The successful completion of these tests marked a significant step forward in readiness for the upcoming Gemini-Titan 3 mission. It underscored the meticulous planning and technical expertise invested in ensuring that astronauts aboard the Gemini spacecraft would have access to vital resources essential for sustaining life during their mission in space.

By January 31st, McDonnell had achieved significant milestones in the production and testing phases for Gemini spacecraft No. 4. This marked the completion of major manufacturing tasks, module tests, and the installation of essential equipment. The meticulous process began with Phase I modular testing on November 30, 1964, paving the way for subsequent assembly and testing phases.

A critical achievement occurred on February 23rd when the spacecraft's reentry and adapter assemblies were successfully mated. This integration phase ensured all components were aligned correctly and functioning cohesively, setting the stage for comprehensive Systems Assurance Tests.

Beginning on February 24th, Systems Assurance Tests commenced, encompassing rigorous evaluations to verify the spacecraft's operational readiness. These tests were designed to assess the functionality of every subsystem and component, ensuring reliability under the demanding conditions of space.

On February 1st, the Manned Spacecraft Center (MSC) achieved a significant milestone with the timely receipt of the first qualification configuration extravehicular life-support system (ELSS) chest pack. This crucial component for supporting astronauts during extravehicular activities (EVA) in space marked a pivotal step forward in advancing manned space exploration capabilities.

At MSC, rigorous testing of the ELSS chest pack and the associated umbilical assembly commenced promptly. These tests were meticulously designed to validate the life-support system's functionality, reliability, and safety under simulated mission conditions.

Meanwhile, AiResearch, a key partner in the ELSS development, diligently prepared for upcoming systems qualification tests. Their contributions were instrumental in ensuring that all components of the ELSS met stringent performance standards necessary for space missions.

Successful zero-gravity flight tests further bolstered the development and testing of the ELSS, demonstrating that astronauts trained in the chest pack could effectively perform egress and ingress maneuvers. These tests provided crucial insights into the operational feasibility of the ELSS in space environments, validating its design and functionality.

As progress continued, each achievement brought NASA closer to realizing the goal of safe and effective extravehicular activities during manned space missions. The advancements in ELSS technology represented a critical stride towards expanding the capabilities and scope of human exploration beyond Earth's confines.

On February 1st, significant developments unfolded in the testing and preparation of Gemini Agena Target Vehicle (GATV) 5001 at Santa Cruz Test Base. Following its removal from the test stand, the vehicle was transported to Sunnyvale, marking a transition to the subsequent rigorous testing and evaluation phase.

Upon arrival, GATV 5001 underwent preliminary testing at systems test complex

C-10, ensuring all systems functioned as designed before proceeding to more specialized evaluations. Subsequently, the vehicle was relocated to the anechoic chamber for electromagnetic interference, radio-frequency interference tests, and critical assessments to safeguard against signal disruptions that could impact mission performance.

However, the timeline for GATV 5001 faced setbacks, with delays totaling 37 calendar days, exacerbated by a critical time-accumulator anomaly discovered during hot-firing tests. This anomaly, responsible for a cumulative delay of 20 days, prompted immediate action. A temporary fix was swiftly implemented to mitigate the issue while Lockheed, the manufacturer, continued its exhaustive efforts to identify the root cause and implement a permanent solution.

As preparations continued, efforts focused on resolving these technical challenges to ensure GATV 5001 met stringent performance standards essential for its role in upcoming Gemini missions.

On February 4th, a significant discussion unfolded regarding spacecraft weight control during a Gemini Management Panel meeting, reflecting ongoing concerns and strategies to manage weight growth within the program. The meeting was prompted by George M. Low, Deputy Director of Manned Spacecraft Center, who expressed keen interest in maintaining vigilance over spacecraft weight as critical missions approached.

Gemini Program Manager Charles W. Mathews reported a modest increase of 12 pounds in spacecraft weight over the past month. He noted a discernible "leveling-off trend" observed over the previous two months, indicating efforts to effectively stabilize and manage weight growth.

Despite these assurances, George M. Low remained cautious about potential unforeseen weight increases as the program progressed through subsequent missions. Walter F. Burke of McDonnell Aircraft Corporation proposed a strategy to eliminate redundant systems once primary systems had been validated, suggesting a pathway to mitigate weight without compromising mission capabilities.

Ernst R. Letsch from Aerospace highlighted concerns over the spacecraft's weight nearing 8,000 pounds, stressing the need for rigorous checks on structural loads to ensure safety and mission success. Responding to these concerns, Low tasked the Air Force Space Systems Division and the Gemini Program Office with a vigilant focus on managing spacecraft weight factors throughout the program's lifecycle.

On February 5th, Gemini spacecraft No. 3 reached a pivotal milestone as it was transported to complex 19 and carefully positioned atop Gemini launch vehicle 3. This significant maneuver began a series of critical test operations to ensure the spacecraft's readiness for launch.

From February 9th to February 13th, preliminary systems tests were conducted, meticulously evaluating the integration of spacecraft systems within the launch complex environment. These tests provided detailed insights into the functionality and compatibility of various systems, particularly those housed within the spacecraft's adapter section.

Following the systems tests, from February 14th to February 16th, a premate Simulated Flight Test was executed. This comprehensive test simulated various flight scenarios to validate the spacecraft's performance under simulated mission conditions. Data gathered during these tests were meticulously compared with results from previous Spacecraft Systems Tests conducted at McDonnell and predelivery acceptance tests conducted at vendors' plants.

The overarching objective of these rigorous tests was to ensure seamless spacecraft integration with the launch vehicle and to thoroughly assess all spacecraft systems before the final mechanical mating to the launch vehicle. This meticulous approach aimed to mitigate risks and optimize operational readiness for the upcoming mission.

On February 5th, significant progress was made with modifications to Gemini launch vehicle five at Martin-Baltimore. Completing these modifications marked a critical step in preparing the vehicle for upcoming missions within the Gemini program.

Following the modifications, Stage I of the launch vehicle was erected in the vertical test facility at Martin-Baltimore on February 5th. This positioning allowed for comprehensive testing and verification of Stage I systems and components.

Subsequently, on February 8th, Stage II of the launch vehicle was also erected, completing the necessary assembly for further testing and evaluation. The integration of both stages set the stage for rigorous testing procedures to ensure the vehicle's readiness for manned space missions.

On February 15th, a significant milestone was achieved as power was applied to Gemini launch vehicle 5 for the first time. This power initiation marked the beginning of Subsystems Functional Verification Tests, a critical phase to validate the functionality and performance of all subsystems within the launch vehicle.

The Subsystems Functional Verification Tests continued until March 8th, encompassing a thorough assessment of various subsystems to ensure they met stringent performance criteria. Following these tests, another modification period commenced, focusing on refining and optimizing the vehicle's systems based on test outcomes and feedback.

Crew Selection for Gemini-Titan 5 Mission

On February 8th, the Manned Spacecraft Center (MSC) made a significant announcement regarding crew assignments for the upcoming Gemini-Titan 5 mission, solidifying key roles crucial to the mission's success.

L. Gordon Cooper, Jr., was selected as the command pilot for the seven-day mission. He brought a wealth of experience and expertise from previous space missions. His leadership and operational skills were essential for overseeing mission objectives and ensuring mission success.

Charles Conrad, Jr., chosen for his exceptional piloting abilities and readiness to support Cooper in all aspects of the mission, joined Cooper as a pilot. Conrad's role as pilot was integral to executing mission tasks and contributing to the overall mission goals.

Additionally, Neil A. Armstrong and Elliot M. See, Jr. were named the backup crew for Gemini-Titan 5, tasked with thorough preparation and readiness to step in if needed. Their role underscored the importance of contingency planning and readiness in NASA's manned space missions.

On February 11th, Atlas standard launch vehicle 5301 underwent critical testing at complex 14, culminating in a successful flight-readiness demonstration. This milestone confirmed the vehicle's readiness for upcoming missions within the Gemini program, demonstrating its capability to meet stringent operational requirements.

After testing, the vehicle was de-erected and transferred to Hanger J for necessary upgrades and maintenance. One of the major tasks involved replacing the sustainer engine, a critical component essential for the vehicle's propulsion and performance during launch.

The sustainer engine was replaced on April 19th, followed by installing a new level sensor and vernier engine on April 21st.

These upgrades enhanced the vehicle's reliability, efficiency, and performance.

After extensive upgrades and maintenance, Atlas standard launch vehicle 5301 was returned to complex 14. On June 18th, the vehicle was erected again, ready to resume testing and preparations for its upcoming missions within the Gemini program.

At the new launch complex, as February arrived, the culmination of months of preparation arrived with a comprehensive Atlas SLV-3 Flight Readiness Demonstration. This milestone was not just a technical checklist but a testament to the dedication and expertise of the entire Gemini team. Every component had been meticulously inspected and validated, from the towering Atlas SLV-3 to the intricate AGE on the ground.

The final verdict was clear on February 11, 1965: the Gemini program's Atlas SLV-3 5301 stood ready. Its journey from design to launch pad had been a testament to human ingenuity and precision engineering. As the countdown commenced, anticipation rippled through the team gathered at Complex 14. They knew that this mission was more than a test of technology; it was a testament to human ambition and the relentless pursuit of exploration.

The launch of Atlas SLV-3 5301 on that historic day marked not just the start of a mission but a pivotal moment in the history of space exploration. It was a testament to the courage of those who dared to push the boundaries of what was possible, paving the way for future triumphs in the Gemini program and beyond.

On February 12th, Director of Flight Operations Christopher C. Kraft, Jr., shared significant updates with the senior staff at the Manned Spacecraft Center regarding the Gemini-Titan (GT) 3 mission and future mission plans.

Kraft indicated that there was a possibility that the GT-3 mission could be conducted between March 22nd and 25th despite its official scheduling for the second quarter of 1965. This tentative scheduling reflected NASA's proactive approach to mission readiness and operational flexibility within the Gemini program.

Moreover, Kraft discussed plans to potentially utilize the Houston control center for the upcoming GT-4 mission. This consideration highlighted NASA's efforts to expand its operational capabilities and infrastructure to support increasing mission complexity and objectives.

On February 15th, the Goddard Space Flight Center made a significant decision by selecting Bendix Field Engineering Corporation, based in Owings Mills, Maryland, for a crucial contract. This contract tasked Bendix with operating, maintaining, and supporting the stations comprising the Manned Space Flight Tracking Network.

The contract, structured as a cost-plus-award-fee agreement, was valued at approximately $36 million and lasted two years. This substantial investment underscored the critical role of the tracking network in supporting NASA's manned spaceflight missions, particularly within the context of the Gemini program.

Bendix Field Engineering Corporation's selection highlighted their capability to manage and sustain the complex infrastructure for tracking and communicating with manned spacecraft during their missions. This included ensuring continuous operation, maintaining high-performance standards, and providing essential support to facilitate seamless communication and data transmission throughout the missions.

February 17 marked a pivotal moment in the Gemini-Titan 3 mission preparations as Gemini launch vehicle 3 and spacecraft No. 3 were successfully mechanically mated at complex 19. This critical step brought together the spacecraft and its launch vehicle, setting the stage for comprehensive testing and validation to ensure mission readiness.

Following the mechanical mating, rigorous testing procedures were carried out to verify the integrated functionality and performance of the combined systems:

The Electrical Interface Integrated Validation Test was completed swiftly by February 19th, focusing on ensuring seamless electrical connectivity and compatibility between the spacecraft and the launch vehicle.

The Joint Guidance and Control Test was conducted on February 22nd. This test evaluated the guidance and control systems' ability to interface and operate effectively in unison.

On February 24th, a pivotal milestone occurred: the Joint Combined Systems Test. This comprehensive test integrated all major systems and assessed their collective performance under simulated mission conditions.

Finally, on March 3rd, the Flight Configuration Mode Test was conducted. This test scrutinized the spacecraft and launch vehicle's configuration in preparation for actual flight, ensuring all systems were configured correctly and ready for launch.

On February 17, crucial live tests for the Gemini personnel parachute system encountered setbacks during high-altitude jumps. The ballute, designed to stabilize descent, failed to deploy due to an aneroid device malfunction. This issue surfaced during a similar high-altitude ejection test on February 12. These failures prompted an immediate design review of the ballute deployment mechanism.

Modifications were swiftly made to the aneroid to address the recurring malfunctions, marking the beginning of a recalibrated qualification test program for the personnel parachute system.

The revised test program substituted 23 scheduled low-altitude live jumps with ten high-altitude dummy drops. These tests, conducted between March 2 and March 5, took place at altitudes ranging from 12,000 to 18,000 feet, and speeds of 130 to 140 knots indicated air speed (KIAS). With the inclusion of the ballute, these tests aimed to verify system functionality under simulated conditions.

All sequences performed as expected during these dummy drops, barring one instance where the ballute remained trapped in its deployment bag. This issue was promptly addressed by eliminating the bag closure pin from the design. Additionally, a procedural adjustment ensured the proper separation of the backboard and egress kit, maintaining system integrity.

Following the successful dummy drops, a crucial phase of five live jumps occurred between March 8 and March 13. These tests were conducted at altitudes ranging from 15,000 to 31,000 feet, maintaining a speed of 130 KIAS.

Despite the meticulous preparations, one test encountered a setback when the ballute again failed to deploy. After a free fall, the jumper initiated a manual override, triggering the personnel parachute at 9,200 feet.

Ultimately, these rigorous tests validated the personnel parachute's functionality and ensured the comprehensive qualification of the entire Gemini escape system.

On February 21, in the Gulf of Mexico, the prime crew of Gemini-Titan 3 embarked on essential egress training from static article No. 5. This critical exercise aimed to simulate post-landing emergency procedures crucial for astronaut safety and mission success.

Following meticulous cockpit checks lasting half an hour, conducted with closed hatches, Astronauts Virgil I. Grissom and John W. Young proceeded with the emergency egress procedures meticulously crafted by the flight crew training staff for the Gemini program.

Both astronauts initiated their egress through the left hatch, designated for the command pilot. Before exiting, they carefully deployed their survival kits into the water, a strategic move to ensure essential supplies were readily available upon egress.

Once in the water, each astronaut practiced boarding a Gemini one-man life raft, an integral part of their emergency response training. To ensure safety throughout the exercise, swimmers remained on standby in a larger raft, ready to assist if needed.

Martin-Denver marked a significant milestone on February 25 by delivering propellant tanks destined for Gemini Launch Vehicle (GLV) 7 to Martin-Baltimore. This delivery culminated efforts that had commenced with the fabrication of these tanks in May 1964.

Upon receiving the tanks, Martin-Baltimore undertook meticulous procedures to ensure their readiness. By April 20, 1965, the tanks underwent thorough recleaning and purging with nitrogen, vital steps to maintain optimal conditions for subsequent operations.

In parallel, flight engines necessary for GLV-7 arrived from Aerojet-General on April 17. The integration process proceeded swiftly, with tank splicing completed by May 6, followed closely by engine installation on May 20.

A comprehensive phase of horizontal testing ensued, culminating in completion by June 14. This period validated the structural integrity and functionality of the assembled components and provided crucial insights for refinement and optimization.

Following the testing phase, a dedicated modification period commenced to implement any necessary adjustments or enhancements identified during testing.

A pivotal milestone in the preparations for Gemini-Titan 3 took place on February 26 with a full-scale rehearsal of the flight crew countdown at the launch site. This exercise was crucial in ensuring that all procedures and personnel involved were fully prepared for the upcoming mission.

The day commenced with the meticulous execution of procedures to transfer the flight crew from their quarters in the Manned Spacecraft Center operations building at Merritt Island to the pilot's ready room at Complex 16 in Cape Kennedy. Here, the crew underwent complete suiting operations, a critical step to ensure their readiness for the mission ahead.

Following suiting, the crew proceeded with the practiced transfer to Complex 19, where the Gemini spacecraft awaited. Here, the crew conducted a simulated ingress into the spacecraft, replicating the precise steps they would take on launch day.

Every step of the countdown, from initial preparations to final ingress, was meticulously rehearsed throughout the rehearsal. The smooth execution of these procedures served as a positive indicator that both equipment and personnel were thoroughly prepared for flight.

In a pivotal development during February, Lockheed launched a "Ten-point Plan for C&C Equipment" to enhance the reliability and performance of the Agena command and communication (C&C) system. This system played a critical role in the Gemini mission, providing essential electronic capabilities for tracking the vehicle, monitoring subsystem performance, and executing orbital commands, crucially supporting rendezvous and docking maneuvers.

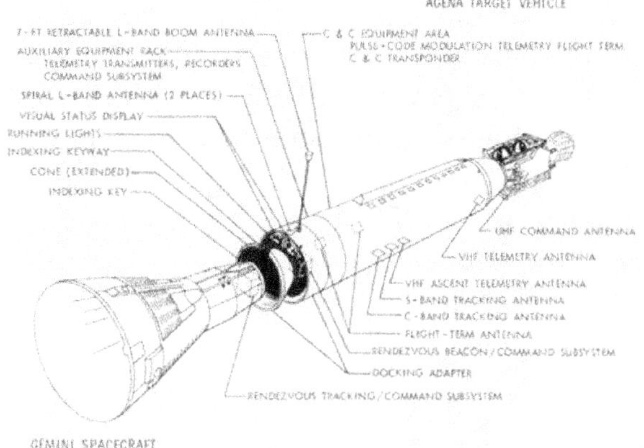

Location of command and communications system equipment on the Agena target vehicle. (Lockheed Photo NP-2-23, June 1, 1965.)

The early stages of manufacturing the C&C system revealed significant challenges, including failures and issues necessitating rework. These difficulties pointed to underlying mechanical and electronic design deficiencies. Assuming technical surveillance responsibilities for the Gemini Agena in the fall of 1964, Aerospace played a key role in highlighting these issues to Air Force and Lockheed leadership.

The comprehensive ten-point initiative yielded several critical improvements:

• Redesigned programmer circuits and packaging changes aimed at enhancing functionality and reliability.

• Enhanced monitoring of vendor work to ensure adherence to specifications and standards.

• Streamlined failure analysis processes to expedite troubleshooting and resolution.

• Improved quality control measures to uphold stringent performance standards.

The implementation of these measures underscored Lockheed's commitment to overcoming initial challenges and optimizing the Gemini Agena C&C system for mission-critical operations.

A pivotal event occurred on March 1-2 as the Office of Manned Space Flight convened the Gemini manned space flight design certification review in Washington. This review brought together chief executives from all major Gemini contractors to assess and certify the readiness of their respective products for manned space flight operations.

During the review, thorough evaluations ensured that all components and systems crucial to the Gemini-Titan 3 mission met stringent safety and operational standards. Chief executives provided assurances regarding the readiness of their products, affirming that they were prepared to support manned space flight missions.

Following the certification review, it was determined that Gemini-Titan 3 was poised for launch, pending the completion of planned test and checkout procedures at Cape Kennedy. This affirmation underscored the confidence in the readiness and reliability of the Gemini spacecraft and associated systems.

On March 2, McDonnell achieved a significant milestone by completing the Systems Assurance Tests for Gemini spacecraft No. 4. These tests represent a crucial phase in ensuring the reliability and functionality of the spacecraft's systems.

From February 27 to March 8, the spacecraft underwent a Simulated Flight Test, a rigorous examination designed to simulate operational conditions and evaluate performance under simulated flight scenarios.

Following the Simulated Flight Test, preparations commenced for altitude chamber testing, a critical step in validating the spacecraft's performance under varying atmospheric conditions. These preparations continued diligently until March 19, underscoring the meticulous approach to preparing Gemini spacecraft No. 4 for its upcoming missions.

AiResearch achieved a significant milestone on March 6 by completing dynamic qualification tests of the Gemini spacecraft's environmental control system (ECS). These tests were essential in validating the ECS's performance and reliability under simulated operational conditions.

The environmental control system was critical in maintaining optimal conditions inside the spacecraft. It regulated temperature, humidity, and air quality to ensure the crew's comfort and safety during missions.

During the dynamic qualification tests, the ECS underwent rigorous evaluations to assess its ability to function effectively under various dynamic conditions that simulate the stresses and demands of spaceflight.

The successful completion of these tests marked a significant advancement in the Gemini program, confirming that the ECS met or exceeded all performance requirements. This achievement underscored AiResearch's dedication to delivering reliable systems essential for the success of manned space missions.

Moving forward, this milestone contributed to the overall readiness of the Gemini spacecraft for its upcoming missions, highlighting the meticulous preparation and testing protocols essential for human space exploration.

On March 8, a pivotal milestone was achieved by successfully conducting the Wet Mock Simulated Launch for Gemini-Titan 3. This exercise simulated crucial aspects of the launch process, providing essential readiness validation for the upcoming mission.

Throughout the Wet Mock Simulated Launch, countdown exercises were meticulously executed to simulate the events leading up to liftoff. These exercises included comprehensive checks and procedures to ensure all systems and personnel were prepared for launch day operations.

The culmination of these preparations occurred on March 18 with the completion of the Simulated Flight Test. This final phase of countdown exercises validated Gemini-Titan 3's readiness for its upcoming mission, affirming that all systems were functioning as expected.

Gemini Agena target vehicle 5001 achieved a critical milestone on March 9 with

the completion of electromagnetic compatibility (EMC) tests conducted in the anechoic chamber at Sunnyvale. These tests were essential to ensure that the vehicle's electronic systems could operate without interference in the electromagnetic environment of space.

Gemini-Titan 3 on pad 19 during final countdown exercises. (NASA Photo No. 65-H-406, released Mar. 19, 196

Following the EMC tests, Gemini Agena target vehicle 5001 remained in the chamber until March 17. Lockheed utilized this extended period to verify corrective actions to resolve issues related to programmer time-accumulator jumps and telemetry synchronization problems identified during testing.

On March 18, the vehicle was transferred to systems test complex C-10 for final Vehicle Systems Tests (VST). These tests represented the last phase of rigorous evaluations to confirm the overall functionality and readiness of Gemini Agena target vehicle 5001 for its intended mission objectives.

On March 10, a significant event occurred at Martin-Baltimore: the official roll-out inspection of Gemini Launch Vehicle (GLV) 4. This inspection marked a crucial step in preparing and accepting the vehicle for its upcoming mission.

Following the roll-out inspection, the Air Force Space Systems Division formally accepted the delivery of GLV-4 on March 21. Immediate preparations commenced to facilitate the vehicle's transfer to Cape Kennedy, where it would undergo final preparations for launch.

GLV-4's first stage arrived at Cape Kennedy on March 22, signaling the beginning of on-site preparations for integration and testing. The second stage followed suit the next day, ensuring that both were promptly available for assembly and final checks.

On March 10, significant decisions regarding the mission planning for Gemini-Titan 6 were made during the Gemini Trajectory and Orbits Panel meeting. These decisions shaped the parameters and constraints that would define the upcoming mission.

During the meeting, the Air Force Space Systems Division reiterated that the nominal plan for Gemini-Titan 6 should not include using the Agena primary propulsion system in orbit. This decision stemmed from concerns that the system had not yet been fully qualified through actual flight tests by the time of this mission.

Simultaneously, the Gemini Program Office announced a critical decision regarding spacecraft No. 6. It was determined that only enough electrical power would be provided to support 22 orbits despite initial plans for a longer mission duration. This constraint and considerations for reentry and recovery necessitated adjustments to the nominal mission plan, limiting it to approximately 15 orbits.

These decisions underscored the careful balance between mission objectives and technical readiness in the Gemini program. By aligning mission plans with operational constraints and system capabilities, stakeholders ensured that Gemini-Titan 6

could proceed safely and effectively within established parameters.

On March 14, McDonnell achieved significant milestones in the production and preparation of Gemini spacecraft No. 5. These milestones marked critical stages in ensuring the spacecraft's readiness for upcoming missions.

McDonnell completed the manufacturing process, including module tests and equipment installation, ensuring all components and systems were integrated and functioning as intended.

The assembly of spacecraft No. 5 was finalized on April 1 with the mating of the reentry and adapter assemblies. This assembly phase combined the spacecraft's key structural and functional elements, preparing it for subsequent testing and validation.

Subsequently, Systems Assurance Tests commenced on April 30, with rigorous evaluations to verify the spacecraft's overall performance and readiness for its designated missions.

On March 18, Gemini Agena target vehicle (GATV) 5001 was transferred from the anechoic chamber to systems test complex C-10 for final systems tests. However, unforeseen challenges surfaced, leading to significant delays in the testing schedule.

Issues arose during the incorporation of filters in the command controller, necessitating substantial redesign efforts. Additionally, alignment problems with the forward auxiliary rack required extensive rectification machining, further complicating the testing timeline.

These unexpected difficulties contributed to a 29-day delay in the GATV 5001 schedule by the end of March, leaving the vehicle significantly behind, with a total slip of 66 calendar days.

Efforts to address these challenges continued, with the machining of the forward auxiliary rack finally completed by April 5. Subsequently, vehicle systems testing commenced on April 9, marking the resumption of critical evaluations to ensure the readiness and functionality of GATV 5001 for its upcoming missions.

Gemini spacecraft No. 4 entering the 14-foot altitude chamber at McDonnell before simulated high-altitude tests. (NASA Photo S-65-3420, Mar. 16, 1965.)

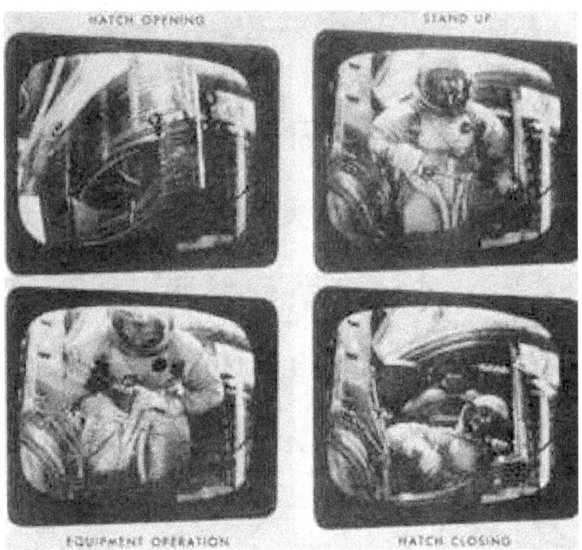

Astronaut Edward H. White II practices standup extravehicular activity at a simulated altitude of 150,000 feet in the McDonnell altitude chamber. (NASA Photo S-65-4896, Mar. 24, 1965.)

On March 20, McDonnell initiated altitude chamber tests for Gemini spacecraft No. 4, marking a critical phase in the spacecraft's preparation for upcoming missions. These tests simulated various flight scenarios to validate the spacecraft's performance under different altitude conditions.

First Run (Unmanned Test): The initial test involved a simulated flight without crew members onboard, serving as a baseline assessment of the spacecraft's systems and responses in the chamber environment.

Second Run (Prime Crew): The prime crew participated in a simulated mission, although the chamber was not evacuated. This run provided insights into crew interactions and operational procedures within simulated flight conditions.

Third Run (Backup Crew): Similar to the second run, the backup crew replaced the prime crew to replicate mission scenarios and verify operational readiness under simulated altitude conditions.

Fourth and Fifth Runs: The prime and backup crews underwent separate flights at simulated altitudes during the fourth and fifth runs, respectively. These tests aimed to confirm the spacecraft's capability to support crew activities and systems functionality under simulated flight conditions.

Altitude chamber testing concluded on March 25, signifying the successful completion of simulated flight scenarios for Gemini spacecraft No. 4. Subsequently, preparations began to transport the spacecraft to Cape Kennedy, where final integration and launch preparations would occur.

Gemini-Titan 3: Pioneering Manned Orbital Flight

On March 23, 1965, at 9:24 a.m. E.S.T., history was made at Cape Kennedy's Complex 19 with the triumphant launch of Gemini-Titan 3 (GT-3), the inaugural manned mission of NASA's Gemini program. Commanded by Astronaut Virgil I. Grissom and pilot Astronaut John W. Young, GT-3 embarked on a pivotal three-orbit journey that would pave the way for extended manned space missions.

The mission's primary goals were manifold: to validate manned orbital flight capabilities within the Gemini spacecraft, thoroughly assess the spacecraft and launch vehicle systems for future long-duration missions, and demonstrate orbital maneuvers using the spacecraft's Orbit Attitude and Maneuver System (OAMS). Additionally, the mission aimed to showcase the OAMS's ability to serve as a backup for retrorockets during critical maneuvers and to master controlled reentry flight paths and landing accuracy.

Astronauts Young and Grissom walk up the ramp leading to the elevator that will carry them to the spacecraft for the first manned Gemini mission. They wear Gemini G3C intravehicular suits. (NASA Photo No. 65-H-438, released Mar. 23, 1965.)

Despite the mission's resounding success in achieving these objectives, the accuracy of the landing point fell short of expectations. GT-3 touched down at 2:16 p.m., approximately 60 nautical miles from its intended target. Prompt recovery operations ensued, with the flight crew exiting the spacecraft shortly after 3:00 p.m. and being swiftly transported by helicopter to the primary recovery vessel, the aircraft carrier Intrepid. The spacecraft's recovery was officially completed at 5:03 p.m., marking the culmination of a meticulously planned mission.

Gemini spacecraft No. 3, wearing a flotation collar, being hoisted aboard the U.S.S. Intrepid after landing. (NASA Photo No. 65-H-462, released Mar. 23, 1965.)

During the flight, Commander Grissom adeptly executed three orbital maneuvers, showcasing the maneuverability and precision of the Gemini spacecraft's systems. Three scientific experiments were among the secondary objectives, two of which were successfully conducted. However, a mechanical malfunction thwarted the third experiment, which aimed to study the effects of zero gravity on sea urchin egg development.

Smuggling Sandwiches and Exploring Biology

Amidst the critical scientific objectives of NASA's Gemini III mission, an unexpected addition found its way into orbit: a corned beef sandwich. This sandwich, not part of NASA's approved provisions, was clandestinely stowed in Pilot John Young's spacesuit pocket before liftoff. Approximately two hours into the nearly five-hour mission, Young surprised Commander Gus Grissom with this unconventional snack.

The exchange over the sandwich was brief but memorable. Initially surprised by Young's gesture, Grissom accepted the rye bread and preserved meat offering. Aware of potential hazards in weightless conditions, Grissom carefully stowed the unfinished sandwich in his own spacesuit pocket to prevent crumbs from interfering with spacecraft instruments or endangering their eyesight.

Meanwhile, the officially sanctioned food for the mission, kept in a nearby box, included practical items like cubed food coated in gel to prevent crumbs—a precaution highlighted by the messiness of the sandwich. The menu also featured rehydrated applesauce, a far cry from the astronauts' whimsical lament about the absence of a pork chop.

Adjacent to the food supplies, another significant cargo lay quietly in a canister: sea urchin eggs, specifically Arbacia punctulata. This payload represented a pioneering moment in NASA's biological research in space—an early step in what would evolve into the field of space biology. This discipline, dedicated to understanding how fundamental life processes like metabolism and growth are impacted by space and spaceflight, had its roots in NASA's earliest missions.

Before Gemini III, NASA's scientific investigations focused on astronaut physiology during the Mercury missions. With Gemini, the agency expanded its scope to include fundamental biology experiments, exploring how life at its most basic functions in the unique environment of space.

Thus, Gemini III not only smuggled a sandwich into orbit but also carried forward the scientific legacy of exploring life beyond Earth's atmosphere. In retrospect, this mission exemplifies both the pioneering spirit and the unexpected quirks that characterize the early days of human space exploration.

On March 23-24, a critical meeting convened representatives from the Air Force Space Systems Division (SSD), Aerospace, Lockheed, and the Gemini Program Office at Sunnyvale. The meeting's focus was the monthly Gemini Agena Target Vehicle (GATV) Management-Technical Review. During discussions, SSD highlighted concerns regarding the oxidizer gas generator solenoid valve on GATV 5001, prompted by a recent failure during 38-day storage tests at Bell Aerosystems.

Following deliberations, SSD recommended removing the current valve configuration from GATV 5001. This decision prompted Lockheed to assemble a specialized team to evaluate and redesign the valve. By July, the redesigned valve had entered qualification testing, marking a proactive step toward ensuring the reliability and functionality of future GATV configurations.

On March 27, significant upgrades were made to Gemini spacecraft No. 4's orbit attitude and maneuver system (OAMS). The existing 25-pound thrusters were replaced with new, long-life engines to enhance reliability and performance in orbital maneuvers.

The new engines were initially slated for installation in spacecraft No. 5, but they were expedited due to their readiness ahead of schedule. Earlier in February, Rocketdyne had completed a substantial portion of the qualification test program for the OAMS and reentry control systems configured for spacecraft Nos. 3, 4, and 5. However, additional testing extended the final qualification phase until mid-April, ensuring all systems met rigorous standards.

By early June, OAMS component qualification for spacecraft configurations from No. 6 onwards was successfully achieved, marking a milestone in the ongoing enhancement of propulsion systems across the Gemini program. The comprehensive ground qualification of all Gemini spacecraft liquid propellant rocket systems concluded in August, culminating with the system qualification of the OAMS in the spacecraft No. 6 configuration.

On March 29, a pivotal moment in space exploration unfolded during an informal meeting in the office of Director Robert R. Gilruth at the Manned Spacecraft Center (MSC). Gilruth and Deputy Director George M. Low attended the meeting, which featured discussions led by Richard S. Johnston from the Crew Systems Division (CSD) and Warren J. North from the Flight Crew Operations Division.

Johnston presented a prototype chestpack for extravehicular activity (EVA) and a mock-up of a handheld maneuvering unit. North expressed confidence in the feasibility of umbilical EVA for the upcoming Gemini-Titan 4 mission. Encouraged by the potential benefits and readiness assessments, Gilruth authorized further exploration of enhanced EVA capabilities.

Subsequently, on April 3, Johnston briefed George E. Mueller, Associate Administrator for Manned Space Flight, in Washington, D.C. Mueller then briefed relevant Headquarters Directorates, initiating preliminary approvals for the expanded EVA plans. This momentum culminated in a visit by Robert C. Seamans, Jr., Associate Administrator of NASA, to MSC on May 14, where he received comprehensive briefings that underscored the program's readiness.

The enthusiastic endorsement brought back by Seamans to Washington swiftly led to final Headquarters approval, solidifying plans to advance Gemini's capabilities through groundbreaking EVA operations. This decision paved the way for innovative spacewalk technologies and reaffirmed NASA's commitment to pushing the boundaries of human exploration in space.

On March 29, Gemini launch vehicle 4 underwent crucial preparations at Complex 19. Following its erection, thorough inspections ensured its readiness for subsequent operations. By March 31, umbilicals were connected to the vehicle, facilitating essential power connections by April 2.

Subsequently, subsystem functional verification tests commenced immediately and proceeded diligently until their completion on April 15. These tests rigorously evaluated the operational integrity and performance of critical systems onboard.

These preparations culminated in the Prespacecraft Mate Combined Systems Test on April 16. This comprehensive test integrated all major systems and components of the spacecraft, ensuring seamless coordination and functionality in anticipation of upcoming mission milestones.

On April 4, McDonnell successfully delivered Gemini spacecraft No. 4 to Cape Kennedy, marking a significant milestone in the Gemini program's ongoing preparations. Following the delivery, a thorough receiving inspection was conducted and concluded by April 6, ensuring the spacecraft met all specified requirements and standards.

Subsequently, various industrial activities in the area commenced, focusing on essential tasks such as pyrotechnic buildup, temporary seat installation, and final preparations necessary for pad testing. These activities were diligently carried out, culminating in completing all necessary tasks by April 14.

Fully prepared, the spacecraft was transferred to Complex 19, the launch site designated for upcoming Gemini missions. This relocation marked the final phase of preparations before the spacecraft underwent comprehensive testing and integration with the launch vehicle.

Selection of Crew for Gemini-Titan 6 Rendezvous and Docking Mission

On April 5, a significant announcement from the Manned Spacecraft Center (MSC) revealed the crew assignments for the upcoming Gemini-Titan 6 mission, slated to pioneer rendezvous and docking capabilities within the Gemini program. Walter M. Schirra, Jr., was designated as the command pilot, while Thomas P. Stafford would serve as the pilot for this historic mission.

Virgil I. Grissom and John W. Young were named as the backup crew, emphasizing NASA's meticulous planning and preparation for every aspect of the mission. This crew assignment underscored the agency's commitment to advancing manned spaceflight capabilities through ambitious objectives such as rendezvous and docking, critical milestones in preparation for future lunar missions under the Apollo program.

The selection of Schirra and Stafford, with Grissom and Young as backups, marked a significant step forward in NASA's pursuit of space exploration goals. It positions the Gemini program at the forefront of technological innovation and operational excellence in space.

On April 13, the Manned Spacecraft Center (MSC) completed and delivered the "Gemini Atlas Agena Target Vehicle Systems Management and Responsibilities Agreement" to the Air Force Space Systems Division (SSD). This agreement, dated April 9, bore the signatures of MSC Director Robert R. Gilruth and Gemini Program Manager Charles W. Mathews.

Before this, Major General Ben I. Funk, Commander of SSD, and Colonel John B. Hudson, SSD Deputy for Launch Vehicles, had signed on behalf of SSD on March 31 and 29, respectively. The agreement, dated back to March 1965, represented the culmination of extensive negotiations and coordination efforts between MSC and SSD.

The agreement clarified and supplemented the "Operational and Management Plan for the Gemini Program," initially drafted on December 29, 1961.

Preparations for Gemini Spacecraft No. 4

On April 14, at Cape Kennedy. Gemini spacecraft No. 4 was carefully hoisted and positioned atop its launch vehicle, marking a crucial step in preparing for an upcoming mission. Following the spacecraft's placement, meticulous cabling for testing purposes was swiftly completed by April 19.

A significant milestone was achieved on April 21 when systems testing concluded successfully. Notably, this phase also marked the first occasion when the Mission Control Center in Houston provided support for operations directly from Kennedy Space Center's launch pad.

The final stages of preparation were marked by the Prespacecraft Mate Simulated Flight Test, conducted diligently over April 22-23. This comprehensive test simulated the conditions and operations expected during the actual flight, ensuring that all systems and procedures were thoroughly vetted and ready for the mission ahead.

Significant progress was made in preparing Gemini Launch Vehicle (GLV) 6 at Martin-Baltimore's vertical test facility from April 14 to 15. GLV-6 was the inaugural vehicle to undergo testing in the newly completed west test cell, which Martin fully installed and checked out earlier in January.

GLV-5 remained in the other test cell during this period, undergoing its own vertical tests. The setup of both vehicles posed logistical challenges as they shared power sources and aerospace ground equipment connections, preventing simultaneous testing. However, Martin's approach allowed for efficient inspection and preparation of one vehicle while the other underwent rigorous testing procedures.

A pivotal moment it occurred on May 13 when power was applied to GLV-6 for the first time, initiating a series of Subsystems Functional Verification Tests that would continue until June 22. These tests were essential for verifying the functionality and integration of various subsystems crucial for the successful launch and mission of GLV-6.

On April 15, Martin-Denver achieved a significant milestone in preparing Gemini Launch Vehicle 8 (GLV-8) by delivering its propellant tanks to Martin-Baltimore. This delivery marked a crucial step in the comprehensive assembly process that began on September 25, 1964, when fabrication of the tanks commenced.

Aerojet-General concurrently played a pivotal role in the project timeline by delivering the Stage I engine on June 16 and the Stage II engine on August 20. The meticulous process of tank splicing was successfully concluded on August 3, ensuring structural integrity and readiness for subsequent stages of assembly.

Further milestones included the completion of engine installation on September 23, followed by the culmination of all horizontal testing by September 27. These achievements reflected the precision and thoroughness of the preparation efforts undertaken at Martin-Baltimore to ensure GLV-8's readiness for upcoming integration and testing phases.

Each step in this process was critical for maintaining the stringent quality standards and operational readiness required for the Gemini program's launch vehicles, contributing to NASA's overall mission success and advancing capabilities in space exploration.

On April 20, McDonnell achieved a significant milestone with the completion of Systems Assurance Tests for Gemini spacecraft No. 5. These tests were crucial for verifying the integrity and functionality of various systems onboard the spacecraft, ensuring they met rigorous operational standards for upcoming missions.

Following this milestone, on April 24, the environmental control system (ECS) was successfully validated. The ECS plays a vital role in regulating temperature, humidity, and air quality within the spacecraft, crucial for maintaining a habitable environment for astronauts during their missions.

However, during the reentry/adapter mating operations on April 16, an issue arose with the spacecraft's fuel cell. Despite this setback, McDonnell promptly addressed the problem, and by April 26, the fuel cell's reinstallation was successfully completed. The fuel cell was essential for providing electrical power to the spacecraft during missions, making its functionality critical for mission success.

These developments underscored McDonnell's commitment to ensuring the reliability and readiness of Gemini spacecraft No. 5, overcoming challenges encountered during the assembly and testing phases to prepare for future missions in NASA's Gemini program.

On April 21, the Combined Systems Acceptance Test (CSAT) of Gemini launch vehicle (GLV) 5 took place at the vertical test facility located at Martin Baltimore. This test was a critical step in verifying the integrated functionality and readiness of GLV-5 for upcoming missions in NASA's Gemini program.

The CSAT process, which had encountered several minor anomalies during earlier attempts from April 15 to April 20, aimed to assess the vehicle's systems and subsystems thoroughly. These tests are essential to ensure that all launch vehicle components operate as expected and can withstand the rigors of spaceflight.

Following completing the CSAT on April 21, the vehicle acceptance team conducted inspections from April 26 to April 30. Their evaluations confirmed that GLV-5 met the necessary standards and was deemed acceptable for flight. Subsequently, the vehicle was removed from the test cell on May 7-8.

On May 15, GLV-5 received formal acceptance from the Air Force, marking a significant milestone in the vehicle's preparation for its mission. Stage I of the launch vehicle arrived at Cape Kennedy on May 17, followed by Stage II on May 19, further advancing the readiness process at the launch site.

Abort Panel Review for Gemini-Titan 4

On April 22, the Abort Panel convened to review the abort criteria specifically for the upcoming Gemini-Titan (GT) 4 mission. The panel's primary task was establishing and finalizing the guidelines and procedures governing potential abort scenarios during the mission.

The panel decided that the abort criteria established for the previous Gemini-Titan 3 (GT-3) mission would suffice for GT-4. This continuity in criteria ensured consistency and familiarity in handling abort situations across successive missions.

The panel also agreed to investigate alternate procedures for delayed mode two abort scenarios. These investigations will be conducted once the Manned Spacecraft Center (MSC) abort trainer becomes available for training related to the GT-5 mission.

The Abort Panel's decisions on April 22 aimed to ensure comprehensive preparedness and operational flexibility in managing potential abort scenarios during the GT-4 mission. By building on previous criteria while exploring enhancements for specific abort modes, the panel enhanced mission safety and readiness for Gemini-Titan missions.

On April 23, in preparation for the Gemini-Titan 4 mission at Complex 19, Gemini launch vehicle (GLV) 4 and spacecraft No. 4 were successfully mechanically mated. This step marked the physical connection of the spacecraft to the launch vehicle, a critical integration milestone.

Following the mechanical mating, the Electrical Interface Integrated Validation and Joint Guidance and Control Test were conducted from April 26 to April 29. These tests were crucial to ensure that all electrical connections between the spacecraft and the launch vehicle were functioning correctly and that the guidance and control systems were integrated seamlessly.

The spacecraft/GLV Joint Combined Systems Test commenced on April 30. This test, which was newly combined for Gemini-Titan 4 onwards, aimed to validate the integrated functionality of all systems between the spacecraft and the launch vehicle in a simulated flight configuration.

The systems testing phase culminated with the Flight Configuration Mode Test, which concluded on May 7. This comprehensive test ensured that all systems operated as expected in the exact configuration they would be in during the actual flight.

On April 26, McDonnell initiated the Simulated Flight Test for Gemini spacecraft No. 5, a critical step in verifying the spacecraft's readiness for space missions. Here are the key events and developments during this testing phase:

The Simulated Flight Test commenced at McDonnell, aiming to simulate various operational conditions the spacecraft would experience during an actual mission.

On April 28, an incident occurred during the test when the ECS was inadvertently overpressurized. The test was halted temporarily to investigate and address the issue with the ECS suit loop.

After resolving the overpressurization issue, the ECS was reinstalled by May 8. Subsequently, from May 9 to May 11, rigorous retesting of the ECS and the guidance and control systems was conducted to ensure proper functionality.

Simulated flight testing resumed on May 11 and continued until May 19. This phase included comprehensive simulations of various mission scenarios to validate the spacecraft's performance under simulated flight conditions.

After simulated flight testing, preparations for altitude chamber testing began and continued until May 25. Altitude chamber testing was crucial for simulating the spacecraft's behavior in low-pressure, high-altitude environments similar to those encountered during space missions.

Throughout late April and May, these activities were essential in verifying the operational readiness and reliability of Gemini spacecraft No. 5, ensuring that it met the stringent requirements for crewed space missions.

On May 4, McDonnell completed the manufacturing process for Gemini spacecraft No. 6. This phase involved fabricating all necessary components and modules that constitute the spacecraft.

Module Tests and Equipment Installation: McDonnell conducted rigorous module tests following manufacturing to ensure all components functioned correctly. They also installed essential equipment and systems within the spacecraft to support its operational capabilities.

The final assembly of Gemini spacecraft No. 6 was completed on May 12. This included mating the reentry and adapter assemblies, marking a crucial step toward integrating all spacecraft components into a cohesive unit.

Post-assembly, the spacecraft underwent cabling installations and preparations for comprehensive testing. These activities ensured that all electrical connections were secure and prepared the spacecraft for subsequent testing phases.

On June 4, Systems Assurance Tests (SAT) commenced. These tests are designed to validate the overall functionality and reliability of the spacecraft's systems and subsystems under simulated operational conditions.

This progression from manufacturing completion to final assembly and initiating Systems Assurance Tests demonstrates McDonnell's systematic approach to ensuring the readiness and quality of Gemini spacecraft No. 6 for upcoming space missions.

Analyzing the Challenges of Gemini

On May 5th, during a pivotal session of the Gemini Management Panel, Charles W. Mathews delved into the complexities surrounding Gemini 3's landing accuracy. The discussion highlighted critical deviations from initial mission projections, shedding light on unforeseen challenges that tested the program's technological limits.

Mathews underscored a crucial discrepancy: the spacecraft's unexpectedly reduced angle of attack, a pivotal parameter influencing its aerodynamic behavior during re-entry. This deviation, coupled with lower-than-anticipated lift capabilities compared to meticulous wind tunnel simulations, presented a significant puzzle for mission engineers and planners alike.

On May 6th, a significant milestone was achieved in the meticulous preparations for Project Gemini when the Gemini Agena Target Vehicle (GATV) 5001 successfully concluded comprehensive vehicle systems testing with a final simulated flight. This pivotal event marked a critical stage in validating the GATV's operational readiness and its ability to support upcoming Gemini missions.

Following the rigorous testing phase, GATV 5001 was meticulously disconnected from the test complex on May 14th, marking the conclusion of data acquisition and initial analysis. The wealth of data collected during these tests provided crucial insights into the vehicle's performance under simulated mission conditions, informing engineers and mission planners alike.

Simultaneously, on May 10th, the First Article Configuration Inspection commenced for GATV 5001. This inspection, essential for verifying the vehicle's structural integrity and adherence to design specifications, underscored NASA's stringent quality control measures in ensuring mission success and astronaut safety.

On May 10th, a collaborative effort involving NASA, the Air Force Space Systems Division, Aerospace, and Lockheed initiated the meticulous First Article Configuration Inspection (FACI) of Gemini Agena Target Vehicle (GATV) 5001 at Sunnyvale. This comprehensive inspection marked a crucial step in validating the vehicle's conformity to stringent engineering standards and operational specifications.

The FACI process encompassed a thorough review and evaluation of all pertinent documentation, including drawings, specifications, test procedures, reports, component and assembly log books, and qualification and certification records specific to GATV 5001. This meticulous scrutiny aimed to identify any discrepancies between the original engineering design and the physical hardware produced, ensuring alignment with mission-critical requirements.

Established as a standard Air Force procedure in June 1962, FACI was a pivotal audit conducted with contractor support to establish a definitive production configuration baseline. This baseline was essential for guiding the manufacturing and delivery of subsequent contract end items, including GATV 5002 and beyond, ensuring consistency and reliability across the Gemini program.

On May 13th, a critical milestone in the preparation for Gemini-Titan (GT) 4 was achieved with the completion of the Wet Mock Simulated Launch (WMSL). This simulated launch test provided essential insights into the readiness of both the spacecraft and launch vehicle, setting the stage for subsequent adjustments and refinements.

Following the WMSL, the spacecraft underwent demating from the launch vehicle to facilitate necessary maintenance tasks. These included replacing batteries within the spacecraft adapter, installing flight seats, and evaluating crew stowage arrangements meticulously. Concurrently, attention turned to the launch vehicle, which underwent a specialized tanking test on May 19th to address loading inaccuracies observed during the WMSL.

Investigations pinpointed the issue to faulty Stage II flowmeters, prompting their replacement on May 21st. A subsequent tanking test focused solely on Stage II on May 27th confirmed the effectiveness of the corrective measures. With these technical challenges addressed, the launch vehicle and spacecraft were successfully remated on May 22nd, marking a pivotal step towards readiness for launch.

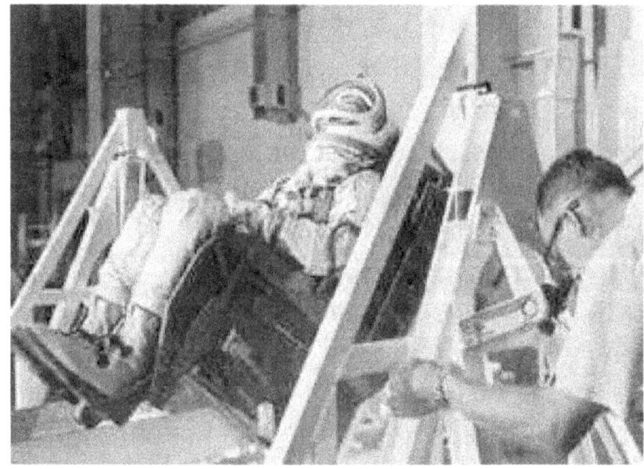

Weight and balance test of Astronaut McDivitt during the Wet Mock Simulated Launch of Gemini-Titan 4. (NASA Photo No. 65-H-797, released May 21, 1965.)

On May 15th, the Gemini program reached a significant milestone by completing the qualification process for the G4C extravehicular suit. This advanced suit, a successor to the G3C model, underwent meticulous enhancements to bolster astronaut safety and operational efficiency during extravehicular activities.

The G4C suit retained many core features of its predecessor while incorporating critical modifications. These included a redundant zipper closure system for enhanced integrity, dual over-visors providing visual clarity and physical protection, and automated ventilation settings to optimize comfort and safety in the harsh space environment. Notably, the suit's outer layer was reinforced with heavier materials to provide thermal insulation and protection against micrometeoroids, essential for safeguarding astronauts during extended missions.

By the end of May, six fully-equipped G4C suits were poised at the launch site, ensuring readiness for the Gemini 4 flight crews. This strategic deployment underscored NASA's proactive approach to mission preparedness, ensuring that astronauts were equipped with state-of-the-art gear capable of meeting the demands of extravehicular operations.

On May 18th, a pivotal stage in the preparation for Project Gemini unfolded as Gemini Agena Target Vehicle (GATV) 5002 completed its final assembly. This milestone marked the culmination of meticulous construction efforts, setting the stage for rigorous testing at systems test complex C-10 in Sunnyvale.

Initially scheduled for transfer on May 5th, the process faced setbacks attributed to parts shortages, engineering complexities, and a substantial backlog of work. A significant hurdle involved rectifying a gap identified between the forward auxiliary rack and the vehicle structure, necessitating precise machining, alignment adjustments, and meticulous refinishing of scraped surfaces. These efforts proved time-intensive but were crucial in ensuring structural integrity and operational readiness.

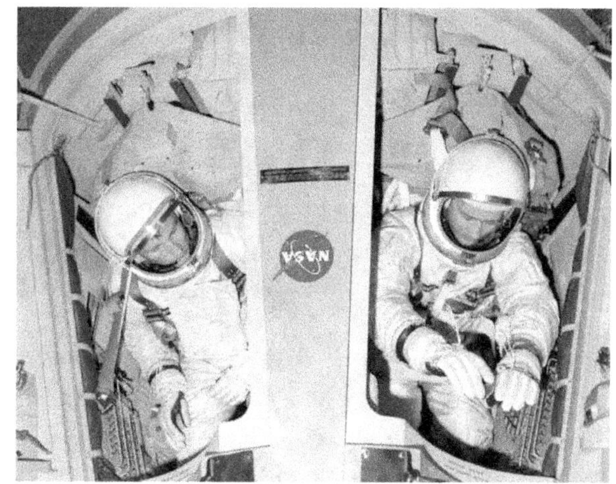

Astronauts Edward H. White II (left) and James A. McDivitt are shown going through tests in a Gemini Crew Simulator at the Cape. NASA

Despite these challenges, GATV 5002 still needed to integrate several critical command equipment items before systems testing could commence. Finally, on May 21st, systems testing officially began, marking a pivotal phase in validating the vehicle's performance under simulated mission conditions.

As preparations for the Gemini 4 mission intensified, May 19th marked a critical milestone with the completion of qualification for all extravehicular equipment planned for the mission. This suite of specialized gear, essential for supporting astronaut activities outside the spacecraft, underwent rigorous testing and validation to ensure reliability and performance in the harsh space environment.

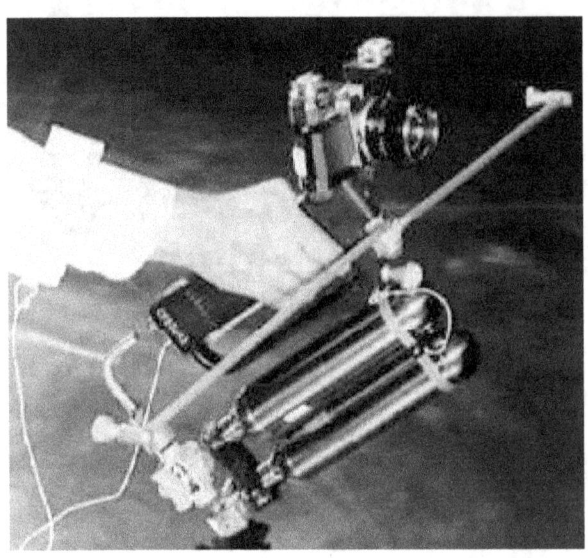

The hand-held maneuvering unit. (NASA Photo S-65-27331, June 2, 1965.)

Gemini spacecraft No. 5 undergoing clean-up prior to being shipped to Cape Kennedy. (NASA Photo S-65-5781, June 2, 1965.)

Key components included the ventilation control module, which regulated air supply within the spacesuit; the extravehicular umbilical assembly, which provided crucial connections for power and communication; and the hand-held maneuvering unit, which enabled controlled movement during spacewalks. Each piece of equipment was meticulously scrutinized to meet stringent operational standards and astronaut safety protocols.

By the end of May, all flight-ready hardware had been successfully transported to the launch site, positioning NASA to proceed confidently with the Gemini 4 mission. This timely readiness underscored NASA's meticulous planning and commitment to equipping astronauts with state-of-the-art technology, essential for executing successful extravehicular activities and advancing human exploration beyond Earth's bounds.

On May 26th, McDonnell initiated altitude chamber tests for Gemini spacecraft No. 5, marking a pivotal phase in the spacecraft's rigorous testing regimen. These tests were essential to simulate the harsh conditions of space and validate the spacecraft's performance under varying atmospheric pressures.

However, the testing process encountered an unexpected setback on June 1st when a fuel cell failure occurred, prompting the replacement of affected fuel sections. This interruption necessitated modifications and thorough preparations before testing could resume.

By June 12th, modifications were completed, and preparations were finalized for the retest, which included an overall systems test to ensure comprehensive functionality, including the repaired fuel cell.

Chapter 10 - Challenges and Commitment

On May 27th, the Air Force Space Systems Division (SSD) critically assessed Gemini Agena Target Vehicle (GATV) 5001, employing standard Air Force acceptance protocols outlined in DD Form 250. The findings from the First Article Configuration Inspection, completed on May 26th, revealed deficiencies that rendered the vehicle non-flightworthy under contractual requirements.

Despite these findings, SSD conditionally accepted the delivery of GATV 5001, stipulating that Lockheed addressed identified deficiencies within the specified timelines outlined in the DD-250 attachments. While some items awaited final documentation, significant components requiring further qualification included the shroud, primary and secondary propulsion systems, the command system, and elements of the electrical power system.

Following conditional acceptance, GATV 5001 was expedited by air to the Eastern Test Range, departing on May 28th and arriving promptly on May 29th. This expedited shipment underscored Lockheed's commitment to rectifying deficiencies swiftly, aligning with stringent timelines to ensure readiness for upcoming mission requirements.

On May 29th, Gemini Agena Target Vehicle 5001 arrived at Cape Kennedy following its conditional acceptance by the Air Force on May 27th. This marked a significant step in the vehicle's journey towards readiness for mission deployment.

Upon arrival, the target vehicle was promptly transferred to the Missile Assembly Building (Hangar E) for comprehensive testing and integration preparations. These crucial steps included meticulous checks and adjustments to ensure the vehicle's operational readiness and compatibility with mission requirements.

On June 18th, Gemini Agena Target Vehicle 5001 reached another milestone as it was successfully mated with target docking adapter No. 1. This integration paved the way for Combined Interface Tests commencing on June 19th. These tests encompassed rigorous evaluations, including functional and static leak checks of the secondary propulsion system (SPS), installation checks, and meticulous preparation of thermal control surfaces.

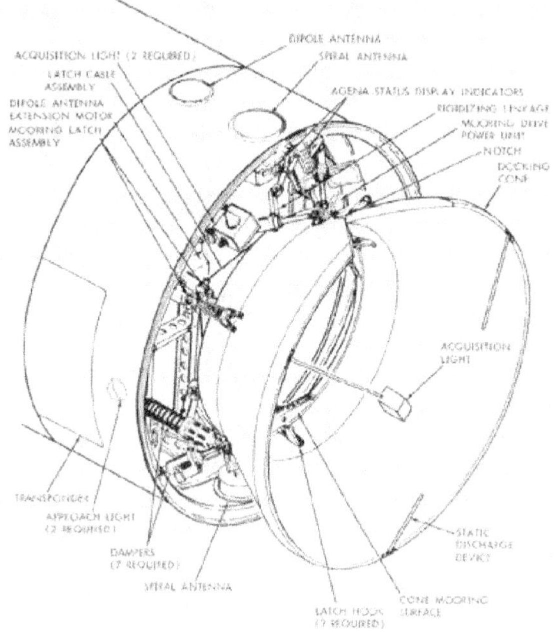

Target Docking Adapter assembly.

Testing concluded on July 8th, confirming the readiness of Gemini Agena Target Vehicle 5001 for its role in upcoming missions. Subsequently, the vehicle was transferred to Complex 14 for final integration with target launch vehicle 5301, marking a crucial phase in preparations for the next stage of the Gemini program.

Gemini 4: Pioneering Long-Duration Space Exploration

Gemini 4, launched from Complex 19 at 10:16 a.m. EST, marked a watershed moment in space exploration as the second manned mission and the first long-duration flight in the Gemini program. Commanded by

Astronaut James A. McDivitt, with Astronaut Edward H. White II as pilot, the mission aimed to validate spacecraft systems during extended spaceflight and assess the physiological effects of prolonged exposure to space.

Launch vehicle erector tower being lowered just prior to launch of Gemini-Titan 4. Difficulty in lowering the erector delayed the launch from the scheduled time of 9:00a.m. to 10:16a.m., e.s.t. (NASA Photo No. 65-H-934, released June 3, 1965.)

The primary objectives included demonstrating the capability of spacecraft systems over four days, conducting extravehicular activity (EVA) in space, and performing intricate maneuvers such as stationkeeping and in-plane/out-of-plane maneuvers—additional goals encompassed testing the Orbit Attitude and Maneuver System (OAMS) and executing 11 scientific experiments.

During the mission, a stationkeeping exercise was cut short due to excessive propellant use, necessitating the cancellation of further rendezvous attempts to safeguard primary objectives. A critical incident occurred during the 48th revolution when computer memory was inadvertently altered, rendering a planned computer-controlled reentry impossible and requiring a ballistic reentry trajectory instead.

Gemini-Titan liftoff. (NASA Photo No. 65-H-934, released June 3, 1965.)

Despite these challenges, all major mission objectives were accomplished. EVA preparations proceeded smoothly, with White performing the first American spacewalk, using a hand-held maneuvering gun for mobility. He re-entered the spacecraft after 20 minutes outside, with the hatch closing after 4 hours and 54 minutes of ground elapsed time.

Gemini 4 maintained a drifting flight pattern for two and a half days to conserve propellant before successfully landing in the Atlantic Ocean, approximately 450 miles east of Cape Kennedy, slightly off its nominal landing point. The crew recovered 34 minutes after landing by helicopter and was transported to the aircraft carrier Wasp, concluding spacecraft recovery by 2:28 p.m. - just over 100 hours after launch.

This historic mission, controlled from Houston's newly established Mission Control

Center, demonstrated NASA's growing expertise in extended space missions and laid crucial groundwork for future endeavors in space exploration. Gemini 4 validated technical capabilities and showcased astronauts' resilience and adaptability in the challenging space environment.

Astronaut Edward H. White II during extravehicular activity on the Gemini-Titan 4 mission. (NASA Photo No. 65-H-1019, released June 3, 1965.)

Launch Preparations of Gemini Launch Vehicle GLV-5

On June 7, 19XX, at Complex 19, a pivotal event unfolded in the meticulous preparations for the upcoming Gemini mission. Dominating the skyline, the Gemini launch vehicle (GLV) 5 was ceremoniously erected, a monumental assembly of cutting-edge technology poised for its journey into the heavens.

Following its dramatic ascent to the launch site, meticulous inspections commenced on June 9. Every facet of GLV-5 underwent rigorous scrutiny to ensure seamless functionality. Vital umbilicals, essential conduits linking power and data, were meticulously affixed to the spacecraft, anchoring it to the operational grid.

With each passing day, anticipation mounted. On June 10, a power surge surged through GLV-5, triggering the intricate ballet of subsystem integration. This pivotal phase marked the inception of a critical testing regime, positioning the vehicle for its ultimate task.

Days later, on June 14, the scene was set for the Subsystems Reverification Tests (SSRT). This stage of the operation was strategically designed to validate the flawless operation of GLV-5's myriad subsystems. Unlike its precursor, the Subsystems Functional Verification Test (SSFVT), conducted at Martin-Baltimore, SSRT streamlined the validation process. It meticulously reaffirmed the meticulous findings of previous tests, ensuring no stone was left unturned in the pursuit of mission readiness.

June 15 At McDonnell, extensive systems assurance testing culminated in the completion of Gemini spacecraft No. 6. Validation of its environmental control system took place from June 16 to 19, paving the way for the initiation of the Simulated Flight Test on June 22.

June 18 Atlas standard launch vehicle 5301 was relocated from Hangar J to Complex 14 and reinstated for operations. By July 9, the Booster Facility Acceptance Composite Test had successfully concluded, marking a critical milestone in readiness assessment.

On June 19, McDonnell delivered Gemini Spacecraft No. 5 to Cape Kennedy, where industrial area activities persisted until June 25. Positioned at Complex 19, the spacecraft was mounted atop the launch vehicle on June 26. Notably, with this spacecraft, the Premate Systems Tests and Premate Simulated Flight Test merged into the comprehensive Premate

Verification Test, executed from June 30 to July 2.

On June 25 Gemini spacecraft No. 6 completed its rigorous Simulated Flight Test at McDonnell. Subsequently, the spacecraft underwent phasing checks and entered the altitude chamber for testing from July 16 to 21, following preparatory activities that concluded on July 15.

Martin-Baltimore completed the Combined Systems Acceptance Test for the Gemini launch vehicle (GLV) 6. After review by the vehicle acceptance team from July 6 to 10, GLV-6 was formally accepted by the Air Force on July 31. Stage I and Stage II were dispatched to Cape Kennedy, with storage preceding the launch of Gemini-Titan 5.

Stage I of Gemini launch vehicle (GLV) 7 was positioned in the east cell of Martin-Baltimore's vertical test facility. Stage II followed on June 28, initiating inspection and testing while GLV-6 underwent vertical tests. On July 26, power was first applied to GLV-7, starting the Subsystems Functional Verification Tests that concluded on August 25 after systems modification and retesting.

On June 28, Gemini launch vehicle (GLV) 7 completed the crucial Subsystems Reverification Tests (SSRT), affirming the readiness of its subsystems for the upcoming mission.

The final stages of manufacturing, module tests, and equipment installation for Gemini spacecraft No. 7 concluded on June 29 at McDonnell. The assembly of reentry and adapter assemblies was finalized on July 26, preceding systems assurance testing, which commenced on August 4.

On June 30, at Sunnyvale, a pivotal milestone was achieved in preparing Gemini Agena target vehicle 5002. Completing rigorous Vehicle Systems Tests underscored the vehicle's readiness for its critical role. Following this, the final acceptance test culminated on a positive note, signifying comprehensive validation of its operational capabilities.

The collaborative effort involving NASA, the Air Force Space Systems Division, Aerospace, and Lockheed proved instrumental in resolving all data discrepancies identified during the final systems tests. This consensus paved the way for the vehicle's disconnection from the test complex on July 13, marking a definitive step toward mission readiness.

On July 1, George E. Mueller, NASA's Associate Administrator for Manned Space Flight, initiated a pivotal development by establishing the "Operations Executive Group." Comprising senior executives from government and contractor organizations involved in manned space flight operations, this group was tasked with a dual mandate.

Firstly, it undertook comprehensive reviews of the status, resource requirements, management approaches, and flight operations of the Gemini and Apollo programs. These assessments provided crucial background information for informed and effective policy decisions at the executive level.

Secondly, the group aimed to foster a cohesive operational environment by facilitating direct collaboration among its members. Regular one-day meetings, scheduled at two to four months, served as platforms for enhancing mutual understanding and ensuring readiness to address urgent, time-critical challenges as they arose.

On July 1, NASA made a significant announcement, revealing the prime flight crew for Gemini VII. Frank Borman and James A. Lovell, Jr. were selected to helm this crucial mission, which was anticipated to extend up to 14 days in space. Edward H. White II and Michael Collins were named as the backup crew for the mission, assuring continuity and preparedness. This selection underscored NASA's strategic approach to crew assignments, ensuring capability and readiness for the challenges ahead in advancing manned space exploration.

NASA made a significant announcement on this day, revealing the prime flight crew for Gemini VII. Frank Borman and James A. Lovell, Jr. were selected to helm this crucial mission, which was anticipated to extend up to 14 days in space. Edward H. White II and Michael Collins were named as the backup crew for the mission, assuring continuity and preparedness. This selection underscored NASA's strategic approach to crew assignments, ensuring both capability and readiness for the challenges ahead in advancing manned space exploration.

Rendezvous evaluation pod installed in the equipment section of Gemini spacecraft No. 5 before launch vehicle mating. (NASA Photo S-65-41884, July 6, 1965.)

On July 8, Gemini Agena target vehicle 5001 achieved a significant milestone with the completion of systems tests conducted in Hangar E. Subsequently, it was relocated to Complex 14 where it underwent a critical phase: mating with the Atlas standard launch vehicle 5301. This integration marked a pivotal step in preparations for the upcoming Simultaneous Launch Demonstration on July 22.

On July 12, NASA Headquarters' Gemini Program Office conveyed a significant decision to the Manned Spacecraft Center: extravehicular activity (EVA) was to be omitted from Gemini missions 5, 6, and 7. This directive marked a strategic shift in mission objectives, focusing the efforts of these missions on other critical goals and objectives within the Gemini program.

On July 22, a pivotal event unfolded in the Gemini program as NASA conducted a Simultaneous Launch Demonstration (SLD), a complex coordination exercise aimed at demonstrating the capability to execute a single countdown for two separate vehicles.

At Complex 14, the Gemini Atlas-Agena target vehicle stood ready, while at Complex 19, Gemini-Titan (GT) 5 awaited its turn. This demonstration was synchronized with the Wet Mock Simulated Launch (WMSL) of GT-5, with the Gemini launch vehicle tanking exercise taking place separately for logistical convenience on July 17.

SLD was a comprehensive dress rehearsal showcasing the intricate coordination required for rendezvous missions. It involved mission control centers in Houston and Cape Kennedy and support facilities across the Eastern Test Range, ensuring seamless operations integration.

During the demonstration, an issue with the Houston computer system transmitted spurious commands to the target vehicle. While some commands were accepted, they primarily consisted of stored program command loads, minimizing any significant impact on the mission.

Following the successful SLD, operations continued on July 26, demating the Atlas and Agena components, marking another milestone in the meticulous preparations for upcoming Gemini missions.

On July 23, a significant milestone was reached as the Air Force Space Systems Division formally accepted the Gemini Agena target vehicle (GATV) 5002 delivery. This acceptance followed a thorough inspection by the vehicle acceptance team, affirming readiness for operational deployment.

Subsequently, GATV 5002 was expedited via air transport to the Eastern Test Range, departing on July 24 and arriving the following day, July 25. Despite its acceptance, several critical items of equipment, including the shroud, secondary and primary propulsion systems, and components of the electrical power and command systems, remained in a "not qualified" status.

On July 23rd, the Gemini-Titan (GT) 5 mission entered a critical phase as engineers at the launch site meticulously demated the spacecraft. This delicate operation followed the completion of the Wet Mock Simulated Launch, a comprehensive simulation designed to verify operational readiness. The primary focus was on replacing the spacecraft's fuel cells and implementing crucial modifications to the coolant bypass system. These enhancements ensured optimal performance and safety during the upcoming mission.

Astronauts Charles Conrad, Jr., and L. Gordon Cooper, Jr., practice procedures for getting into their spacecraft in the Gemini 5 Wet Mock Simulated Launch. (NASA Photo S-65-41895, July 22, 1965.)

After meticulous adjustments and system upgrades, the spacecraft and its launch vehicle were remated on August 5th. This marked a significant milestone in the pre-launch preparations, signaling the integration of the GT-5 mission components in readiness for the next stages of testing and validation.

With the spacecraft and launch vehicle securely integrated, rigorous testing procedures commenced. On August 6th, the Modified Electrical Interface Integrated Validation tests were conducted to verify seamless communication and functionality across all electrical systems. Simultaneously, the Joint Guidance and Control Tests were performed to ensure precise navigation and control capabilities crucial for mission success.

Standard Agena D 108 being delivered to final assembly area.
(NASA Photo S-65-8066, July 23, 1965.)

As the launch date approached, the intensity of testing increased. From August 9th to 10th, the spacecraft scrutinized the Final Systems Test, meticulously assessing every onboard system and subsystem for operational integrity. This comprehensive evaluation aimed to identify and resolve any potential issues that could jeopardize the mission's objectives.

On August 13th, the culmination of pre-launch preparations arrived with the Simulated Flight Test. This final trial run simulated the actual mission conditions, allowing mission controllers and engineers to validate operational procedures and readiness under realistic scenarios.

Simultaneously, critical preparations were underway for the Gemini Agena target vehicle 5003. Stored since its completion in June, the Agena D (AD-108) underwent meticulous modifications and final assembly at Building 104 in Sunnyvale. These adjustments included removing and adapting several components to align with the Gemini mission's specific requirements, ensuring optimal functionality and compatibility.

As the team meticulously finalized these preparations, the countdown to the GT-5 mission launch on August 19th approached with excitement and precision. Every step, from demating and modification to final testing and assembly, underscored the meticulous planning and engineering prowess required to propel humanity further into the frontier of space exploration.

Preparing for Gemini VI: Rendezvous Mission

On July 26th, following the successful Simultaneous Launch Demonstration on July 22nd, significant operations unfolded at Complex 14. The Atlas standard launch vehicle 5301 and Gemini Agena target vehicle (GATV) 5001 underwent demating procedures. This meticulous process involved separating the components to facilitate necessary adjustments and preparations.

Following demating, GATV 5001 was carefully transported and stored at Hangar E, where it assumed the role of the backup vehicle for GATV 5002. This strategic decision ensured redundancy and readiness in the event of unforeseen challenges or adjustments to mission plans.

Meanwhile, on August 18th, GATV 5002 was officially designated as the primary target vehicle for the upcoming Gemini VI mission. This mission, slated to be the first rendezvous mission of the Gemini program, required meticulous planning and preparation to achieve a groundbreaking milestone in space exploration.

Concurrently, GATV 5001 remained in flight-ready condition at Hangar E, poised to intervene if any issues arose with GATV 5002. This dual-vehicle strategy underscored NASA's commitment to mission success through comprehensive contingency planning.

In anticipation of the upcoming mission, Atlas 5301, temporarily housed at Hangar J after demating, was relocated back to Complex 14 on August 16th. This relocation positioned Atlas 5301 as the designated launch vehicle for GATV 5002, aligning all components for the imminent mission.

On July 27th, Charles W. Mathews, the Gemini Program Manager, introduced a pioneering initiative—the spacecraft manager program. This strategic move involved assigning dedicated engineers to oversee the development and preparation of specific Gemini spacecraft, starting with No. 5 and No. 6. This marked a significant departure from previous practices in the Mercury program and early Gemini missions.

Under this innovative program, each designated engineer would meticulously shepherd their assigned spacecraft through every stage of its lifecycle—from manufacturing and assembly to rigorous testing and eventual launch. This approach ensured that a singular individual remained intimately familiar with every detail and nuance of their spacecraft, serving as a critical source of real-time information and expertise.

The establishment of the spacecraft manager program was built upon established precedents set by Martin, McDonnell, and Aerojet-General in previous space missions. By adopting this model, NASA aimed to enhance efficiency, streamline communication, and proactively address potential challenges or technical issues that could arise during the spacecraft's preparation and mission readiness phases.

Initially implemented for Gemini spacecrafts Nos. 5 and 6, the spacecraft manager program was poised to expand to encompass other spacecraft in subsequent phases of the Gemini program. This expansion would further solidify NASA's commitment to meticulous oversight and quality control throughout the Gemini missions.

On August 4th, McDonnell achieved a significant milestone by delivering Gemini spacecraft No. 6 to Cape Kennedy. This marked the beginning of a crucial phase in the mission's preparation, signaling the start of intensive activities in the industrial area over the ensuing three weeks.

Following its delivery, Gemini spacecraft No. 6 underwent meticulous preparations and modifications in the industrial area at Cape Kennedy. These activities focused on integrating essential pyrotechnics and conducting crucial spacecraft modifications critical for mission readiness.

By the last week of August, the spacecraft was relocated to the Merritt Island Launch Area, where it underwent Plan X integrated tests with the designated target vehicle. These rigorous tests were crucial in verifying seamless communication, integration, and operational readiness between the spacecraft and its accompanying components.

Arrival of Atlas Launch Vehicle 5302 at Cape Kennedy

On August 5th, Atlas standard launch vehicle 5302 embarked on a crucial journey from San Diego to Cape Kennedy, underscoring a pivotal moment in the Gemini program's logistical preparations. The vehicle, having completed its production phase and subsequent testing milestones, arrived at its destination on August 11th after a meticulous transport operation.

Production and Assembly Timeline

Atlas 5302 had been meticulously assembled and prepared for its role in the Gemini program:

April 2: The vehicle came off the production line and was officially designated for the Gemini program.

May 25: Final assembly of the launch vehicle was completed, marking a significant milestone in its preparation.

June 3: Installation of flight equipment and Gemini-specific components, essential for mission compatibility, was finalized.

July 22: Factory testing was successfully conducted, ensuring all systems operated as intended under simulated mission conditions.

The Air Force Space Systems Division formally accepted Atlas 5302 on July 29th, signifying its readiness for operational deployment to support upcoming Gemini missions. This acceptance validated the vehicle's adherence to stringent quality and performance standards set forth by both NASA and the Air Force, crucial for ensuring mission success.

On August 12th, McDonnell completed systems assurance testing for Gemini spacecraft No. 7, marking a critical milestone in its preparation for upcoming missions. This rigorous testing phase ensured that all onboard systems and subsystems operated seamlessly and met stringent performance criteria.

Following the systems assurance testing, validation of the spacecraft's environmental control system was concluded on August 19th. This essential system was pivotal in maintaining a habitable and safe environment for astronauts during their missions in space.

With foundational testing completed, preparations swiftly commenced for the Simulated Flight Test scheduled for August 26th. This comprehensive test simulated mission conditions, allowing engineers and mission controllers to assess the spacecraft's readiness and performance under realistic scenarios.

On August 12th, the Gemini Program Office decided to incorporate the new lightweight G5C space suit for the upcoming Gemini VII mission. This decision followed extensive testing and evaluation by the Crew Systems Division, which demonstrated significant advancements in comfort and mobility while maintaining essential pressure integrity and crew safety standards.

The G5C space suit, weighing approximately nine pounds, represented a notable evolution from its predecessor, the G4C suit. Key design modifications included eliminating the restraint layer, introducing a soft helmet design with an integrated visor, and removing the traditional neckring. These enhancements aimed to enhance astronaut comfort and mobility during extended missions in space.

Notably, discussions were underway regarding the potential for astronauts to remove their suits during the Gemini VII mission. This consideration reflected ongoing efforts to optimize operational flexibility while ensuring crew safety and mission success under evolving NASA directives.

Earlier directives from NASA Headquarters on July 2nd stipulated that the flight crew for Gemini VII should not utilize full pressure suits during their mission. This directive underscored NASA's strategic approach to mission planning and operational protocols, balancing technological advancements with safety considerations in human spaceflight.

On August 16th, Martin-Baltimore received the propellant tanks destined for the Gemini Launch Vehicle (GLV) 9 from Martin-Denver. This marked a significant logistical operation as these were the first GLV tanks transported by rail from Denver to Baltimore. Previous tanks had all been transported by air, but rail transport was deemed necessary for this shipment due to a shortage of suitable aircraft.

The fabrication of these propellant tanks commenced on February 25th at Martin-Denver. Following their completion, the tanks were shipped out on August 9th, embarking on their journey to Martin Baltimore for integration into GLV-9.

In addition to the propellant tanks, Aerojet-General played a crucial role by delivering the Stage I engine for GLV-9 on August 20th and the Stage II engine on September 22nd. These engines were integral components required for the launch vehicle's operational readiness.

Upon arrival at Martin-Baltimore, the process of tank splicing—a critical assembly procedure—was completed by October 21st. Subsequently, the installation of the engines took place on November 10th, marking significant progress in the vehicle assembly process. Horizontal testing, aimed at validating the integrated functionality and readiness of GLV-9, concluded on November 23rd.

On August 19th, just 10 minutes before the scheduled launch of Gemini-Titan 5, a critical malfunction in the spacecraft computer prompted an immediate hold in the countdown process. This unforeseen issue necessitated thorough investigation and troubleshooting by mission controllers and engineers.

As efforts to resolve the computer malfunction intensified, adverse weather conditions, including approaching thunderstorms in the Cape Kennedy area, further complicated the situation. The deteriorating weather posed additional risks and safety concerns, influencing the decision-making process for mission control.

Ultimately, due to the unresolved spacecraft computer issue and the worsening weather conditions, the mission was officially scrubbed on August 19th. NASA promptly rescheduled the launch for August 21st, initiating a comprehensive recycling process that involved unloading propellants from the launch vehicle to ensure safety and readiness for the upcoming attempt.

Following the scrubbing of the mission, recycling procedures focused on meticulously resetting and re-verifying all systems, including the spacecraft and launch vehicle, in preparation for the rescheduled launch. This comprehensive approach aimed to address any underlying issues, mitigate risks, and optimize conditions for a successful mission.

From August 19th to August 24th, Lockheed conducted critical shroud separation tests at its Rye Canyon Research Center. These tests were designed to simulate conditions at various altitudes to ensure the shroud's successful separation during actual flight scenarios.

Lockheed executed a series of four separation tests, each meticulously crafted to replicate the environmental conditions and stresses anticipated during the launch and ascent phases of a mission. The objective was to verify the shroud's ability to reliably and safely separate from the spacecraft at critical stages without compromising mission integrity.

All four separation tests successfully validated the shroud's design and functionality under simulated flight conditions. The tests demonstrated that the shroud could withstand the rigors of launch and separation processes, confirming its flightworthiness for upcoming missions.

After completing the tests, Lockheed's engineering teams analyzed the data collected to assess the shroud's performance and structural integrity under varying altitudes and stress scenarios. Based on thorough analysis and evaluation, the shroud was deemed ready for flight, meeting Lockheed and NASA's stringent safety and reliability standards.

Gemini 5: Pushing the Limits of Long-Duration Spaceflight

August 21 marked the launch of Gemini 5 from complex 19 at 9:00 a.m. EST, featuring Command Pilot Astronaut L. Gordon Cooper, Jr., and Pilot Astronaut Charles Conrad, Jr. This eight-day mission aimed primarily to assess the performance of the rendezvous guidance and navigation system using a rendezvous evaluation pod (REP) and evaluate the physiological effects of extended space exposure on the crew.

Secondary objectives included demonstrating controlled reentry guidance, evaluating fuel cell performance, and testing the guidance and control systems essential for rendezvous missions. The mission also included conducting 17 experiments to gather crucial data for future spaceflights.

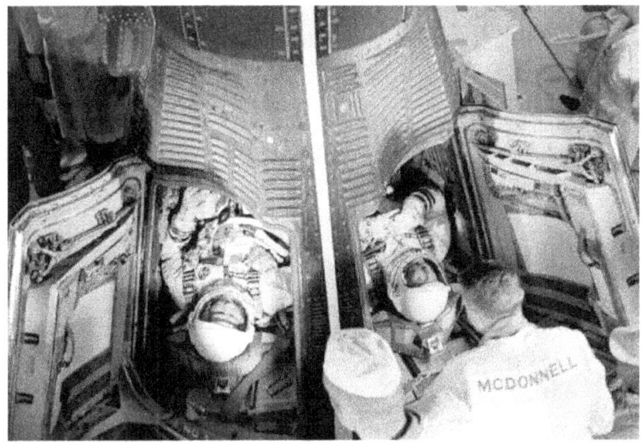

ug. 1965) — Astronauts L. Gordon Cooper Jr. (left) and Charles Conrad Jr. are seen in the Gemini-5 spacecraft in the white room at Pad 19 just after insertion. NASA

Initial operations proceeded smoothly through the first two orbits and the REP's ejection. However, a significant issue arose about 36 minutes into the rendezvous system evaluation when the oxygen supply tank pressure in the fuel cell system began to decline rapidly. Despite stabilizing at 70 psia after 4 hours and 22 minutes, this forced the crew to power down the spacecraft and abort the REP exercise.

Christopher C. Kraft, Jr., Robert R. Gilruth, and George M. Low in the Houston Mission Control Center when falling pressure in the oxygen supply tank of the fuel cell threatened the Gemini V mission. (NASA Photo S-65-28691, Aug. 22, 1965.)

Ground experts swiftly diagnosed the issue by the seventh revolution, implementing a power-up procedure that enabled real-time scheduling of mission activities. Four rendezvous radar tests were successfully conducted, demonstrating the radar's

capability to track ground transponders at Cape Kennedy during the second day.

Photograph of the Florida peninsula taken from the Gemini 5 spacecraft, looking south along the east coast, with Cape Kennedy in the foreground projecting into the Atlantic Ocean. (NASA Photo S-65-45388, Aug. 21-29, 1965.)

On the third day, a simulated Agena rendezvous was executed under full electrical load, encompassing critical maneuvers such as apogee adjust, phase adjust, plane change, and coelliptical maneuvers using the OAMS. However, by the fifth day, OAMS operations became sluggish, with thrusters No. 7 and 8 eventually failing, complicating mission operations.

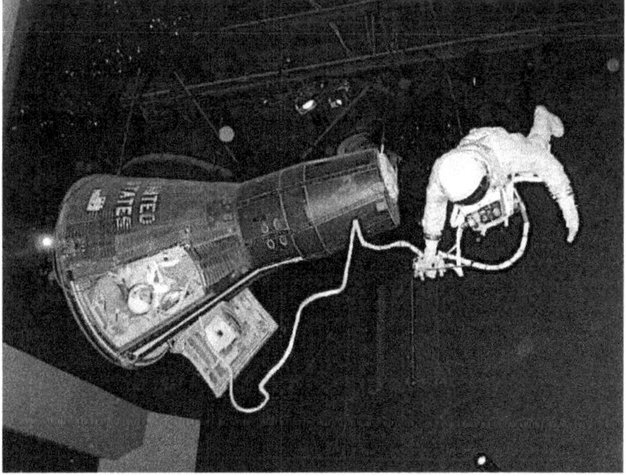

Despite these challenges, retrofire was initiated on the eighth day, one revolution earlier than planned due to adverse weather in the intended recovery area. The spacecraft reentered and landed successfully at 7:56 a.m. on August 29, 190 hours and 55 minutes after launch, albeit 89 miles short of the intended landing point due to incorrect navigation coordinates transmitted to the spacecraft computer.

Astronauts Cooper and Conrad were recovered by the aircraft carrier Lake Champlain at 9:25 a.m. the spacecraft recovery was completed by 11:51 a.m. This mission underscored NASA's resilience in managing in-flight challenges and gathering valuable data to advance space exploration capabilities in the Gemini program.

L. Gordon Cooper, Jr.

Colonel L. Gordon Cooper, Jr., an esteemed pilot and astronaut, was chosen as one of the original Mercury astronauts in April 1959. His historic journey aboard the "Faith 7" spacecraft marked the successful conclusion of Project Mercury's operational phase. Cooper's pivotal role continued as he assumed command pilot duties for Gemini 5, a mission to advance NASA's capabilities in long-duration spaceflight and rendezvous maneuvers. His expertise and dedication were further showcased as he served as backup command pilot for Gemini 12 and backup commander for Apollo 10, contributing significantly to the preparation and success of these missions. Throughout his illustrious career, Colonel Cooper logged 222 hours in space. He retired from the Air Force and NASA in 1970, leaving behind a legacy of exploration and achievement in human spaceflight.

Charles "Pete" Conrad, Jr.

Astronaut Pete Conrad, Jr. began his journey into space exploration after joining the U.S. Navy in 1953, following his graduation from Princeton University. Trained extensively as a test pilot and flight instructor, Conrad's exceptional skills led to his selection as part of NASA's second group of astronauts in 1962. His career in spaceflight was marked by notable achievements, including his roles as pilot on two Gemini missions and commander of Apollo 12, the mission that achieved the second successful landing on the Moon. Conrad's leadership and ingenuity were further demonstrated during his mission to the Skylab space station in 1973, where he and his crewmates performed critical repairs that enabled them to conduct groundbreaking experimental work aboard the orbiting workshop. Pete Conrad's contributions to space exploration spanned decades and left an indelible mark on NASA's quest to explore the frontiers of space.

Plan X Testing for Gemini Agena Target Vehicle 5002

On August 23, 1965, the Gemini Agena target vehicle 5002 underwent preliminary systems testing at Hangar E before being transported to Merritt Island Launch Area. This marked a critical phase as it was paired with spacecraft No. 6 for Plan X testing.

The Plan X tests commenced on August 25 following meticulous ground equipment checks. These tests were crucial to ensuring the compatibility and operational readiness of the spacecraft and its associated systems. Fortunately, no significant interference issues were identified during the testing phase, allowing the operations to proceed smoothly.

On August 30, 1965, Stage I of the Gemini launch vehicle (GLV) 6 was erected at Complex 19, a critical step in preparing for an upcoming mission. Stage II was erected swiftly on a subsequent day, ensuring that both stages were in place for the next preparation phases.

By September 1, umbilicals—essential connections for fueling and data exchange—were meticulously attached and inspected, marking a crucial milestone in the readiness checks. These connections were vital for maintaining communication and power during the launch sequence.

Gemini spacecraft No. 7 in final shakedown in the cleanroom at McDonnell. (NASA Photo S-65-54127, Sept. 29, 1965.)

Subsystems Reverification Tests commenced on September 2, focusing on reconfirming the functionality and integration of various critical systems within the GLV-6. These rigorous tests spanned until September 15, ensuring all components operated seamlessly under simulated launch conditions.

These preparations culminated in the Prespacecraft Mate Verification Test of GLV-6, conducted on September 16. This final test validated the spacecraft's successful integration with the launch vehicle, confirming readiness for the upcoming mission.

On August 30, 1965, the Simulated Flight Test of Gemini spacecraft No. 7 successfully concluded at McDonnell, a critical milestone in the spacecraft's preparation for its upcoming mission. Following the completion of this test phase, the spacecraft underwent thorough cleaning procedures in preparation for further testing.

Subsequently, on September 9, Gemini spacecraft No. 7 was relocated to the altitude chamber, where it underwent a series of phasing checks on September 10-11. These checks ensured the spacecraft's systems were properly synchronized and calibrated for altitude chamber testing.

Altitude chamber tests commenced on September 13 and continued until September 17. They focused on simulating the environmental conditions and stresses the spacecraft would encounter during its mission. These tests were crucial in verifying the spacecraft's performance under extreme conditions and ensuring its readiness for the challenges of spaceflight.

Upon completion of the altitude chamber tests, the spacecraft underwent deservicing, updating, and additional retesting procedures to address any findings from the tests. These comprehensive measures were essential to guaranteeing the spacecraft's optimal performance and safety.

Following these rigorous preparations and tests, Gemini spacecraft No. 7 was deemed ready for shipment to Cape Kennedy (now Cape Canaveral), where it would be further integrated with its launch vehicle and prepared for its mission into space.

On August 31, 1965, the Gemini Program Office reported a significant advancement in the development of skin-tracking procedures. This development marked a critical enhancement in the ability to track Gemini spacecraft during missions, even under challenging conditions.

During the missions of Gemini 4 and 5, which preceded this report, the effectiveness of skin-tracking procedures had been successfully demonstrated. Utilizing C-band radars, the spacecraft could be tracked in both beacon and skin-track modes. This capability proved instrumental, enabling tracking data to be obtained even when the spacecraft was powered down and its tracking beacons were not operational.

Successfully integrating skin-tracking procedures into the network support infrastructure was a pivotal decision following the validation of its effectiveness. This integration ensured that all subsequent Gemini missions would benefit from enhanced tracking capabilities, improving the program's overall reliability and operational flexibility.

On September 1, 1965, final troubleshooting activities were completed on the Gemini Agena target vehicle (GATV) 5002 following the conclusion of Plan X testing at Merritt Island Launch Area (MILA). Plan X testing had been conducted to ensure the compatibility and functionality of GATV 5002 with spacecraft operations.

GATV 5002 was transported back to Hanger E from MILA on the following day. At Hanger E, a series of comprehensive tests commenced to verify the operational readiness of all vehicle systems. These tests were crucial to ensure that GATV 5002 was fully prepared for the next phase: erection and mating with the launch vehicle.

On September 8, 1965, representatives from the Air Force Space Systems Division, Aerospace, and Lockheed convened for a technical review. The focus was on the flight verification test program for the oxidizer gas generator solenoid valve, a critical component of the Agena primary propulsion system.

The review marked a significant milestone as it represented the final qualification stage for this particular component. The oxidizer gas generator solenoid valve was essential for controlling the oxidizer flow in the propulsion system and was crucial for maneuvering and rendezvous operations in space.

The valve had been tested earlier by August 26, followed by thorough disassembly, inspection, and evaluation processes that concluded on September 3. The consensus among the attending experts was positive, confirming that the successful test program had effectively demonstrated the flightworthiness of this configuration.

On September 9, 1965, Gemini Spacecraft No. 6 underwent significant preparations and testing activities at Complex 19. The spacecraft was moved to the launch complex and hoisted to the top of the launch vehicle, marking a crucial step toward its integration and readiness for an upcoming mission.

Originally scheduled for September 2, the move had to be postponed due to Hurricane Betsy, which posed risks to operations near Cape Kennedy from September 3 to 8. Once the weather conditions improved and it was safe to proceed, the spacecraft was successfully transferred to its designated launch site on September 9.

The Prespacecraft Mate Verification Test occurred following the September 13 to 16 relocation. This rigorous test phase ensured that all systems on the spacecraft were functioning correctly and ready for integration with the launch vehicle. It included comprehensive checks and verifications to confirm the spacecraft's operational readiness and compatibility with the upcoming mission requirements.

On September 16, 1965, Martin-Denver encountered an issue during the shipment of propellant tanks intended for Gemini launch vehicle (GLV) 10 to Martin-Baltimore. The tanks, being transported by rail, suffered from leaking battery acid that corroded the dome of the stage II fuel tank during the journey.

Upon arrival at Martin-Baltimore on September 21, it was determined that the stage II fuel tank was damaged beyond repair due to the corrosion caused by the leaking battery acid. As a result, the tank was rejected and returned to Martin-Denver for further assessment and replacement.

In response to this setback, the stage II fuel tank originally intended for GLV-11 was identified as a suitable replacement. This tank had completed its final assembly by September 25 and underwent inspection and certification before being shipped to Baltimore. The replacement tank arrived at Martin-Baltimore on November 3, ensuring the continuity of the assembly process for GLV-10.

The fabrication of the propellant tanks for GLV-10 had commenced back in April, highlighting the meticulous planning and production timelines involved in the Gemini launch vehicle program. Despite the setback caused by the leaking battery acid, swift measures were taken to replace the damaged tank with a certified alternative, demonstrating the resilience and adaptability of the project team in overcoming unexpected challenges during the spacecraft manufacturing and assembly phases.

Milestones in Gemini-Titan 6 Systems Testing

On September 17, 1965, significant progress was made in the Gemini-Titan 6 mission preparations:

Mechanical Mating: Gemini launch vehicle (GLV) 6 and spacecraft No. 6 were mechanically mated at complex 19. This critical step involved physically connecting the spacecraft to the launch vehicle, ensuring they functioned as a single integrated unit.

Following the mechanical mating, the Electrical Interface Integrated Validation and Joint Guidance and Control Test were completed by September 21. These tests verified the electrical connections between the spacecraft and the launch vehicle and tested the joint guidance and control systems.

On September 23, the spacecraft/GLV Joint Combined Systems Test was conducted. This comprehensive test evaluated the integrated systems of the spacecraft and the launch vehicle, ensuring they operated correctly as a unified system.

On September 29, the Gemini launch vehicle underwent a tanking test. This test involved filling the launch vehicle's propellant tanks to simulate launch conditions and checking for leaks or operational issues.

On October 1, the Flight Configuration Mode Test was conducted, marking the completion of systems testing for Gemini-Titan 6. This final test ensured that all systems were configured correctly for the upcoming mission and were ready for flight.

On September 20, 1965, McDonnell achieved a significant milestone in the preparation of spacecraft No. 8 for the Gemini program:

McDonnell completed mating the reentry module and adapter assemblies of spacecraft No. 8. This involved physically connecting these components, ensuring they were properly aligned and integrated.

Gemini spacecraft No. 8 in cleanroom at McDonnell for systems validation testing. (NASA Photo S-65-54125, Sept. 29, 1965.)

After mating the assemblies, the complete spacecraft underwent alignment and adjustment processes. This step ensured that all components within the spacecraft were correctly positioned and aligned according to specifications.

Systems Assurance Tests began on September 30. These tests are critical to verify the functionality and reliability of various systems within the spacecraft. They cover a range of tests to ensure that all systems perform as expected under simulated conditions.

On September 20, 1965, significant progress was made in the Gemini program with the completion of the Combined Systems Acceptance Test for Gemini Launch Vehicle (GLV) 7 at Martin-Baltimore. Here are the key details of the events surrounding GLV-7 during this period:

The Gemini launch vehicle GLV-7 underwent a rigorous Combined Systems Acceptance Test at Martin-Baltimore's vertical test facility. This comprehensive test evaluated the launch vehicle's integrated systems to ensure they functioned correctly and reliably together.

After completing the acceptance test, the vehicle acceptance team thoroughly inspected GLV-7. This inspection period began on September 27 and concluded on October 1. The team confirmed that GLV-7 met all required standards and was deemed acceptable for its intended mission.

GLV-7 was deerected from the vertical test facility on October 5. It was formally accepted by the Air Force on October 15, signifying its readiness for operational use.

Stage I of GLV-7 was airlifted to Cape Kennedy on October 16, followed by Stage II on October 18. Both stages were then placed in storage at Cape Kennedy, awaiting the launch of the Gemini VI mission.

Gemini VIII Crew Announcement

On September 20, 1965, the Manned Spacecraft Center made a significant announcement regarding the crew assignment for the upcoming Gemini VIII mission. Here are the details:

Crew Assignment: Command Pilot: Neil A. Armstrong, Pilot: David R. Scott

Backup Crew: Command Pilot: Charles Conrad, Jr., Pilot: Richard F. Gordon, Jr.

Gemini VIII was slated to include extensive practice in rendezvous and docking maneuvers.

The mission would also involve a spacewalk, planned to last up to one Earth orbit, approximately 95 minutes.

This announcement marked the formal assignment of Neil Armstrong and David Scott as the primary crew members for Gemini VIII, with Charles Conrad, Jr., and Richard F. Gordon, Jr., designated as their backups. Gemini VIII's focus on critical maneuvers such as rendezvous, docking, and spacewalking reflected NASA's ongoing efforts to develop and refine techniques crucial for future space missions, including the Apollo program aimed at landing astronauts on the Moon.

On September 28, 1965, significant progress was made in the testing and preparation of the Gemini launch vehicle (GLV) 8 at Martin-Baltimore's vertical test facility. Here are the details:

GLV-8 was erected in the west cell of the vertical test facility at Martin-Baltimore on September 28. Following the deerection (disassembly) of GLV-7, power was applied to the vehicle on October 13. These tests were completed on November 4.

The Subsystems Functional Verification Tests aimed to ensure that all critical systems and subsystems of the Gemini launch vehicle (GLV-8) were functioning correctly and meeting performance specifications.

On October 1, 1965, significant activities were conducted regarding the Gemini Agena Target Vehicle (GATV) 5002 at complex 14. GATV 5002 was transported to complex 14 and successfully mated to the target launch vehicle 5301. Preliminary checks were conducted following the mating to ensure the integrated system's readiness and compatibility. On October 4, the Joint Flight Acceptance Composite Test (J-FACT) took place.

J-FACT involved a comprehensive simulation that checked all contractors, the range, vehicles, and aerospace ground equipment. It simulated a full countdown and flight scenario, although propellants and high-pressure gases were not loaded, and the gantry was not removed during this test. Completed on October 7, the Simultaneous Launch Demonstration likely tested coordination and readiness across multiple systems and operations required for a launch scenario.

On October 6, a pivotal milestone in the Gemini program unfolded as engineers convened for the final design review of the Gemini Atlas-Agena target vehicle ascent guidance equations. This critical meeting marked the culmination of meticulous testing and analysis to ensure the precision and reliability of the spacecraft's ascent trajectory.

The equations under scrutiny encompassed the intricate interplay between the Gemini spacecraft and its Agena target vehicle, integrating precise pitch and yaw steering controls of the Atlas launch vehicle with the nodal steering capabilities of the Gemini Agena target vehicle. Rigorous testing protocols had rigorously vetted these equations, affirming their adherence to stringent accuracy standards mandated for manned spaceflight.

After comprehensive scrutiny and deliberation, the assembled experts reached a consensus: the ascent guidance equations were deemed operationally sound and ready for deployment in upcoming missions. This approval signified a technical achievement and a testament to the collaborative efforts of the engineers and scientists who dedicated themselves to advancing the frontiers of space exploration through the Gemini program.

Following the WMSL, the spacecraft and its launch vehicle underwent essential maintenance procedures. One critical task was replacing the spacecraft battery, ensuring an optimal power supply for the mission ahead. This period also involved carefully remating the spacecraft with the launch vehicle, a meticulous process from October 8 to October 13.

On October 15, the rigorous evaluation of spacecraft systems was completed. This comprehensive Systems Test ensured that all onboard systems, from propulsion to communication, functioned seamlessly in preparation for the demanding challenges of spaceflight.

The culmination of pre-launch testing occurred on October 20 with the Simulated Flight Test. This final assessment simulated the entire mission sequence, validating the integrated performance of spacecraft and ground systems under simulated flight conditions. Each phase of the mission was meticulously rehearsed to anticipate and mitigate any potential challenges, affirming readiness for the upcoming launch.

Preparation of Gemini Spacecraft No. 7 for Launch

On October 9, McDonnell delivered Gemini spacecraft No. 7 to Cape Kennedy, marking the beginning of a sequence of meticulous preparations for an upcoming mission. Upon arrival, intensive industrial activities commenced, preparing the spacecraft for its critical role in the Gemini program.

Over the ensuing weeks, a dedicated team undertook a series of tasks essential for mission readiness. These included the meticulous buildup of pyrotechnics, crucial for various spacecraft operations, and the precise installation of fuel cells to ensure sustained power supply throughout the mission. Additionally, modifications were made to the water management system, vital for maintaining environmental control within the spacecraft's interior.

By October 29, all industrial area activities were completed, and Gemini spacecraft No. 7 was transported to Launch Complex 19. Here, it was meticulously hoisted atop its designated launch vehicle, a pivotal step towards final integration and readiness for the mission ahead.

From November 1 to November 5, the spacecraft underwent rigorous verification testing known as the Prespacecraft Mate Verification Test. This comprehensive assessment included critical procedures such as activating and deactivating the fuel cell system, ensuring all systems functioned flawlessly under operational conditions.

On October 14, Gemini Agena target vehicle 5003 achieved a significant milestone. Following its final assembly on October 9, it was transferred to the Vehicle Systems Test, marking the initiation of a critical phase in the meticulous testing and verification process essential for its upcoming mission.

Testing of Gemini Agena target vehicle 5003 commenced on October 18. The tests focused on ensuring the integrated performance and functionality of all systems. They encompassed a rigorous evaluation of propulsion systems, guidance controls, and communication interfaces to confirm operational readiness for its role in the Gemini program.

On October 20, a critical milestone was achieved at Complex 14. Systems testing of the Gemini Atlas-Agena target vehicle for Gemini VI concluded with a launch readiness demonstration. This marked the culmination of rigorous evaluations to ensure the vehicle's readiness for its upcoming mission.

After successful systems testing, final preparations for launch commenced on October 21. These meticulous tasks included the comprehensive closeout of all vehicle systems and payloads, ensuring every component was securely integrated and functioning optimally. These efforts were crucial in preparing for the final countdown, which commenced on October 25, marking the last phase before liftoff.

On October 22, McDonnell achieved significant milestones in preparing Gemini spacecraft No. 8 for its upcoming mission. Systems Assurance Tests were completed, rigorously verifying the spacecraft's operational readiness and reliability under simulated conditions. Concurrently, validating the spacecraft's environmental control system ensured optimal conditions for crew comfort and operational efficiency in space.

From October 26 to November 4, Gemini spacecraft No. 8 underwent a comprehensive Simulated Flight Simulation. This critical phase replicated the entire mission profile in a controlled environment, testing the integration and functionality of all spacecraft systems. The simulation provided essential insights into the spacecraft's performance under simulated flight conditions, enabling engineers to fine-tune operations and mitigate potential risks ahead of the actual mission.

Incident and Investigation of Gemini VI Mission Cancellation

On October 25, the Gemini VI mission faced an unexpected setback when the Gemini Agena target vehicle (GATV) 5002 experienced what appeared to be a catastrophic failure shortly after separating from the Atlas launch vehicle. The mission had commenced from Complex 14 at 10:00 a.m. EST, with initial vehicles separating at 10:05 a.m. Initially, all telemetry signals indicated normal operation. However, telemetry was lost approximately 375 seconds into the flight, and subsequent attempts to regain communication were unsuccessful.

At 10:54 a.m., the decision was made to halt the Gemini VI countdown and cancel the mission due to the target vehicle's failure to achieve orbit. Per protocols established by the Air Force Space Systems Division (SSD) and NASA management directives, which mandate investigation following such incidents, Major General Ben I. Funk, Commander of SSD, reconvened the Agena Flight Safety Review Board. Concurrently, NASA instituted a dedicated GATV Review Board to examine the circumstances surrounding the failure thoroughly.

This unforeseen event underscored space exploration's inherent risks and complexities, prompting a rigorous investigation to identify the failure's root cause and implement corrective measures to safeguard future missions. The coordinated efforts of both military and civilian aerospace authorities reflected their commitment to ensuring the safety, reliability, and success of future Gemini missions amidst the challenges of space exploration.

On October 27, in response to the catastrophic anomaly of Gemini Agena target vehicle (GATV) 5002 on October 25, NASA Associate Administrator Robert C. Seamans, Jr., officially classified the event as a mission failure. In a directive to George E. Mueller, Associate Administrator for Manned Space Flight, Seamans requested establishing a comprehensive GATV Review Board. This board was tasked with investigating all aspects of the Agena failure, encompassing both technical and managerial dimensions.

NASA's Manned Spacecraft Center Director, Robert R. Gilruth, and Major General O. J. Ritland, Deputy Commander for Space at Air Force Systems Command, were appointed as co-chairmen of the review board. The primary responsibility for determining the cause of the failure rested with the Air Force Space Systems Division (SSD), which committed to providing its findings to the board for comprehensive analysis.

On October 28, the White House announced a bold plan by NASA to attempt a dual mission involving Gemini VI and Gemini VII. This follows the cancellation of the original Gemini VI mission due to the catastrophic failure of its target vehicle on October 25. NASA Administrator James E. Webb conveyed in a memorandum to the President the feasibility of reerecting the Gemini VI spacecraft and launch vehicle shortly after the launch of Gemini VII.

Webb outlined that much of the prelaunch checkout procedures for Gemini VI would not need to be repeated, facilitating a swift turnaround for a potential launch. This strategy aimed to capitalize on the 14-day duration of the Gemini VII mission, allowing Gemini VI to rendezvous with its counterpart in orbit if the launch pad sustained minimal damage from Gemini VII's liftoff.

Following the October 25 failure, NASA officials, guided by suggestions from Walter F. Burke and John F. Yardley of McDonnell, immediately began discussions on the feasibility and logistics of executing a dual mission. This initiative was built upon six months of prior discussions and preliminary planning involving NASA, Air Force, Martin, and McDonnell personnel for rapid manned flight launch demonstrations.

Following October 28 and 29, Gemini spacecraft No. 6 and its associated launch vehicle components underwent careful handling and storage procedures at Cape Kennedy. These actions were prompted by the cancellation of the original mission due to the catastrophic failure of the Gemini Agena target vehicle on October 25.

On October 28, Gemini spacecraft No. 6 and the second stage of the Gemini launch vehicle (GLV-6) were carefully removed from Complex 19. GLV-6's first stage was removed from the complex the following day. To ensure their preservation, GLV-6 was transported and placed in storage at the Satellite Checkout Building, where strict environmental controls for temperature and humidity were maintained under guard. This bonded storage method aimed to preserve the integrity of previously conducted tests, minimizing the need for repeated testing.

Gemini spacecraft No. 6 was stored at the Pyrotechnics Installation Building within the Merritt Island Launch Area. This facility provided secure storage for the spacecraft, protecting it from environmental factors while awaiting potential future deployment or further evaluation.

On October 28, significant progress was reported regarding the First Article Configuration Inspection (FACI) of Gemini Agena target vehicle 5001, which commenced in June. At this stage, most of the 819 discrepancies identified during the inspection were successfully resolved. Notably, 128 discrepancies remained unresolved, as they had not yet been reconciled against the acceptance document (DD-250).

Throughout October, efforts focused on closing out all remaining FACI discrepancies, including those pertaining to various subsystems critical for the vehicle's operational readiness. This meticulous process ensured all components met stringent standards and specifications before deployment.

From October 29 to October 30, Gemini launch vehicle (GLV) 7 was erected at Complex 19 following the deerection of GLV-6. This marked a crucial step in preparing for an upcoming mission, signaling the beginning of intensive testing and verification processes.

On October 31, power was applied to GLV-7, initiating the Subsystems Reverification Tests (SSRT) without delay. These tests, designed to verify the integrated performance of critical subsystems, commenced immediately after power-up.

SSRT continued rigorously until November 9, ensuring all systems operated flawlessly under simulated flight conditions. The thorough evaluation of subsystems confirmed readiness for subsequent mission phases.

On November 10, the Prespacecraft Mate Verification Test was conducted. This comprehensive test included dropping all umbilicals, which obviated the need for a Flight Configuration Mode Test (FCMT). Consequently, no FCMT was performed on GLV-7 or any subsequent vehicles.

On November 1, a pivotal meeting of the subpanel for Gemini VI within the Agena Flight Safety Review Board took place at Lockheed. Chaired by Colonel John B. Hudson, Deputy Commander for Launch Vehicles at Air Force Space Systems Division, the subpanel convened to review Lockheed's comprehensive flight safety analysis of the failure of Gemini Agena target vehicle (GATV) 5002 on October 25.

The subpanel endorsed Lockheed's analysis, attributing the catastrophic anomaly to a "hard start" of the Agena's main engine. This anomaly was likely caused by a fuel, rather than oxidizer, lead into the thrust chamber before ignition. Unlike previous standard Agenas, GATV 5002 had explicitly been sequenced to prioritize a fuel lead to conserve oxidizer for multiple planned restarts during its mission.

Following deliberations, the subpanel approved Lockheed's findings and conclusions, emphasizing the criticality of the sequencing anomaly in the failure event. These findings were reported to the parent Agena Flight Safety Review Board on November 3, contributing crucial insights for ongoing investigations and future mission planning.

On November 3, Martin-Baltimore received the propellant tanks for Gemini launch vehicle (GLV) 11 from Martin-Denver. These tanks were fabricated on June 28 and shipped by rail on October 27. This marked a crucial step in the assembly process of GLV-11, ensuring the timely receipt of essential components.

The stage II fuel tank originally intended for GLV-11 was previously utilized in GLV-

10. Similarly, the stage II fuel tank from GLV-12 was reassigned to GLV-11, arriving by air from Martin-Denver on January 16, 1966. Aerojet-General delivered the engines for GLV-11 on December 14, 1965, setting the stage for subsequent assembly phases.

By March 31, the stage I tank splicing and engine installation for GLV-11 had been completed. Subsequently, horizontal tests for stage I were concluded on April 12, followed by the completion of stage II horizontal tests by April 25. These milestones reflected the meticulous preparation and assembly efforts to ensure the readiness and reliability of GLV-11 for upcoming space missions.

On November 3, the Agena Flight Safety Review Board convened at Lockheed to continue its investigation into the failure of Gemini Agena target vehicle 5002 on October 25. Chaired by George E. Mueller, NASA Associate Administrator of Manned Space Flight, the board reviewed the findings of the subpanel for Gemini VI. It concurred that the failure resulted from a "hard start" likely caused by a fuel lead issue.

The following day, November 4, the board formulated and presented its recommendations to the Air Force Space Systems Division. These recommendations proposed a contractual change to modify the design of the Model 8247 main rocket engine to revert to oxidizer lead. The plan included subsequent design verification testing to ensure the efficacy of the modifications.

As part of the initiative, existing engines would be recycled through Bell Aerosystems to incorporate the design modifications identified through the investigation. Two existing engines would undergo design verification testing, necessitating the procurement of two new engines to replace those used in testing.

On November 8, Martin-Baltimore hosted the Combined Systems Acceptance Test of Gemini launch vehicle (GLV) 8. This critical test evaluated the integrated performance of all systems to ensure readiness for upcoming missions.

From November 16 to November 19, the vehicle acceptance team conducted thorough inspections of GLV-8. Their assessment concluded that the vehicle met all required specifications and achieved an excellent rating for its readiness and reliability.

GLV-8 was deerected on December 13-14, marking the completion of all required tests and inspections. Subsequently, on December 23, the Air Force formally accepted GLV-8, signifying its approval for operational deployment.

In January 1966, stage I of GLV-8 was airlifted to Cape Kennedy on January 4, followed by stage II on January 6. Both stages were carefully placed in storage facilities, ensuring their preservation and readiness for future missions.

Preparations for Gemini VII

On November 11, at Complex 19, the Gemini launch vehicle (GLV) 7 and spacecraft No. 7 were meticulously integrated. Suspended approximately six feet above stage II, the spacecraft was electrically connected to the GLV via an interface jumper cable. This milestone marked the beginning of a series of crucial tests and validations.

Unlike previous missions, Gemini VII skipped the traditional Wet Mock Simulated Launch (WMSL) in favor of the Simultaneous Launch Demonstration (SLD) and a separate tanking test. This decision streamlined the testing process and allowed for greater efficiency in preparation.

With the elimination of WMSL, the erector lowering step was deferred until later in the testing sequence. This adjustment enabled continuous access to the spacecraft adapter without repeated demating and remating of the spacecraft and launch vehicle. Integrated testing proceeded swiftly, effectively shortening the overall test schedule.

Critical milestones followed quickly: the Electrical Interface Integrated Validation and Joint Guidance and Control Test were successfully conducted by November 13. The Joint Combined Systems Test further validated the integrated systems on November 15. The only countdown exercise was the GLV tanking test on November 16, ensuring readiness for fueling procedures.

By November 20, the spacecraft's Final Systems Test confirmed the operational readiness of all onboard systems. Mechanical mating of the spacecraft and launch vehicle was completed on November 22, marking a pivotal stage in the preparation process.

The comprehensive Simulated Flight Test concluded on November 27, solidifying the spacecraft's readiness for its upcoming mission.

From November 12 to 13, Lockheed hosted a pivotal symposium focused on hypergolic rocket ignition at altitude. The gathering was prompted by insufficient diagnostic data from the Gemini Agena Target Vehicle (GATV) 5002 flight, which failed to identify the cause of a hard start issue conclusively. This unresolved challenge cast uncertainty over proposed modifications to prevent similar malfunctions, particularly the reintroduction of oxidizer lead.

Sixteen propulsion experts representing government, industrial, and academic sectors convened to dissect the hard-start phenomenon and isolate potential mechanisms within the Agena engine. Their discussions aimed at defining robust test programs to validate proposed solutions, building upon earlier findings to refine their understanding.

Concurrently, delays in Agena Target Vehicle production necessitated adjustments in crew rotations. As a result, the slated initially Schirra and Young mission for Gemini 6 was rescheduled, with the astronauts reassigned as backups to Shepard and Stafford. Grissom and Borman, in turn, were assigned to the upcoming Gemini 5 mission, which entailed an extended-duration flight to further NASA's exploration objectives.

On November 15, Lockheed Martin presented a crucial proposal to Colonel A. J. Gardner, the Gemini Target Vehicle Program Director at the Air Force Space Systems Division (SSD). This proposal, aimed at enhancing the reliability of the Gemini Agena Target Vehicle (GATV) engine, marked a significant step in addressing previous technical challenges.

Immediately following the presentation, a specialized team comprising B. A. Hohmann from Aerospace, Colonel J. B. Hudson, Deputy Commander for Launch Vehicles at SSD, and L. E. Root from Lockheed undertook a detailed review. By November 18, they finalized a comprehensive plan that outlined three key initiatives:

The Agena engine would undergo modifications to incorporate oxidizer lead during the start sequence to mitigate potential hard-start issues.

Bell Aerosystems would conduct rigorous sea-level tests to validate the engine's operational readiness under simulated flight conditions.

An extensive altitude test program at the Arnold Engineering Development Center would assess the engine's performance in varying atmospheric conditions, which was crucial for mission success.

These efforts culminated in a refined proposal, which was subsequently presented to the GATV Review Board at the Manned Spacecraft Center on November 20.

On November 19, Aerojet-General achieved a critical milestone by delivering the stage II engine for the Gemini Launch Vehicle (GLV) 10 to Martin-Baltimore. This delivery followed the arrival of the stage I engine on August 23. The subsequent integration process at Martin-Baltimore proceeded swiftly to ensure readiness for upcoming missions.

By January 12, 1966, Martin Baltimore completed the splicing of stage I, integrating various components to form a cohesive unit. The splicing of stage II, utilizing the fuel tank reassigned from GLV-11, was finalized by February 2. This meticulous process ensured that both stages were prepared for subsequent installation and testing phases.

Engine installation for stage I was successfully completed by February 7, followed by horizontal tests on February 11 to validate its operational capabilities. Similarly, stage II underwent comprehensive horizontal testing, concluding on March 2. These rigorous tests were crucial in verifying the performance and reliability of both stages of the GLV 10, reinforcing NASA's commitment to ensuring mission success and astronaut safety in the demanding environment of space exploration.

On November 19, the Air Force Space Systems Division (SSD) instructed Lockheed to return Gemini Agena Target Vehicle (GATV) 5001 to Sunnyvale. Currently stored at Hanger E, Eastern Test Range, the GATV was undergoing preparation for future missions, albeit with its main engine already dispatched to Bell Aerosystems on November 9 for essential modifications.

Despite SSD and NASA's considerations to potentially use GATV 5001 as the second flight vehicle, significant refurbishment, repairs, and updates were deemed necessary—tasks feasible only at Lockheed's facilities. A dummy engine was promptly installed onboard the vehicle to simulate weight and center of gravity.

On November 20, GATV 5001 commenced its journey from Cape Canaveral via commercial van, ultimately arriving at Sunnyvale on November 24. This meticulous process highlighted NASA's commitment to ensuring its spacecraft's operational readiness and reliability, paving the way for future

missions within the Gemini program amidst evolving technological advancements and operational requirements.

On November 24, Lockheed took a decisive step by submitting an engineering change proposal to the Air Force Space Systems Division (SSD) for Project Surefire. This initiative, part of the Gemini Agena Target Vehicle (GATV) Modification and Test Program, aimed to rectify the malfunction that led to the failure of GATV 5002 on October 25. The proposal marked a pivotal moment in addressing critical issues affecting mission reliability.

Recognizing the situation's urgency, SSD provisionally approved Project Surefire on November 27, designating it as an emergency priority to ensure swift implementation. Concurrently, Lockheed established a dedicated Project Surefire Engine Development Task Force on the same day. This team was entrusted with executing the program's objectives swiftly and effectively, aligning efforts with the scheduled launch of GATV 5003 for Gemini VIII.

As part of the program, systems testing for GATV 5003 was temporarily suspended to prioritize modification efforts. The main engine, removed on November 23, was promptly sent to Bell Aerosystems for necessary modifications, underscoring Lockheed's commitment to enhancing vehicle performance and reliability.

Furthermore, adjustments were made to the schedule for GATV 5004, facilitating its final assembly with a modified engine configuration. These proactive measures demonstrated NASA and SSD's collaborative approach to overcoming challenges, ensuring the continued success of the Gemini program and advancing capabilities in space exploration.

On November 26, McDonnell proposed a contingency plan for Gemini rendezvous missions by suggesting the development of a backup target vehicle known as the

Augmented Target Docking Adapter (ATDA). This initiative arose amidst efforts to address challenges with the Gemini Agena Target Vehicle (GATV), which had caused the mission to abort on October 25. The ATDA, designed to be compatible with Gemini spacecraft and operations, served as an alternative in case modifications to the GATV did not align with the schedule for launching Gemini VIII.

Mockup of the augmented target docking adapter at McDonnell, along with a spacecraft mockup. (NASA Photo S-65-62180, Dec. 12, 1965.)

The ATDA was essentially an enhanced version of the Target Docking Adapter (TDA), incorporating modifications to enhance stability and facilitate spacecraft acquisition and docking. Key additions included a comprehensive communications system encompassing tracking, telemetry transmission, and command subsystems. Instrumentation, a guidance and control system featuring a target stabilization system and rendezvous radar transponder, and electrical and reaction control systems identical to those on the Gemini spacecraft were also integrated.

Robert C. Seamans, Jr., NASA Associate Administrator, endorsed the procurement of the ATDA on December 9, signaling official approval for the project. McDonnell promptly initiated assembly operations on December 14, underscoring NASA's proactive approach to mission preparedness and ensuring operational flexibility within the Gemini program.

On November 29, Director Robert R. Gilruth of the Manned Spacecraft Center sought NASA Headquarters' approval for adjustments to suit protocol during orbital flight for Gemini VII. Astronauts preferred to doff their G5C pressure suits after the second sleep period, reserving them solely for dynamic maneuvers like rendezvous and reentry. This request prioritized crew comfort and aimed to prevent potential performance degradation during the mission's extended duration.

The Gemini Program Office had previously certified the G5C suit for intravehicular use in the Gemini spacecraft on November 19, validating its suitability for extended orbital missions.

When Gemini VII launched on December 4, the original mission plan mandated that at least one astronaut always remained suited. However, on December 12, NASA Headquarters authorized both crew members to remove their suits simultaneously—a decision reflective of ongoing evaluations of crew safety, comfort, and operational needs during the mission.

On December 3, McDonnell initiated crucial altitude chamber and Extravehicular Support Package (ESP) tests for Spacecraft No. 8. These rigorous evaluations aimed to validate the spacecraft's performance under simulated space conditions and assess the functionality of its extravehicular support systems. The tests concluded successfully on December 13, marking a significant milestone in readiness assessment.

Throughout the remainder of December, McDonnell proceeded with necessary updates and additional rounds of testing to further refine Spacecraft No. 8. These efforts were essential to ensuring compliance with stringent safety and operational standards required for manned space missions.

By January 8, 1966, the spacecraft underwent final preparations and was shipped to Cape Kennedy.

Chapter 11 - Pioneering Long-Duration Orbital Flight

Gemini VII

On December 4, Gemini VII, the fourth manned mission of the Gemini program, launched from Complex 19 at 8:30 p.m. EST. Commanded by Astronaut Frank Borman, with Pilot Astronaut James A. Lovell, Jr., the mission's primary objectives were to demonstrate manned orbital flight for approximately 14 days and evaluate the physiological effects of prolonged spaceflight on the crew.

Astronauts Frank Borman and James A. Lovell, Jr., walking up the ramp to the elevator at pad 19 prior to their Gemini VII flight. They are wearing the new lightweight G5C suits. (NASA Photo S-65-44290, Dec. 4, 1965.)

Secondary objectives included serving as a rendezvous target for Gemini VI-A, stationkeeping with the launch vehicle's second stage and spacecraft No. 6, conducting 20 experiments, testing lightweight pressure suits, and evaluating spacecraft reentry guidance capabilities. Despite equipment failures that caused the loss of two experiments, all primary objectives were successfully achieved.

Immediately after separating from the launch vehicle, the crew maneuvered Gemini VII to within 60 feet of the second stage, performing a 15-minute stationkeeping exercise before powering down for the extended mission. Five maneuvers were executed throughout the mission to extend orbital duration and position the spacecraft for a successful rendezvous with Gemini VI-A on the 11th day.

Astronauts Borman (right) and Lovell on the deck of the U.S.S. Wasp after completing their 14-day mission. (NASA Photo No. 65-H-2323, released Dec. 18, 1965.)

Lovell removed his pressure suit about 45 hours into the mission, later donning it again at 148 hours, while Borman followed suit some 20 hours later. Both astronauts flew the remainder of the mission without suits, except during critical phases like rendezvous and reentry. Despite minor issues, such as a malfunctioning telemetry recorder and degraded fuel cell stacks, the spacecraft and its systems performed nominally until reentry.

Retrofire occurred precisely on schedule, and reentry and landing were nominal, with Gemini VII touching down only 6.4 miles from the intended landing point at 9:05 a.m. on December 18. Half an hour later, the aircraft carrier Wasp recovered the crew, followed shortly by spacecraft recovery.

Gemini VII's successful completion of extended orbital flight and rendezvous operations not only demonstrated NASA's growing capabilities in manned spaceflight but also provided invaluable insights into human endurance and spacecraft performance in long-duration missions, paving the way for future lunar exploration missions in the Apollo program.

December 4 marked a pivotal day in the meticulous preparations for Gemini VII's launch, underscoring NASA's operational efficiency and dedication to mission readiness.

Gemini spacecraft No. 6 after removal from storage, being hoisted to the top of the launch pad at complex 19. (NASA Photo No. 65-H-1906, released Dec. 5, 1965.)

Following the launch of Gemini VII, both stages of the Gemini Launch Vehicle (GLV-6) were promptly retrieved from storage and transported to Complex 19. Despite a two-hour delay, Stage No. 6 of the spacecraft was returned to Complex 19 on December 5,

ensuring swift readiness for the upcoming mission.

Astronauts Thomas Stafford and Walter Schirra NASA

Within an impressive 24-hour timeframe post-Gemini VII launch, both stages of GLV-6 were meticulously erected, and the spacecraft was seamlessly mated to the launch vehicle. Electrical power was swiftly applied, and crucial subsystem verification tests were initiated by December 8. Despite encountering a notable issue with the spacecraft's computer memory, which required replacement and subsequent verification on December 7-8, NASA's team effectively navigated these challenges.

On December 8-9, the Simulated Flight Test was conducted, marking the final prelaunch evaluation phase. This comprehensive test ensured all systems were fully operational and prepared for the impending launch.

Initially slated for December 13, GLV-6's launch date was rescheduled to December 12, reflecting NASA's commitment to optimal launch conditions and mission success.

From December 8 to 10, significant milestones were achieved in preparing Gemini Launch Vehicle 9 (GLV-9), highlighting the meticulous process and dedication of the teams involved at Martin Baltimore.

Erection at Vertical Test Facility: From December 8 to 10, Gemini Launch Vehicle 9 was erected in the east cell of the vertical test facility at Martin Baltimore. This phase involved carefully positioning the launch vehicle in preparation for subsequent testing and integration procedures.

Power Application and Testing: On December 22, following the erection process, power was applied to Gemini Launch Vehicle 9 for the first time. This critical step allowed engineers to begin assessing the functionality and operational readiness of the vehicle's various subsystems.

Subsystem Functional Verification Tests: From December 22 to January 20, 1966, rigorous Subsystem Functional Verification Tests were conducted over the ensuing weeks. These tests comprehensively evaluated each subsystem's performance and integration within the launch vehicle, ensuring all components operated seamlessly and reliably under simulated mission conditions.

December 12 marked a pivotal moment in the Gemini VI-A mission, highlighting NASA's adept handling of unexpected challenges and its commitment to ensuring mission success through meticulous troubleshooting and preparation.

The scheduled launch of Gemini VI-A encountered an abrupt halt when the Master Operations Control Set automatically shut down the Gemini launch vehicle mere seconds after engine ignition. This unexpected shutdown was traced to a premature separation of an electrical umbilical connector. The launch was officially canceled at 9:54 a.m. EST, initiating immediate emergency procedures.

Following the shutdown, emergency procedures necessitated delaying the raising of the erector until 11:28 a.m. The crew was subsequently removed from the spacecraft at 11:33 a.m., ensuring their safety amidst the abortive launch sequence.

Routine analysis of the engine data revealed critical insights into the cause of the shutdown. Engineers discovered decaying thrust in one of the first-stage engine subassemblies, attributed to a restriction in the gas generator circuit. This obstruction was later pinpointed to a protective dust cap inadvertently left in place within the gas generator oxidizer injector inlet port. This oversight could delay shutdown by approximately one second, correlating with the umbilical disconnect event.

Following this diagnosis, prompt corrective measures were implemented. Anomalies were swiftly rectified, and the launch team seamlessly executed long-prepared contingency plans. These efforts ensured that the Gemini VI-A mission could proceed without incident, culminating in the successful rescheduled launch on December 15.

Attempted launch and the shutdown of Gemini VI-A. (NASA Photo No. 65-H-1944, released Dec. 12, 1965.)

On December 14, a significant milestone occurred in the preparation and refurbishment efforts for the Gemini Agena target vehicle 5001, reflecting meticulous inspection protocols and refurbishment procedures authorized by the Air Force Space Systems Division.

The Air Force Space Systems Division granted Lockheed authorization to commence the disassembly and inspection of Gemini Agena target vehicle 5001. This critical decision marked the beginning of thorough evaluations to determine the extent of refurbishment required to ensure the vehicle's readiness for upcoming missions.

Lockheed initiated the disassembly process by systematically stripping down Gemini Agena target vehicle 5001 to its major structural components. This meticulous approach was essential to expose all areas potentially affected by contamination or wear, enabling comprehensive assessments of the vehicle's condition.

The primary objective of this disassembly and inspection was to assess the vehicle's structural integrity, functionality, and performance capabilities. By scrutinizing each component and subsystem, engineers aimed to identify any signs of wear, damage, or contamination that could compromise future mission success.

Upon completion of the inspection phase, Lockheed prepared detailed reports outlining findings and recommendations for refurbishment. These insights would guide subsequent efforts to refurbish and optimize Gemini Agena target vehicle 5001, ensuring it met stringent operational standards and reliability criteria.

Gemini VI-A: First Rendezvous in Space

On December 15, at 8:37 a.m. Eastern Standard Time, Gemini VI-A marked a pivotal moment in space exploration as the fifth manned mission and the first rendezvous mission of NASA's Gemini program. Commanded by Astronaut Walter M. Schirra, Jr., with Astronaut Thomas P. Stafford as pilot, the primary objective of this historic flight was to rendezvous with spacecraft No. 7.

After a flawless liftoff, the spacecraft entered a precise 87 by 140 nautical mile orbit from launch complex 19. A series of meticulously planned maneuvers followed, designed to bring Gemini VI-A into proximity with its target. Approximately ninety minutes after insertion, the crew initiated the first of several critical adjustments to align their orbit with the target spacecraft. These maneuvers included height adjustments, phase adjustments, a plane change, and crucial radar activations to gauge distance and positioning.

The Mission Control Center at Houston just after the announcement from the orbiting spacecraft that Gemini VI-A and VII had achieved rendezvous. (NASA Photo No. S-65-62720, Dec. 15, 1965.)

Three hours into the mission, the onboard radar made initial contact, confirming a distance of 246 miles between the two spacecraft. This milestone was followed by a coelliptic maneuver and subsequent midcourse corrections, all leading to the final approach phase. At approximately 5 hours and 50 minutes into the flight, the crew achieved technical rendezvous, halting relative motion with spacecraft No. 7 just 120 feet away.

Over the next three and a half orbits, the astronauts maintained stationkeeping, demonstrating the capability to hold positions within distances ranging from 1 to 300 feet. Following this successful phase, Gemini VI-A executed a controlled separation and moved to a safe distance of approximately 30 miles from its companion.

Despite a minor setback with the telemetry tape recorder later in the mission, which caused a loss of delayed-time telemetry data, the mission concluded with a textbook reentry and landing in the West Atlantic. The spacecraft splashed down just 7 miles from the intended target point at 10:29 a.m. on December 16. An hour later, the prime recovery ship, the aircraft carrier Wasp, retrieved the capsule, ensuring the safe return of Astronauts Schirra and Stafford after their groundbreaking mission.

Gemini VI-A achieved the first rendezvous in space and validated crucial techniques and technologies essential for future space missions, setting the stage for NASA's ambitious goals in manned space exploration.

On December 17, the Air Force accepted the main rocket engine for the Gemini Agena Target Vehicle (GATV) 5003, a significant milestone in the Gemini program. This critical engine, which underwent modifications under Project Surefire by Bell Aerosystems, was promptly shipped and arrived at Lockheed the following day, December 18.

U.S. Navy swimmers attaching the cable to the Gemini VI-A spacecraft, containing the astronauts, to haul it aboard the U.S.S. Wasp. The crew remained in the spacecraft during recovery. (NASA Photo No. 65-H-2294, released Dec. 16, 1965.)

Lockheed wasted no time in reinstalling the engine, completing the task by December 20. This swift turnaround allowed for systems retesting on GATV 5003 starting December 27, following the installation of additional equipment modifications.

On December 27, preparations for the upcoming Gemini IX mission took a crucial step forward with the acceptance meeting for Atlas 5303, the designated launch vehicle. Held in San Diego, the meeting initially faced a setback due to an unresolved issue concerning a liquid oxygen tank

pressurization duct. However, subsequent investigations confirmed the duct's satisfactory condition, clearing the way for formal acceptance.

Following this resolution, Atlas 5303 was prepared for transport, departing San Diego by truck on February 4. It completed its journey to Cape Kennedy on February 13, 1966, where final preparations and integration for the Gemini IX mission would occur.

Preparing for Gemini VIII: Atlas 5302

On January 5, 1966, at Cape Canaveral's Complex 14, the monumental Atlas 5302 rocket stood tall, poised to serve as the launch vehicle for Gemini VIII. This pivotal mission followed the setbacks of the Agena failure just months earlier, prompting rigorous efforts by the Air Force Space Systems Division and General Dynamics/Convair to ensure the rocket's flight readiness.

In response to the October 25, 1965, Agena mishap, engineers underwent a comprehensive overhaul of Atlas 5302. This involved implementing a series of procedural and design modifications aimed squarely at bolstering the vehicle's reliability. Notably, out of the 20 engineering change proposals introduced between Atlas 5301's launch and Atlas 5302's preparation, all but one had been rigorously validated through prior Atlas missions. The remaining adjustment, a novel destruct unit, underwent its inaugural flight test aboard Atlas 5302.

Verification of the booster's subsystems continued unabated until February 23. This meticulous testing phase was integral to ensuring that every facet of the rocket's operation met exacting performance and safety standards.

On January 8, 1966, McDonnell Aircraft Corporation delivered spacecraft No. 8 to Cape Kennedy, marking a significant milestone in the lead-up to the Gemini VIII mission. This spacecraft underwent an intensive two-week process of meticulous checks and installations. Key tasks included installing fuel cells, rigorous resistance checks on heaters, and the careful buildup of pyrotechnic systems.

Following these preparations, the spacecraft was relocated to Merritt Island Launch Area for critical tests. From January 26 to 28, it underwent an integrated (Plan X) test alongside the designated launch vehicle, Atlas 5302. This comprehensive evaluation aimed to ensure seamless coordination and operational readiness between the spacecraft and its accompanying rocket.

Subsequently, on January 29, another pivotal assessment took place: an extravehicular equipment compatibility test. This rigorous examination scrutinized the spacecraft's compatibility with essential extravehicular activity gear, validating that all systems could function harmoniously in demanding space conditions.

On January 8, 1966, the Gemini Agena target vehicle (GATV) 5003 completed its final acceptance tests at Sunnyvale, culminating a series of rigorous evaluations to ensure its readiness for the upcoming Gemini VIII mission. Earlier challenges, including an elusive command system issue, necessitated a rerun of the final systems test on January 4. This rerun revealed no discrepancies, affirming the vehicle's robustness and reliability.

Formal acceptance by the Air Force Space Systems Division was followed on January 18 after a thorough inspection by the vehicle acceptance team. The same day, GATV 5003 commenced its journey to the Eastern Test Range, poised to play a pivotal role as the designated target vehicle for Gemini VIII. However, adverse weather conditions delayed its delivery until January 21, underscoring the logistical complexities of preparing for space missions.

Preparing for Launch: Gemini Launch Vehicle (GLV) 8

On January 13, 1966, the Gemini Launch Vehicle (GLV) 8 was erected at complex 19, setting the stage for critical preparations ahead of its scheduled mission. After meticulous inspections and the connection of umbilicals, power was applied on January 19, initiating the next phase of readiness checks.

Subsystems Reverification Tests commenced promptly the following day and continued rigorously until January 31, ensuring that every component of GLV-8 met stringent operational standards. The Prespacecraft Mate Verification on February 1 marked a significant milestone, confirming the seamless integration between the launch vehicle and the spacecraft it would carry.

A comprehensive launch test-procedure review was conducted on February 2-3 to finalize preparations. However, during leak checks of the stage II engine on February 7, engineers discovered small cracks in the thrust chamber manifold. X-ray inspections revealed that these cracks were localized to the weld. Swift action addressed the issue through rewelding, successfully eliminating the problem by February 9.

Meanwhile, on January 16, another crucial preparation phase began at Bell Aerosystems with Project Surefire verification testing. This initiative focused on demonstrating the sea-level flightworthiness of the modified Agena main engine, pivotal for the upcoming mission. Bell Aerosystems completed testing on March 4 with a full 180-second simulation firing, affirming the engine's readiness for the mission ahead.

Planning Future Gemini Missions

During a pivotal NASA-McDonnell Management Panel meeting on January 17, 1966, W. B. Evans from the Gemini Program Office outlined ambitious plans for upcoming missions beyond Gemini VIII. Detailed discussions focused on mission activities, emphasizing the complex maneuvers and technological advancements planned for future Gemini missions.

Gemini VIII was slated to feature three periods of extravehicular activity (EVA), strategically scheduled to include both daylight and nighttime operations. During these EVAs, the astronaut would undock from the spacecraft with the right hatch snubbed against the umbilical guide, secured into the adapter section. Each redocking maneuver would precede with an orbit of stationkeeping to ensure precise alignment.

Key tasks during EVA included retrieving the emulsion pack from the adapter, initiating the S-10 Micrometeorite Collection experiment on the Agena, and employing a power tool for necessary tasks. The astronaut would utilize the extravehicular support pack, maneuver with the handheld maneuvering unit, and test various tether lengths to optimize mobility and safety.

Operational plans also included intricate maneuvers between the spacecraft and the Agena, incorporating burns from the secondary propulsion system to refine orbital trajectories. Gemini VIII was slated for a challenging three-day mission duration, setting the stage for increasingly complex space operations.

Looking ahead, Gemini IX and Gemini X were envisioned as three-day missions with even more sophisticated objectives. Gemini IX would simulate lunar module rendezvous scenarios, conduct primary propulsion system burns with docked Agenas, and execute multiple rendezvous maneuvers, including a simulated lunar module abort. EVAs would integrate the modular maneuvering unit for enhanced mobility and operational flexibility.

Gemini X, meanwhile, would pioneer dual rendezvous operations involving both a parked and a newly docked Agena. This mission would culminate in retrieving the S-10 experiment during EVA after undocking with the new Agena, pushing the boundaries of extravehicular activity and mission complexity.

January 20, 1966, marked a significant milestone in the preparations for Gemini launch vehicle (GLV) 12. Martin-Denver delivered propellant tanks crucial for the mission to Martin-Baltimore via air transport. This delivery was part of a complex reshuffling effort necessitated by earlier technical challenges.

Initially, the GLV-12 stage II fuel tank had been reassigned from GLV-11 due to operational needs. In turn, GLV-12 utilized the stage II fuel tank originally designated for GLV-10, following extensive reworking to address a damaged dome issue. The reworked tank arrived on March 12, ensuring the readiness and reliability of GLV-12's propulsion systems.

Aerojet-General had previously delivered the stage I engine on December 13, 1965, with the stage II engine arriving on January 20, completing crucial propulsion components for the mission. Subsequent splicing of the stage I tank was finalized on April 25, followed by stage II on May 4. Engine installations for both stages were completed by May 19, paving the way for comprehensive horizontal testing.

Testing concluded on June 1 for stage I and June 22 for stage II, marking successful readiness assessments for GLV-12's propulsion systems.

On January 21, McDonnell achieved a critical milestone by completing the final assembly of the augmented target docking adapter (ATDA). A series of rigorous tests followed, including Voltage Standing Wave Ratio Tests on January 21-22 and Systems Assurance Tests by January 25, culminating in vibration tests on January 27. Additional simulated flight and phasing tests from January 30 to February 1 validated the adapter's operational integrity.

The culmination of these tests affirmed the ATDA's readiness, and it was subsequently shipped to Cape Kennedy on February 4.

On January 21, 1966, significant progress was made in preparation for the Gemini VIII mission by completing qualification testing for the freon-14 extravehicular propulsion system. Earlier tests had encountered freezing issues, prompting a meticulous review of gas loading procedures. By refining drying protocols during gas loading, engineers successfully mitigated the freezing problem, ensuring subsequent tests were executed without issue. This advancement marked a critical step toward enhancing the reliability of extravehicular activities in space, following Astronaut Edward H. White II's pioneering use of oxygen for propulsion during Gemini IV.

The following day, January 22, saw the integration of the Gemini Agena target vehicle (GATV) 5003 with the target docking adapter (TDA) 3. McDonnell had delivered TDA-3 to Cape Kennedy on January 8, laying the groundwork for crucial integration tests. The GATV/TDA interface functional test, completed by January 24, validated seamless operational compatibility between the two components.

Subsequently, GATV 5003 was relocated to Merritt Island Launch Area for integrated tests with spacecraft No. 8 and extravehicular equipment, culminating in successful completion on January 28. These integrated tests were pivotal in ensuring cohesive functionality across all systems, setting the stage for the forthcoming Gemini VIII mission.

Crew Selection and GATV 5004 Preparations

On January 25, 1966, Astronaut John W. Young was appointed as the command pilot for the upcoming Gemini X mission, marking a pivotal moment in crew selection for NASA's ambitious space exploration program. Astronaut Michael Collins would join him as the pilot, forming the primary crew for this mission. The backup crew comprised James A. Lovell, Jr., as command pilot and Edwin E. Aldrin, Jr., as pilot, highlighting NASA's meticulous approach to mission and crew readiness.

The following day, January 26, preparations continued with the Gemini Agena target vehicle (GATV) 5004, which was transferred to Sunnyvale's vehicle systems test area. Essential upgrades included the installation of a modified main engine, delivered promptly by Bell Aerosystems on January 12 and installed by January 20. Despite these advancements, delays were encountered due to prioritizing GATV 5003's final acceptance tests. Critical electronic assemblies, including the command system, were temporarily relocated from GATV 5004 to expedite GATV 5003's readiness, resulting in an eight-day setback from its initially scheduled transfer date of January 18.

January 28, 1966, marked a pivotal moment in the preparations for Gemini missions with the return of the Gemini Agena target vehicle (GATV) 5003 to Hangar E, following the completion of Plan X tests at

Merritt Island Launch Area. These tests validated critical systems and operational readiness, setting the stage for comprehensive Systems Verification and Combined Interface Tests that continued through February 18. Functional checks of both primary and secondary propulsion systems further ensured the vehicle's reliability. The testing phase concluded on February 28, after which GATV 5003 was transferred to complex 14, marking another milestone in the meticulous preparations for upcoming missions.

On January 31, Gemini Spacecraft No. 8 was transported to Complex 19 and positioned atop its designated launch vehicle. Detailed preparations ensued with the connection of essential cables for testing on February 1-2, followed by the Prespacecraft Mate Verification Tests conducted from February 3-8. Activation of fuel cells on February 8, followed by deactivation the next day, marked critical steps in verifying spacecraft readiness.

Integrated tests between the spacecraft and launch vehicle commenced on February 10, consolidating efforts to ensure seamless functionality and operational compatibility. These meticulous procedures underscored NASA's commitment to meticulous planning and technical proficiency in preparing for groundbreaking manned space missions.

On February 2, 1966, a pivotal mission planning meeting for Gemini flights IX through XII occurred at McDonnell, gathering key Gemini Program Office and Flight Operations Division members. The meeting concluded with a significant proposal from McDonnell regarding the Gemini spacecraft's capabilities in achieving high elliptic orbits and ensuring safe reentry under specific conditions.

McDonnell emphasized the spacecraft's potential to reach a relatively high elliptic orbit with an apogee ranging from 500 to 700 nautical miles. This concept involved utilizing the Agena primary propulsion system to maneuver into the higher orbit and return to a nominal 161-mile circular orbit for controlled reentry.

This proposal highlighted innovative strategies to expand the operational scope of Gemini missions, potentially enabling extended mission durations and enhanced capabilities for scientific research and exploration. The discussion underscored NASA's forward-thinking approach in leveraging existing spacecraft capabilities to maximize mission flexibility and scientific yield in Earth's orbit.

On February 2, 1966, Agena D (AD-129) achieved a significant milestone as it was accepted by the Air Force for delivery to the Gemini program. Subsequently, this pivotal component was transferred to the final assembly area at Sunnyvale to undergo modifications, transforming into Gemini Agena target vehicle 5005. This adaptation process aimed to enhance its capabilities and readiness for upcoming missions within the Gemini program.

Two days later, on February 4, the augmented target docking adapter (ATDA) arrived at Cape Kennedy, marking another critical step in mission preparedness. Following its arrival, rigorous modifications, testing, and troubleshooting efforts were diligently carried out, culminating in completion by March 4. Designed to serve as a backup for the Gemini Agena target vehicle (GATV), the ATDA was strategically placed in storage from March 8 onwards.

Events took a pivotal turn on May 17, when the failure of target launch vehicle 5303 prevented GATV 5004 from achieving orbit as intended. The ATDA was swiftly repurposed to become the primary target for the subsequent Gemini IX-A mission, highlighting its critical role in maintaining mission flexibility and continuity within the Gemini program.

February 9, 1966 marked a significant achievement with completing the Combined

Systems Acceptance Test for Gemini launch vehicle (GLV) 9 at Martin-Baltimore's vertical test facility. This critical milestone validated the integrated functionality of GLV-9, affirming its readiness for upcoming missions within the Gemini program. Following the test, the vehicle acceptance team convened on February 14 and concluded their comprehensive review by formally accepting GLV-9 on February 17. Subsequent activities included dismantling GLV-9, completed by February 25, and its formal acceptance by the Air Force on March 8. The stage I components arrived at Cape Kennedy on March 9, with stage II following on March 10, ensuring timely readiness for mission preparations.

On February 10, pivotal advancements continued as Gemini launch vehicle 8 and spacecraft 8 were electrically mated, marking a crucial step in their integration process. This milestone facilitated the completion of the Electrical Interface Integrated Validation and Joint Guidance and Control Test by February 14. Following a thorough review of test data on February 15, the Joint Combined Systems Test was successfully conducted on February 16, further validating the seamless operational compatibility and performance of the integrated spacecraft and launch vehicle systems.

Final Preparations and Testing for Gemini Missions

February 17, 1966, marked a crucial phase in the preparation for Gemini missions with the conduct of the tanking test for Gemini launch vehicle (GLV) 8. This test, aimed at validating the fueling procedures and systems of GLV-8, was completed, ensuring readiness for upcoming launch operations. Concurrently, spacecraft No. 8 underwent its Final Systems Test, culminating in completion on February 23, further solidifying spacecraft readiness for integration with the launch vehicle.

On February 23, a significant milestone was achieved with successfully completing the Booster Flight Acceptance Composite Test (B-FACT) for target launch vehicle 5302. This test, which encompassed subsystem testing, addressed component challenges that had previously delayed the completion of vehicle pad tests. Despite encountering difficulties in propellant tanking tests, the successful completion of B-FACT signified progress in overcoming technical hurdles and advancing preparations for upcoming Gemini missions.

The Astronaut Maneuvering Unit

On February 23, 1966, preparations for the Gemini IX mission took a critical step forward, when the astronaut maneuvering unit (AMU) was delivered to Cape Kennedy. This specialized equipment, essential for extravehicular activities (EVA), underwent a receiving inspection that revealed issues: nitrogen leaks in the propulsion system and oxygen leaks in the oxygen supply system.

Prompt corrective actions were initiated, and by March 11, meticulous reworking efforts successfully eliminated the leakage problems in both systems. Subsequent systems tests were conducted to ensure the AMU's operational integrity and readiness for its pivotal role in supporting EVA during the Gemini IX mission.

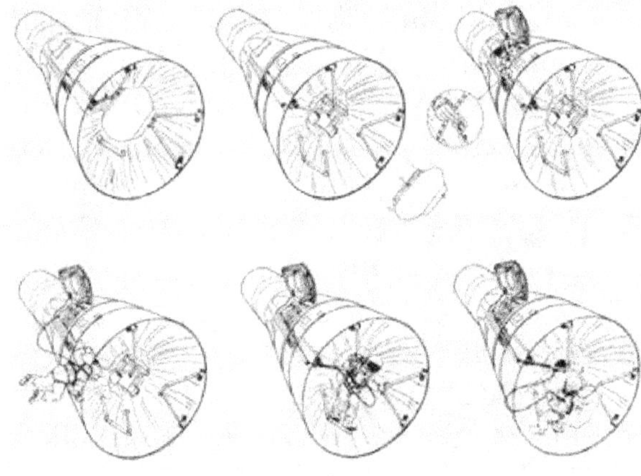

Method of donning the astronaut maneuvering unit, carried in the adapter section. (NASA Photo S-66-24197, Mar. 16, 1966.)

On February 25, GLV-8 and spacecraft No. 8 were temporarily mated for an erector-cycling test, a critical step in verifying the operational functionality and compatibility of the integrated systems. Throughout this period, from February 26 to March 5, meticulous checks and installations were conducted on the spacecraft, including verifying and installing the extravehicular support package and life support systems.

Meanwhile, from February 28 to March 3, modifications and revalidation efforts were underway for GLV systems, ensuring optimal performance and reliability. These comprehensive activities underscored NASA's commitment to meticulous planning and technical proficiency to mitigate risks and enhance mission success in manned spaceflight endeavors.

From February 23 to 24, a pivotal event unfolded at the Manned Spacecraft Center: the Gemini Midprogram Conference. This gathering drew over 600 delegates from governmental agencies and industry leaders deeply involved in Project Gemini. The conference was a critical platform for sharing insights and advancements in spacecraft development, launch vehicles, and operational strategies.

Throughout the event, the attendees were treated to a comprehensive array of 44 technical papers. These papers delved into every facet of Gemini's progress, including detailed analyses of the first seven Gemini missions. Attendees eagerly absorbed findings from the experiments conducted during these missions, illuminating new frontiers in space exploration.

Gemini Program: Triumphs and Tragedy

On February 27, a significant milestone marked the advancement of Project Gemini as the Gemini Agena Target Vehicle (GATV) 5004 completed rigorous systems testing in Sunnyvale. This pivotal achievement culminated in its formal acceptance by the Air Force on March 11, following a meticulous inspection by the vehicle acceptance team. Just a day later, on March 12, GATV 5004 embarked on its journey to the Eastern Test Range, arriving promptly on March 14, poised to play a crucial role in upcoming Gemini missions.

On February 28, 1966, tragedy struck the NASA community with the untimely deaths of Elliot M. See Jr. and Charles A. Bassett II, prime crew members of the upcoming Gemini IX mission. They were aboard a T-38 Talon aircraft en route from Houston to St. Louis for a rendezvous simulation at McDonnell Aircraft Corporation, the manufacturer of the Gemini spacecraft.

During an instrument landing attempt at Lambert International Airport amidst marginal weather conditions, their aircraft tragically struck the roof of a McDonnell building before crashing into an outside storage area. See and Bassett lost their lives instantly, while several McDonnell employees sustained minor injuries. Miraculously, their Gemini spacecraft, prepared for shipment, escaped unscathed.

Robert R. Gilruth, director of the Manned Spacecraft Center (MSC), later known as NASA's Johnson Space Center in Houston, expressed profound sadness at their passing, describing both men as outstanding professionals and test pilots. He remarked, "Both of these men were fine persons and excellent professional test pilots. We will miss them more than I can say."

In the wake of this devastating loss, Thomas P. Stafford and Eugene A. Cernan, the backup crew members flying separately in another T-38, landed safely. Their readiness and the resilience of the NASA team ensured that the Gemini program continued to progress, honoring the legacy of See and Bassett as pioneers in space exploration.

An investigative panel chaired by Alan B. Shepard, Chief of the Astronaut Office at MSC, considered multiple factors in the cause of the crash, such as the inclement weather that included rain, snow, and fog that limited visibility, necessitating an instrument approach. In the final analysis, the panel ruled pilot error as the primary cause due to the pilot's inability to maintain a visual reference for a landing. Since the spacecraft escaped damage, McDonnell shipped it to NASA's Kennedy Space Center on schedule two days after the accident. The deaths of See and Bassett resulted in a shuffling of astronaut flight assignments but did not delay the remainder of the Gemini missions. Stafford and Cernan replaced See and Bassett as the prime crew for Gemini-IX and flew the mission in June 1966. See and Bassett were buried at Arlington National Cemetery on March 4.

In the aftermath of the accident, the Gemini IX backup crew, Thomas P. Stafford and Eugene A. Cernan, who were scheduled for simulator training at McDonnell, landed safely. Despite the tragedy, NASA swiftly appointed Alan B. Shepard, Jr., to lead a thorough investigation into the incident. Determined to honor their fallen colleagues' commitment to space exploration, NASA Headquarters affirmed that Stafford and Cernan would proceed with the Gemini IX mission as planned.

In a critical step forward for the Gemini program, Stage I of Gemini launch vehicle 10 was meticulously erected in the east cell of Martin-Baltimore's vertical test facility on February 28. Following comprehensive horizontal testing, which concluded on March 3, Stage II was assembled on March 7. The vehicle saw its first application of power on March 14, marking a pivotal moment in its readiness assessment. By April 13, Subsystems Functional Verification Tests were successfully completed, affirming its operational readiness for upcoming missions.

On March 1, significant progress continued as Gemini Agena target vehicle 5003 was expertly mated to target launch vehicle 5302 at Complex 14. Rigorous ground equipment compatibility tests were meticulously executed, paving the way for the successful Joint Flight Acceptance Composite Test on March 7. Demonstrating the program's thoroughness, a Simultaneous Launch Demonstration spanning March 8-9 effectively validated the Gemini Atlas-Agena target vehicle systems in preparation for the impending Gemini VIII mission, scheduled for launch on March 15.

On March 2, a crucial phase commenced as Spacecraft No. 9, and Target Docking Adapter No. 5 arrived at Cape Kennedy after manufacturing at McDonnell. The integration process began swiftly, with the installation of spacecraft fuel cells completed by March 4. Over the following days, meticulous preparations ensued, including pyrotechnics buildup and additional installations, culminating in comprehensive testing preparations that extended until March 18.

Subsequently, the spacecraft was relocated to Merritt Island Launch Area, where from March 22 to 24, it underwent rigorous Plan X integrated tests. These tests encompassed critical evaluations of its compatibility with the target vehicle and extravehicular systems, setting the stage for upcoming missions.

Meanwhile, on March 6, at Complex 19, another pivotal moment unfolded as Gemini Launch Vehicle 8 and Spacecraft No. 8 were successfully mated in preparation for their upcoming flight. This milestone marked a significant step forward in the meticulous preparations for the mission. A Simultaneous Launch Demonstration with the Gemini Atlas-Agena target vehicle at Complex 14 was executed seamlessly by March 9, underscoring the readiness of the launch systems. The Final Simulated Flight Test, conducted on March 10, conclusively validated all prelaunch procedures and readiness, ensuring a comprehensive readiness assessment for the impending mission.

A setback occurred on March 13 when the fuel tank of Target Launch Vehicle 5302 was inadvertently overfilled during propellant loading. This incident necessitated urgent maintenance, including replacing the fuel tank regulator and fuel relief valve, which was promptly completed by the following day. Due to these necessary repairs and safety checks, the scheduled launch, originally set for March 15, was rescheduled to March 16. This delay highlighted the meticulous attention to detail and safety protocols essential to the success of every mission within the Gemini program.

From March 14 to March 18, the AMU underwent comprehensive installation procedures within spacecraft No. 9, culminating in intensive preparations to equip the spacecraft with the necessary tools for upcoming mission objectives. These efforts highlighted NASA's commitment to meticulous engineering and readiness protocols, ensuring astronaut safety and mission success amidst the challenges and complexities of manned space exploration.

Gemini VIII: Pioneering Space Docking

The Gemini VIII mission commenced with the launch of the Gemini Atlas-Agena target vehicle (GATV) from complex 14 at 9:00 a.m. EST, followed by the launch of the Gemini space vehicle with Command Pilot Astronaut Neil A. Armstrong and Pilot Astronaut David R. Scott from complex 19 at 10:41 a.m. The primary goals of this three-day mission included rendezvous and docking with the GATV and conducting extravehicular activities (EVA). Secondary objectives encompassed various maneuvers and experiments crucial for advancing space exploration capabilities.

After insertion, the GATV settled into a 161-nautical mile circular orbit, while the Gemini spacecraft adopted an elliptical orbit ranging from 86 to 147 nautical miles. Nine maneuvers were executed over the following six hours to facilitate rendezvous, culminating in the spacecraft reaching a stationary position just 150 feet from the GATV. Docking was successfully achieved at 6 hours 33 minutes into the mission.

However, a critical issue arose shortly after docking when an Orbit Attitude and Maneuver System (OAMS) thruster malfunctioned, causing the spacecraft to rotate rapidly. The crew swiftly undocked from the GATV to regain control and utilized the Reentry Control System (RCS). Despite stabilizing the spacecraft, premature RCS usage necessitated an early mission termination.

Retrofire was initiated during the seventh revolution, followed by a nominal reentry and landing in the western Pacific Ocean at 10:22 p.m., less than seven miles from the intended landing point. The destroyer Leonard Mason recovered both crew and spacecraft three hours later.

Although the mission concluded prematurely, it achieved one primary objective - rendezvous and docking - and several secondary objectives, including evaluating systems like the auxiliary tape memory unit and demonstrating controlled reentry. While two experiments were only partially completed, Gemini VIII demonstrated NASA's ability to manage in-flight challenges and laid essential groundwork for future space missions, particularly in mastering orbital rendezvous and docking techniques.

Astronauts Neil A. Armstrong (right), command pilot; and David R. Scott, pilot, were the prime crew of the Gemini-8 space mission.NASA

Following the premature termination of the Gemini VIII mission, the Gemini Agena target vehicle (GATV) 5003 continued its operations in orbit, demonstrating resilience and achieving several significant milestones.

GATV 5003's systems were put through extensive tests and operations while in orbit: The main engine of GATV 5003 was fired nine times, surpassing the contractual requirement of five firings. This demonstrated the robustness and reliability of the propulsion system under operational conditions.

The launch of the Gemini Atlas-Agena target vehicle for the Gemini VIII mission from complex 14. (NASA Photo No. 66-H-296, released Mar. 16, 1966.)

The command and communications system of GATV 5003 processed and executed 5000 commands during its mission. This far exceeded the contractual requirement of 1000 commands, showcasing the vehicle's capability to handle complex operational tasks in orbit.

After ten days in orbit, GATV 5003's electrical power reserves were depleted, leading to the vehicle's loss of control. Before this, all remaining attitude control gas was vented to prevent any potential malfunction of thrusters.

As part of its mission objectives following Gemini VIII, GATV 5003 was maneuvered into a 220-nautical mile circular decay orbit. This deliberate orbit was planned to gradually lower the vehicle's altitude over time, ensuring it would be within reach during subsequent Gemini missions for astronaut inspection.

250

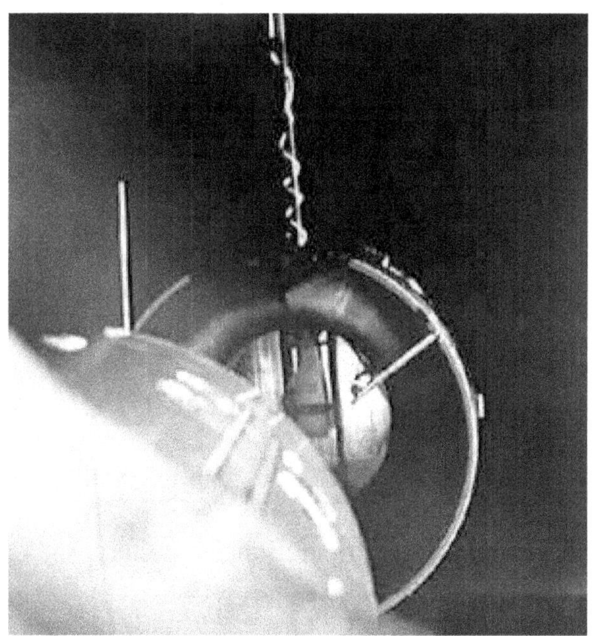

The Gemini VIII spacecraft approaching the Gemini Agena target vehicle in the final stage of rendezvous (the distance between the two craft is approximately two feet). (NASA Photo No. 66-H-225 [66-HC-191], released Mar. 16, 1966.)

The docked Gemini and Agena. (NASA Photo No. 66-H-226 [66-HC-192], released Mar. 16, 1966.)

Crisis and Resolution in Space

Shortly after the historic docking at 5:41 p.m., Gemini VIII encountered a sudden and perilous situation when the combined spacecraft began to undergo a violent yaw and tumble. Recognizing the severity of the situation, Astronaut Neil Armstrong promptly disengaged the Gemini capsule from the Agena Target Vehicle (GATV). However, the tumbling persisted and even intensified, reaching potentially hazardous rates of over one revolution per second.

Armstrong and David Scott acted decisively, using quick thinking and exceptional skill. They deactivated the Orbit Attitude and Maneuver System (OAMS) and initiated a desperate attempt to stabilize the spacecraft using all 16 reentry control system (RCS) thrusters. This critical maneuver, executed at 6:06:30 p.m., halted the violent tumbling and restored stability to Gemini VIII.

Upon investigation, it was determined that one of the 25-pound OAMS roll thrusters (roll thruster no. 8) aboard Gemini VIII had malfunctioned. The thruster had short-circuited during its operation to maneuver the Gemini-GATV combination, causing it to remain stuck in an open position and continuously fire. This malfunction was identified as the primary cause of the spacecraft's sudden and dangerous gyrations.

Despite the harrowing ordeal, Armstrong and Scott's skillful handling of the emergency and successful stabilization of Gemini VIII demonstrated NASA's preparedness and the effectiveness of astronaut training and decision-making under extreme conditions. The incident also highlighted the critical importance of rigorous technical checks and redundancy in space systems, influencing future mission planning and spacecraft design to enhance crew safety and mission success.

Due to the critical depletion of reentry control system (RCS) fuel caused by the earlier emergency, Gemini VIII was compelled to execute an immediate landing under NASA safety protocols. As a result, planned activities such as the ExtraVehicular Activity (EVA) and other experiments were regrettably canceled. Retrofire operations

251

commenced during the 7th revolution at 9:45:49 p.m. on March 16, a little over 10 hours after liftoff.

Gemini VIII safely splashed down in the western Pacific Ocean approximately 800 km west of Okinawa, near coordinates 25.22 N, 136.00 E, at 10:22:28 p.m. EST (during daylight at the landing site). Within minutes of touchdown, United States Air Force (USAF) frogmen descended from a C-54 rescue plane, swiftly attaching a flotation collar around the spacecraft. The crew was subsequently retrieved by the recovery ship USS Mason three hours later, at 1:28 a.m. EST on March 17. The spacecraft itself was recovered shortly after that, at 1:37 a.m., and the total elapsed mission time from launch to recovery amounted to 10 hours, 41 minutes, and 26 seconds.

Despite the premature conclusion, Gemini VIII accomplished significant milestones. The successful rendezvous and historic docking with the Agena Target Vehicle validated crucial space maneuvering techniques. Evaluating the auxiliary tape memory unit and demonstrating controlled reentry further advanced NASA's understanding and capabilities in manned spaceflight.

Of the six scientific experiments planned, only the Agena micrometeorite collection yielded definitive results. The others, including zodiacal light photography, frog egg growth, synoptic terrain photography, nuclear emulsions, and spectrophotography of clouds, were unfortunately incomplete due to the mission's early termination.

Following Gemini VIII's mission, ground maneuvers successfully placed the Agena Target Vehicle into a stable circular orbit, ensuring ongoing utilization and operational success for subsequent missions. The mission's challenges and achievements underscored the resilience and adaptability of NASA's astronauts and ground teams, contributing valuable lessons to the evolving field of space exploration.

In mid-March, a pivotal delivery arrived at Cape Kennedy, marking a critical juncture in the Gemini spacecraft program. The Extravehicular Life Support System (ELSS) designated for Gemini spacecraft No. 9 made its way to the launch site, signaling the next rigorous testing and preparation phase.

The ELSS, a cornerstone of astronaut safety and mission success, underwent meticulous compatibility assessments. These tests, which commenced promptly upon its delivery, integrated the ELSS with the astronaut maneuvering unit and the spacecraft. The goal was to ensure seamless functionality and operational readiness under the demanding space conditions.

By March 24th, after days of intensive testing, engineers and astronauts alike breathed a cautious sigh of relief as compatibility trials concluded successfully. This milestone marked a significant step forward in the Gemini program, validating the intricate interplay between humans and machines in the harsh environment beyond Earth's atmosphere.

However, the journey towards perfection was not without its challenges. Following these initial tests, the ELSS was temporarily returned to its contractors on April 6th for essential modifications. This refinement phase underscored the program's commitment to continuous improvement and the uncompromising pursuit of excellence in aerospace engineering.

As engineers diligently worked to enhance the ELSS, anticipation mounted among the space exploration community. Each adjustment aimed to meet and exceed the stringent standards required for the impending missions of Gemini spacecraft No. 9.

Gemini XI Mission Announced

On March 19th, NASA made a pivotal announcement that reverberated through the halls of space exploration. The astronaut assignments for the upcoming Gemini XI mission were unveiled, solidifying the roster of pioneers who would navigate space challenges.

At the helm of the mission stood command pilot Charles Conrad, Jr., joined by pilot Richard F. Gordon, Jr., comprising the esteemed prime crew. This selection underscored NASA's confidence in their expertise and readiness to command the Gemini XI spacecraft through the depths of space.

In a strategic move reflecting NASA's commitment to contingency planning, Neil A. Armstrong, a seasoned astronaut, assumed the role of command pilot for the backup crew, supported by pilot William A. Anders. Their assignment as backup crew for Gemini XI highlighted NASA's meticulous approach to mission readiness and crew preparedness.

Meanwhile, James A. Lovell, Jr., and Edwin E. Aldrin, Jr., originally slated for the Gemini X mission, received a reassignment as backup crew for Gemini IX. This reallocation of talent underscored NASA's flexibility and strategic maneuvering to optimize crew dynamics and readiness across missions.

Adding to the intrigue of crew realignment, Alan L. Bean and Clifton C. Williams, Jr., were appointed as the fresh backup crew for Gemini X. Their inclusion showcased NASA's commitment to cultivating a deep bench of talent, ensuring robust backup support for critical missions in the Gemini program.

On March 21st, a pivotal collaboration unfolded at the Merritt Island Launch Area Radar Range, marking a significant milestone in the Gemini program. Gemini Agena target vehicle 5004 and spacecraft No. 9 embarked on rigorous Plan X compatibility tests, a crucial phase to ensure seamless integration and operational harmony between the two essential components.

The intricate dance of technology and precision engineering lay at the heart of these tests. Engineers meticulously orchestrated the interaction between the Gemini Agena target vehicle 5004 and spacecraft No. 9, scrutinizing every detail to guarantee flawless functionality under demanding space conditions.

The Merritt Island Launch Area Radar Range was the perfect backdrop for these trials, providing a controlled environment where engineers could monitor and adjust parameters in real time. This collaborative effort between NASA and its partners underscored the program's commitment to precision and safety in all phases of mission preparation.

As Plan X compatibility tests unfolded, anticipation mounted within the aerospace community. Each successful maneuver and data point gathered brought Gemini closer to its ultimate goal: unlocking the mysteries of space and advancing humankind's quest for exploration and discovery.

On March 22nd, a pivotal chapter unfolded in the preparation for the Gemini program as Agena D (AD-130) was formally accepted by the Air Force. This crucial milestone marked its transformation into Gemini Agena target vehicle 5006, signaling a phase of meticulous modification and final assembly.

Following its formal acceptance, Agena D (AD-130) was swiftly transported to Building 104 at Sunnyvale, a hub of aerospace innovation and precision engineering. Here, a dedicated team of engineers and technicians embarked on the intricate modification process, fine-tuning every aspect to meet the exacting standards of the Gemini program.

At Sunnyvale, the transformation of Agena D (AD-130) into Gemini Agena target vehicle 5006 represented more than mere technical refurbishment. It symbolized a collaborative effort between the Air Force and NASA, blending military expertise with civilian ingenuity to advance the frontiers of space exploration.

As modifications progressed, anticipation grew within the aerospace community. Each adjustment and enhancement brought Gemini Agena target vehicle 5006 closer to its pivotal role in supporting upcoming missions, reinforcing its critical role in enabling rendezvous and docking maneuvers essential for the Gemini program's objectives.

March 24th marked a pivotal moment in the Gemini program as Gemini launch vehicle 9 emerged from storage and was meticulously erected at Complex 19. This event signaled the beginning of intensive preparations for an upcoming mission characterized by meticulously orchestrated tests and verifications.

The vehicle was inspected thoroughly to ensure every component met stringent safety and operational standards following its assembly. By March 28th, umbilicals essential for power and data transfer were meticulously connected, setting the stage for the next phases of readiness testing.

On March 29th, a critical milestone was reached as power was applied to the Gemini launch vehicle 9, initiating a series of comprehensive tests known as the Subsystems Reverification Test (SSRT). Engineers meticulously scrutinized every subsystem from March 30th to April 11th, ensuring seamless integration and functionality under simulated mission conditions.

The conclusion of SSRT on April 11th marked a significant achievement, validating the readiness of Gemini launch vehicle 9 for its upcoming mission. This success paved the way for the next phase: the Prespacecraft Mate Verification Combined Systems Test, which commenced promptly and concluded on April 12th.

On March 24th, the Air Force Space Systems Division and Lockheed decided to maintain momentum in the Project Surefire test program despite the stellar performance of Gemini Agena target vehicle (GATV) 5003 during the Gemini VIII mission. This commitment underscored their dedication to ensuring robustness and reliability in future missions.

The final phase of Project Surefire commenced on March 28th at the Arnold Engineering Development Center, marking a critical juncture in testing protocols. This phase focused on rigorous evaluations, including low-temperature starts and deliberately induced malfunctions, designed to simulate and mitigate potential operational challenges.

Engineers meticulously executed two firings throughout testing, culminating on April 4th with a planned fuel lead test. As anticipated, an engine hard start occurred during this phase, mirroring previous findings and validating critical hypotheses related to engine performance under varying conditions.

The analysis of engine damage provided invaluable insights, correlating closely with data from previous GATV 5002 failures. This correlation strengthened the hypothesis, attributing previous failures to hard starts caused by fuel preceding oxidizer into the thrust chamber during ignition—a crucial discovery informing future design and operational protocols.

On March 28th, the preparations for an upcoming mission reached a critical phase as Gemini spacecraft No. 9 was carefully transferred to Complex 19 and positioned atop its designated launch vehicle. This pivotal maneuver set the stage for meticulous tests and verifications to ensure the spacecraft's readiness for space exploration.

Over the following two days, engineers and technicians diligently cabled Gemini Spacecraft No. 9 for testing, meticulously connecting essential systems to the launch vehicle. This preparatory phase laid the foundation for the next crucial step: premate verification, which commenced on March 31st and concluded on April 6th.

During premate verification, every subsystem and component of the spacecraft underwent rigorous testing to validate functionality and operational readiness. This comprehensive evaluation process was essential to mitigate potential risks and ensure optimal performance during the upcoming mission.

Following the successful activation and deactivation of the fuel cells, preparations swiftly transitioned to the spacecraft/launch vehicle integrated tests, scheduled to commence on April 11th. These integrated tests represented a pivotal milestone, consolidating the seamless interaction between Gemini spacecraft No. 9 and its launch vehicle under simulated mission conditions.

As engineers meticulously scrutinized each test result and fine-tuned systems, anticipation mounted within the aerospace community. Every successful milestone achieved during these preparatory phases brought Gemini spacecraft No. 9 closer to its historic mission, embodying NASA's commitment to precision, safety, and the relentless pursuit of scientific discovery in space.

On March 31st, the Atlas target launch vehicle (TLV) 5304 encountered a setback during the San Diego acceptance meeting for the Gemini program due to an unmet contractual requirement. Despite completing systems testing on March 23rd, procedural details needed to be resolved before formal acceptance.

After addressing these technicalities, the Air Force formally accepted TLV-5304 on April 14th, signaling a significant step forward in its journey toward supporting upcoming Gemini missions. The vehicle embarked on its next phase, transported by truck to Cape Kennedy, poised to play a crucial role in advancing space exploration.

However, en route to its destination, TLV-5304 encountered an unforeseen challenge. This accident damaged the skirt on booster engine No. 1. Prompt inspection and rigorous analysis by contractors determined that the dented tubes resulting from the accident could be safely utilized without requiring repair.

Despite this setback, TLV-5304 persevered, arriving at Cape Kennedy on May 8th after a resilient nine-day road trip. Upon arrival, the vehicle underwent thorough receiving inspection protocols to ensure its operational integrity and readiness for future missions.

On May 11th, TLV-5304 was securely placed in storage, marking the culmination of its tumultuous yet ultimately triumphant journey to readiness within the Gemini program. This resilient response to challenges highlighted the program's commitment to adaptability, safety, and the unwavering pursuit of excellence in space exploration.

Atlas 5303 for Gemini IX

On April 4th, Atlas 5303, designated as the target launch vehicle for Gemini IX, achieved a pivotal milestone as it was meticulously erected at Launch Complex 14. This began extensive preparations to ensure the vehicle's readiness for its critical role in the upcoming mission.

By April 11th, electrical power was successfully applied to Atlas 5303, initiating a series of comprehensive tests to validate its systems and readiness for flight. Among these tests, the Booster Flight Acceptance Composite Test stood out as a crucial evaluation of the vehicle's integrated capabilities.

Engineers meticulously tested and analyzed every aspect of Atlas 5303's performance under simulated mission conditions. This rigorous evaluation aimed to confirm operational integrity and mitigate potential risks associated with the upcoming mission.

These efforts culminated on April 27th with completing the Booster Flight

Acceptance Composite Test. This milestone affirmed Atlas 5303's readiness to support the Gemini IX mission, marking a significant achievement in the meticulous preparation required for human spaceflight.

On April 12th, a pivotal milestone was reached in the Gemini program as Gemini Agena target vehicle 5005 underwent final assembly following the installation of critical electrical and electronic components. This phase included integrating essential systems such as the guidance module, flight control junction box, and flight electronics package, which were crucial for its operational readiness.

With these components in place, Gemini Agena target vehicle 5005 was transferred to test complex C-10 at Sunnyvale, marking the commencement of rigorous Vehicle Systems Tests. Engineers conducted preliminary test tasks to verify all systems' seamless integration and functionality.

By April 23rd, preliminary test tasks were completed, demonstrating the vehicle's initial readiness for advanced evaluations. This progress paved the way for a meticulous preliminary inspection on April 26th and 27th, ensuring every component met stringent safety and operational standards.

Throughout these critical phases, meticulous attention to detail and rigorous testing protocols underscored NASA's commitment to precision and safety in space exploration. Each milestone achieved with Gemini Agena target vehicle 5005 brought the program closer to achieving complex rendezvous and docking maneuvers in orbit, pushing the boundaries of human exploration and scientific discovery.

Simultaneously, on April 12th, Gemini Agena target vehicle 5004 embarked on a crucial phase in its preparation for upcoming missions by initiating the Combined Interface Test (CIT) at Hangar E, Eastern Test Range. This milestone followed the successful completion of Plan X tests on March 24th, highlighting a seamless transition to comprehensive integration testing.

The CIT, a pivotal evaluation conducted at Hangar E, focused on validating the harmonious interaction and functionality between Gemini Agena target vehicle 5004 and associated spacecraft systems. Engineers meticulously scrutinized every interface to ensure optimal performance under simulated mission conditions.

Over ten intensive days, the CIT progressed with precision, culminating on April 22nd with successful outcomes that affirmed the readiness of integrated systems. Subsequently, rigorous engine functional tests were conducted on both primary and secondary propulsion systems, confirming operational reliability.

By May 1st, all testing at Hangar E was completed, marking a significant milestone in the meticulous preparation of Gemini Agena target vehicle 5004 for its critical role in supporting upcoming Gemini missions. These efforts underscored NASA's commitment to excellence in space exploration and laid the groundwork for future successes in advancing human capabilities beyond Earth's orbit.

A pivotal phase in the Gemini program began on April 13th. The Electrical Interface Integrated Validation and Joint Guidance and Control Test began following the electrical mating of Gemini launch vehicle 9 and spacecraft No. 9. These critical activities, aimed at ensuring seamless integration and operational readiness, were successfully concluded by April 15th.

The Joint Combined Systems Test, a comprehensive evaluation of integrated functionalities, was conducted on April 19th. This crucial test assessed the collective performance of both the launch vehicle and spacecraft systems under simulated mission conditions, marking a significant milestone in the meticulous preparation for upcoming missions.

Engineers and technicians meticulously validated every interface and system component throughout these rigorous testing phases to mitigate potential risks and ensure optimal performance. These efforts underscored NASA's commitment to precision and safety in human spaceflight, paving the way for successfully executing complex maneuvers and scientific objectives in orbit.

April 14th marked a significant milestone in the Gemini program, as the Combined Systems Acceptance Test (CSAT) for Gemini Launch Vehicle (GLV) 10 was completed at Martin-Baltimore. This rigorous evaluation validated GLV-10's integrated systems and operational readiness, setting the stage for critical performance assessments.

Following the CSAT, a thorough performance data review was conducted, culminating in a comprehensive analysis completed by April 19th. This detailed review provided crucial insights into the vehicle's capabilities and readiness for upcoming missions.

April 18th marked a crucial milestone in the Gemini program as Stage I of Gemini Launch Vehicle 11 was meticulously erected in the west cell of the vertical test facility at Martin-Baltimore. This pivotal step initiated a series of meticulous preparations to ensure the vehicle's readiness for upcoming missions.

Following the completion of horizontal tests on April 25th, Stage II of Gemini Launch Vehicle 11 was swiftly erected by April 29th, setting the stage for comprehensive integrated testing. On May 9th, a significant milestone was achieved as power was applied to the vehicle for the first time, marking the commencement of Subsystems Functional Verification Tests.

Over the ensuing weeks, engineers and technicians meticulously conducted Subsystems Functional Verification Tests to validate the performance and integration of critical systems. This rigorous evaluation was completed on June 8th, demonstrating the vehicle's operational readiness and capability under simulated mission conditions.

Demonstration of the astronaut maneuvering unit. (NASA Photo S-66-32550, May 12, 1966.)

On April 18, the extravehicular life support system (ELSS) designated for Gemini spacecraft No. 9 arrived at Cape Kennedy. Following its return, rigorous testing ensued to ensure seamless integration with the astronaut maneuvering unit (AMU). The initial phase involved an exhaustive electrical compatibility test, setting the stage for subsequent evaluations.

The critical ELSS/AMU Joint Combined System Test commenced promptly the following day, probing every aspect of their interplay in simulated space conditions. This pivotal test was challenging, prompting a rerun on April 21 to refine system synchronicity and operational reliability.

By April 22, the ELSS journeyed to the Manned Spacecraft Center, where meticulous tests awaited to scrutinize its functionality under varying operational scenarios. Simultaneously, preparations intensified for the installation of the AMU into the spacecraft adapter, ensuring all components were meticulously aligned for optimal performance.

Returning to Cape Kennedy on April 26 after comprehensive evaluations, the ELSS entered its final readiness phase. May 7 marked the successful completion of the AMU Final Systems Test, a critical milestone before its definitive integration into the spacecraft configuration. With meticulous attention to detail, the ELSS was serviced and secured into place by May 16, poised and ready for the impending mission ahead.

On April 26th, the vehicle acceptance team convened to assess the findings and formally accepted GLV-10 on April 29th, marking a pivotal achievement in the vehicle's journey toward deployment. Subsequently, from May 2nd to May 4th, stage I of GLV-10 was carefully transported to Cape Kennedy, followed by stage II on May 20th.

Upon arrival at Hanger L, both stages underwent meticulous preparations, including purging and pressurization with dry nitrogen, ensuring optimal conditions for their subsequent storage in controlled-access facilities. These meticulous procedures underscored NASA's commitment to safety and precision in every aspect of the Gemini program.

On April 20, a crucial milestone in the Gemini program was reached with the tanking test of Gemini Launch Vehicle (GLV) 9. This test, essential for evaluating fueling procedures, set the stage for meticulous checks and preparations over the following weeks.

Immediately following the tanking test, attention turned to the spacecraft's systems.

From April 21 to 22, rigorous retesting of the spacecraft computer and extravehicular systems was conducted, ensuring their readiness for the demanding mission ahead. Concurrently, pyrotechnics crucial to mission success were meticulously installed in the spacecraft on April 25.

At the end of April, intensive efforts were made to finalize spacecraft readiness. From April 27 to 28, comprehensive spacecraft final systems tests were executed, scrutinizing every subsystem for optimal functionality. Crew stowage arrangements were meticulously reviewed on April 29, ensuring all essential supplies and equipment were securely stored for the mission.

The astronaut maneuvering unit, a pivotal component for extravehicular activities, underwent meticulous reverification from April 30 to May 2, ensuring it would perform flawlessly in the vacuum of space.

May 3 marked a critical operational test as the spacecraft and launch vehicle were temporarily mated for an erector-cycling test, validating their integration and operational readiness. Subsequent revalidation of GLV systems paved the way for the Simultaneous Launch Demonstration (SLD), a pivotal event scheduled for May 10.

As preparations intensified, spacecraft extravehicular equipment underwent meticulous rework and revalidation to ensure seamless operation during the mission. On May 8, the culmination of weeks of preparation arrived as the spacecraft and GLV were successfully mated for flight, marking a significant step toward the upcoming mission.

Announcement of Gemini X, XI, and XII Launch Dates

On April 22, a pivotal announcement from Gemini Program Manager Charles W. Mathews set the trajectory for upcoming missions in the Gemini program. Mathews confirmed tentative launch dates for three significant missions, providing a glimpse into the future of American space exploration.

Gemini X was scheduled for launch on July 18, marking a crucial step in advancing capabilities for manned spaceflight. Following closely, Gemini XI was slated for September 7, emphasizing the program's rapid cadence and ambitious objectives. Lastly, Gemini XII was set for October 31, 1966, highlighting meticulous planning to achieve mission milestones within a tight timeframe.

On May 2, a significant milestone was reached in the Gemini program as Gemini Agena target vehicle 5004 was transferred to Complex 14 and meticulously mated to Atlas target launch vehicle 5303. This crucial step began a series of tests and demonstrations to ensure mission readiness.

By May 6, the joint effort culminated in completing the Joint Flight Acceptance Composite Test, a comprehensive evaluation validating the integrated functionality of both the Gemini Agena target vehicle and the Atlas target launch vehicle. This test underscored the seamless coordination required for successful orbital operations.

The culmination of these preparations came on May 10 with the Simultaneous Launch Demonstration, a pivotal event validating the synchronized launch procedures and operational readiness of both vehicles. This demonstration served as a final confirmation of readiness before the mission's actual launch.

On May 3, Lockheed achieved a significant milestone in the Gemini program by completing the Combined Systems Acceptance Test on Gemini Agena target vehicle 5005. This rigorous test occurred at Test Complex C-10 in Sunnyvale, marking a crucial step toward ensuring the vehicle's operational readiness.

The Air Force formally accepted Gemini Agena target vehicle 5005 after successful testing on May 14. This acceptance underscored Lockheed's adherence to stringent quality standards and readiness criteria mandated for space missions.

By May 16, the vehicle was delivered to the Eastern Test Range, where it would undergo final preparations for its upcoming mission. These meticulous steps in testing, acceptance, and delivery highlighted the precision and dedication of Lockheed and the broader Gemini program team in advancing capabilities for orbital rendezvous and docking missions.

On May 8, Lockheed embarked on a pivotal initiative within the Gemini program by establishing a dedicated task force to refurbish the Gemini Agena Target Vehicle (GATV) 5001. This strategic decision underscored Lockheed's commitment to ensuring the vehicle's readiness for its crucial role in the upcoming Gemini XII mission.

Central to this effort was the announcement of the GATV 5001 Reassembly Plan, meticulously designed to guide the vehicle's refurbishment process. The plan outlined a detailed operational baseline and provided stringent guidelines for reassembling the vehicle, which had been completely disassembled down to the level of riveted or welded parts.

The task force's primary objective was clear: to restore GATV 5001 to a flightworthy condition promptly and economically, aligning with the stringent timelines of the Gemini XII mission. Scheduled for acceptance on September 20, GATV 5001 represented a vital component in advancing the capabilities of orbital rendezvous and docking.

On May 13, McDonnell achieved a significant milestone in the Gemini program by delivering Gemini spacecraft No. 10 to Cape Kennedy. This marked the beginning of a meticulous series of preparations to ensure the spacecraft's readiness for its upcoming mission.

By May 18, the installation of crucial fuel cells was completed, laying the groundwork for subsequent operational checks and tests. The integration of pyrotechnics into the spacecraft followed on May 25, essential for critical mission events such as separation maneuvers and docking procedures.

Preparations intensified with the completion of Plan X testing on June 1. This comprehensive evaluation ensured all systems and subsystems functioned seamlessly under simulated mission conditions. These tests were pivotal in validating the spacecraft's readiness for the challenges of spaceflight.

On June 3, the spacecraft was relocated to Merritt Island Launch Area, marking the final phase of preparations before the mission. This relocation underscored the meticulous planning and coordination required to execute a successful launch and mission in the demanding space environment.

Incident and Investigation of Gemini IX Launch Postponement

On May 17, the scheduled launch of Gemini IX faced a critical setback when target launch vehicle 5303 malfunctioned. Consequently, Gemini Agena target vehicle 5004 failed to achieve orbit, leading to an extensive investigation into the incident.

Initially, the launch proceeded normally until approximately 120 seconds after liftoff, just 10 seconds before booster engine cutoff. At this juncture, booster engine No. 2 unexpectedly gimbaled to a full pitchdown position. Efforts to automatically correct this deviation proved ineffective.

Stabilization was eventually achieved after booster separation but not before the vehicle executed a drastic 216-degree pitchdown maneuver. During this unplanned maneuver, the vehicle inadvertently pointed toward Cape Kennedy, ascending at an angle of about 13 degrees above the horizontal.

Compounding the issue, ground guidance systems were lost temporarily, further complicating the trajectory adjustments. Despite these challenges, the vehicle continued through its sequences, including the vernier engine cutoff, albeit on a significantly altered trajectory.

Ultimately, while the Agena separated as planned, it could not achieve the intended orbit and descended into the Atlantic Ocean approximately 90 miles off the Florida coast, approximately seven and a half minutes after launch.

Subsequent investigation revealed that the root cause of the failure was traced to a short in the servo control circuit. This critical finding underscored the importance of rigorous system checks and redundancy in ensuring the reliability of space missions.

Following the cancellation of the Gemini IX mission on May 17 due to technical issues, immediate recycling operations were initiated to prepare for the redesigned mission, now designated Gemini IX-A. These operations encompassed meticulous procedures to ensure the readiness of the spacecraft and the launch vehicle.

Propellants were carefully unloaded, and ordnance, along with pyrotechnics, were methodically removed from both the launch vehicle and the spacecraft on May 17. The following day, on May 18, the spacecraft and launch vehicle were demated to facilitate detailed inspections and servicing.

After thorough checks and necessary maintenance, the spacecraft and launch vehicle were remated on May 24. This crucial step was followed by the Electrical Interface Integrated Validation on May 24, ensuring all systems were fully integrated and operational.

On May 26, the comprehensive Simulated Flight Test was conducted, marking the final retesting phase to confirm readiness for the redesigned Gemini IX-A mission. This test validated mission procedures and ensured all systems were prepared to meet the demands of spaceflight.

On May 18, in response to the failure of Atlas target launch vehicle (TLV) 5303 and the subsequent loss of Gemini Agena target vehicle 5004, NASA made the pivotal decision to launch the augmented target docking adapter (ATDA). This decision initiated a rapid series of modifications and preparations to ready TLV-5304 as the launch vehicle for the ATDA mission.

Initially designed for placing an Agena into a specified coast ellipse, the Atlas standard launch vehicle (SLV-3) required substantial modifications for the ATDA mission. Unlike the standard mission, the ATDA required the SLV-3 to place the target directly into an Earth orbit via direct ascent, necessitating numerous adaptations to the launch vehicle.

Anticipating such scenarios, preparations for the ATDA program began following the Agena failure on October 25, 1965. By March 1, 1966, ATDA modification kits were strategically positioned at Cape Kennedy, ready to be deployed within 18 days of approval. Remarkably, these modifications were completed in just 14 days, underscoring NASA's agility and preparedness in response to mission contingencies.

Following the completion of modifications on May 20, TLV-5304 was erected at Complex 14 on May 21. Subsequent steps included mating TLV and ATDA on May 25, swiftly followed by comprehensive launch preparations culminating by May 30.

On May 25, a crucial step in the Gemini program was achieved as Gemini Agena target vehicle 5005 was successfully mated to the target docking adapter (TDA) in Hangar E at Cape Kennedy. This milestone marked the beginning of meticulous preparations to ensure the integration's operational readiness for upcoming mission objectives.

McDonnell had delivered the TDA earlier, on May 4, paving the way for the seamless integration process. Following the mating, comprehensive interface functional tests were conducted from May 25 to May 27. These tests were essential to verify the seamless communication and operational compatibility between Gemini Agena target vehicle 5005 and the TDA.

Subsequently, preparations commenced for Plan X testing with spacecraft No. 10 at Merritt Island Launch Area. This critical testing phase aimed to validate the integrated functionality of the spacecraft, target vehicle, and docking adapter under simulated mission conditions.

On June 1, at 10:00 a.m. EST, the augmented target docking adapter (ATDA) was successfully launched from Complex 14. This marked a significant achievement as the ATDA promptly achieved a near-circular orbit, with an apogee of 161.5 nautical miles and a perigee of 158.5 nautical miles.

However, the scheduled launch of Gemini IX-A, which was set to follow the ATDA launch, encountered an unforeseen setback. One hour and 40 minutes later, the launch was postponed due to a ground equipment failure. This failure hindered the transfer of crucial updating information from the Cape Kennedy Mission Control Center to the spacecraft computer.

In response to this delay, a meticulously prepared 48-hour recycling plan was swiftly implemented. This plan addressed the technical issue and ensured that all systems were operating optimally for the next launch attempt.

Subsequently, the Gemini IX-A mission was rescheduled for launch on June 3, allowing sufficient time for the necessary adjustments and verifications to be completed. This incident underscored the meticulous planning and readiness protocols essential for the success of manned space missions, emphasizing NASA's commitment to safety and operational excellence in space exploration endeavors.

On June 1, Gemini Agena target vehicle 5005 underwent preliminary testing at Hanger E, Eastern Test Range, marking a critical phase in its preparation for upcoming missions. Subsequently, the vehicle was relocated to Merritt Island Launch Area to conduct Plan X tests with spacecraft No. 10.

Originally scheduled for May 23, the Plan X tests were rescheduled to June 2-3 to accommodate operational adjustments. To mitigate scheduling impacts, several activities were typically conducted after Plan X was advanced, including the fit check and alignment of secondary propulsion system (SPS) modules, SPS heatshield fit check, and booster adapter fit check.

Despite these efforts, further rescheduling of the vehicle work plan ensued, delaying the commencement of Plan X until June 7. The primary cause for these adjustments was the postponement of the Gemini IX-A launch, necessitating flexible adaptation of operational timelines.

Upon successfully completing Plan X on June 8, Gemini Agena target vehicle 5005 was returned to Hanger E for subsequent systems verification tests, which commenced on June 9. These tests were integral in ensuring the vehicle's readiness and operational capability for future mission objectives.

Chapter 12 -Rendezvous and EVA

Gemini IX-A, NASA's seventh manned mission and third rendezvous mission within the Gemini program, lifted off from Complex 19 at 8:39 a.m. EST. Commanded by Astronaut Thomas P. Stafford with Astronaut Eugene A. Cernan as pilot, the mission aimed primarily to rendezvous and dock with the Augmented Target Docking Adapter (ATDA) and to conduct extravehicular activities (EVA), marking crucial steps in advancing space exploration capabilities.

Gemini IX crew Thomas Stafford and Eugene Cernan. NASA

Early in the mission, the crew achieved rendezvous during the third revolution, a secondary objective. However, attempts to dock were thwarted when the ATDA shroud failed to separate, preventing both primary and secondary docking objectives. Despite this setback, Stafford and Cernan completed other secondary objectives, including an equiperiod rendezvous using onboard optical techniques and a rendezvous from above, simulating future Apollo missions.

EVA plans were initially postponed due to crew fatigue, with the second day dedicated to experiments. Finally, at 49 hours 23 minutes into the mission, Cernan embarked on the EVA. While successful overall, challenges arose when Cernan's visor fogged up, impeding the evaluation of the Astronaut Maneuvering Unit (AMU). This issue was traced to excess moisture in the extravehicular life support system, complicating tasks during the spacewalk.

The augmented target docking adapter with shroud partly open and still attached, as seen from the Gemini IX-A spacecraft in orbit. Shroud's failure to separate precluded docking. (NASA Photo No. 66-H-725, released June 7, 1966.)

Following the EVA, the crew focused on experiments, preparing for retrofire during the 45th revolution. The spacecraft landed remarkably close to the primary recovery ship, the aircraft carrier Wasp, and was swiftly recovered just 53 minutes after landing.

Despite achieving partial success with its objectives, Gemini IX-A underscored the complexities of space operations and provided invaluable lessons for future missions. The mission's challenges with docking and EVA highlighted the need for continued advancements in space technology and astronaut training as NASA pushed forward toward its lunar exploration goals.

Thomas P. Stafford: From Backup to Prime Crew

Thomas P. Stafford's journey in NASA's space program began in September 1962 when he was selected as part of the second group of astronauts for Projects Gemini and Apollo. His career took a pivotal turn following the tragic loss of Elliot M. See Jr. and Charles A. Bassett II, the original prime crew of Gemini IX.

In December 1965, Stafford piloted Gemini VI, achieving the historic first rendezvous in space. His expertise and contributions were instrumental in developing and proving the essential techniques for space rendezvous, marking a significant advancement in space exploration capabilities.

Eugene A. Cernan: A Legacy of Exploration

Eugene A. Cernan was initially the backup pilot for Gemini IX and was elevated to the prime crew after losing the original crew. A Captain in the U.S. Navy and one of the fourteen astronauts selected by NASA in October 1963, Cernan left an indelible mark on human exploration.

He made history by flying into space three times, including two missions to the moon. Cernan became the second American to perform a spacewalk, traversing the globe twice in just over two and a half hours. As commander of Apollo 17, the final mission to the moon, he was the last human to walk on its surface, leaving his footprints as a testament to human ingenuity and courage.

On June 6, Gemini Agena target vehicle 5006 reached a significant milestone as it completed modification and final assembly, marking a crucial step in its preparation for upcoming missions. Subsequently, the vehicle was transferred to Vehicle Systems Test (VST) at Sunnyvale for comprehensive testing and validation.

Upon arrival at VST, testing commenced promptly despite the initial lack of the flight control electronics package and guidance module. The guidance module was received on June 7, swiftly followed by the flight control electronics package on June 9, enabling full-scale testing to proceed without delay.

Preliminary Vehicle Systems Test (VST) was successfully concluded on June 17, underscoring the vehicle's readiness for further evaluation. Subsequently, the Air Force Plant Representative Office at Sunnyvale authorized the commencement of final acceptance tests, scheduled to begin on June 20. These tests represented the last phase of rigorous evaluations to ensure the vehicle's operational integrity and readiness for its designated mission objectives.

TLV 5305: Preparing for Gemini's Flight

On June 7, 1965, a crucial acceptance meeting convened at General Dynamics/Convair in San Diego, marking a pivotal moment in Project Gemini's timeline. This meeting centered on the final approval of the target launch vehicle (TLV) 5305, a meticulously engineered craft explicitly tailored for the Gemini missions.

Initially completing its systems test on March 25, TLV 5305 underwent extensive refinements over the subsequent two months to align with the latest flight configurations. From May 26 to June 1, rigorous systems tests were rerun, ensuring every component met exacting standards. A composite test conducted on June 2-3 further validated the vehicle's readiness for the demanding journey ahead.

Following meticulous scrutiny and comprehensive testing, TLV 5305 was deemed flight-ready. A notable milestone was achieved as the vehicle departed for Cape Kennedy on June 9, transported swiftly by air—a first for such a craft destined for the Cape. Remarkably, TLV 5305 arrived the very same day, underscoring the precision and efficiency of Project Gemini's logistical operations.

On June 7, 1965, a pivotal operation unfolded at Complex 19 as Gemini launch vehicle 10 was carefully removed from storage and meticulously erected. This marked the commencement of a meticulously orchestrated series of events essential for its readiness.

By June 9, the vehicle's umbilicals were securely connected, and power surged through its intricate systems, initiating a critical testing and verification phase. The Subsystems Reverification Tests (SSRT) commenced immediately, scrutinizing each subsystem precisely to ensure optimal functionality and reliability.

Over the following week, from June 9 to June 16, engineers and technicians meticulously verified and validated each component during the SSRT phase. The culmination of this rigorous testing regimen occurred on June 17 with the execution of the Prespacecraft Mate Verification Combined Systems Test—a comprehensive assessment designed to simulate the operational conditions the spacecraft would face during its forthcoming mission.

On June 9, 1965, a pivotal operation unfolded as Gemini spacecraft No. 10 was meticulously transported to Complex 19 and carefully hoisted atop its designated launch vehicle. This began an intricate series of preparatory steps crucial for its upcoming mission.

By June 13, the spacecraft's cabling for testing purposes was meticulously completed, ensuring all connections were secure and ready for rigorous evaluation. Over the ensuing days, meticulous checks and verifications continued. By June 16, key milestones were achieved with the completion of premate verification procedures, alongside the critical activation and subsequent deactivation of its fuel cells—an essential step in ensuring the spacecraft's operational readiness.

The meticulous preparation culminated on June 17 with the seamless integration of Gemini spacecraft No. 10 with its launch vehicle, setting the stage for comprehensive integrated tests. This phase aimed to validate the seamless interaction between the spacecraft and its propulsion system, ensuring optimal performance and reliability during the mission ahead.

On June 9, 1966, a pivotal chapter unfolded in the history of the Gemini program as the launch vehicle acceptance test for Gemini Launch Vehicle (GLV) 11 commenced. This marked the beginning of a meticulous process aimed at ensuring the vehicle's readiness for its upcoming mission.

Following the initial test, the vehicle acceptance team convened on June 20 to scrutinize every aspect of GLV-11's performance and functionality. By June 24, their rigorous evaluations culminated in the formal acceptance of the vehicle, signifying its readiness for deployment.

The first and second stages of Gemini launch vehicle 11 arriving at complex 19. (NASA Photo No. 66-H-1045, released July 23, 1966.)

On June 29, GLV-11 was carefully de-erected, setting the stage for its next preparation phase. A significant milestone followed on July 11, as the United States Air Force formally accepted the vehicle, underscoring its military significance and operational capabilities. On the same day, Stage I of the vehicle was swiftly transported by air to Cape Kennedy, followed quickly by Stage II on July 13.

Upon arrival at Hanger U, both stages underwent meticulous procedures, including tank purging and pressurization, ensuring optimal conditions for their subsequent integration and deployment. The stages remained in controlled-access storage throughout this period, safeguarding their integrity until the launch pad revalidation process commenced.

The final validation step occurred on July 21, following the successful launch of Gemini X, ensuring that all systems and components were primed for GLV-11's imminent mission. This sequence of events not only highlighted the meticulous planning and execution prowess of NASA and its partners but also underscored America's commitment to advancing manned space exploration during the Gemini era.

On June 13, 1965, a pivotal phase commenced in the Gemini program as Combined Interface Tests (CIT) of the Gemini Agena target vehicle (GATV) 5005 began. This marked the initiation of rigorous testing procedures crucial for ensuring the vehicle's readiness and compatibility for upcoming missions.

The CIT proceeded smoothly over nine days, from June 13 to June 22. Metricious checks confirmed seamless interfaces between the Gemini spacecraft and the Agena target vehicle. Notably, no significant anomalies were detected during this critical testing phase, underscoring GATV 5005's robust design and meticulous preparation.

After completing CIT, attention turned to the primary and secondary propulsion systems. By June 30, comprehensive functional checks of these propulsion systems were successfully conducted, further validating their reliability and operational readiness for the demanding tasks ahead.

With all systems verified and confirmed, GATV 5005 was subsequently moved to Complex 14, positioning it for the next integration and mission preparation stages. This period of meticulous testing and preparation highlighted the technical precision and thoroughness of NASA's operations. It set a solid foundation for the successful execution of Gemini missions, advancing America's capabilities in manned space exploration during this pivotal era.

Atlas 5305: Preparing for Gemini X

On June 15, 1966, a critical phase began at Launch Complex 14 as Atlas 5305, designated as the target launch vehicle for Gemini X, was carefully erected. This marked the start of a meticulous series of preparations for successfully executing the upcoming mission.

Two days later, on June 17, electrical power was applied to Atlas 5305, initiating the crucial phase of subsystem testing. Over the next ten days, engineers and technicians diligently evaluated each subsystem to ensure flawless performance during the mission.

However, a significant issue arose during the meticulous checks of the propellant system. A leak was discovered in the fuel start tank, necessitating immediate attention and repair. Rectifying the leak required intricate procedures, including removing the sustainer engine and the fuel tank apex cone to gain access to the affected area.

This unexpected challenge underscored the meticulous nature of spaceflight preparations, where even minor issues could have significant implications. Despite this setback, the launch team's dedication and expertise ensured that repairs were conducted swiftly and effectively, minimizing any potential delays to the mission timeline.

Announcement of Gemini XII Crew and Mission Schedule

On June 17, 1966, a significant announcement marked the culmination of the Gemini IX-A mission's post-launch activities. During a press conference attended by Astronauts Thomas P. Stafford and Eugene A. Cernan, Director Robert R. Gilruth of the Manned Spacecraft Center revealed pivotal details regarding the next and final Gemini mission, Gemini XII.

Director Gilruth announced that Astronauts James A. Lovell, Jr., and Edwin E. Aldrin, Jr., would serve as the prime crew for Gemini XII, the last scheduled mission of the Gemini program. This selection underscored their expertise and readiness to undertake the challenges of this historic mission.

Additionally, Astronauts L. Gordon Cooper, Jr., and Eugene A. Cernan were designated as the backup crew, standing ready to support the prime crew and step in if needed.

The mission was scheduled to launch in late October or early November, setting a definitive timeline for the culmination of the Gemini program—a crucial stepping stone in America's ambitious journey towards lunar exploration and beyond.

On June 19, 1966, NASA made a significant announcement outlining the details of the upcoming Gemini X mission, setting the stage for another milestone in America's manned space exploration efforts.

The mission, scheduled for no earlier than July 18, would be led by Command Pilot John W. Young and Pilot Michael Collins, appointed as the prime crew. Command Pilot Alan L. Bean and Pilot Clifton C. Williams served as their backups, highlighting NASA's meticulous crew rotation and readiness protocols.

Central to the mission objectives were ambitious plans for rendezvous, docking, and extravehicular activity (EVA), showcasing advancements in orbital maneuvering and astronaut capabilities. A key mission highlight would involve the Gemini X spacecraft rendezvousing and docking with an Agena target vehicle, slated for launch on the same day—an intricate maneuver critical for future lunar mission planning.

If feasible, Gemini X aimed to rendezvous with the Agena target vehicle launched during the March 16 Gemini VIII mission, demonstrating NASA's commitment to refining and mastering complex spaceflight techniques.

June 20, 1965, marked a critical milestone at Complex 19 as Gemini launch vehicle 10 and spacecraft No. 10 were meticulously electrically mated. This pivotal step initiated a series of rigorous tests and validations to ensure the seamless operation of the spacecraft and launch vehicle as a unified system.

Over the following days, from June 20 to June 21, engineers conducted the Electrical Interface Integrated Validation, meticulously verifying the electrical connections and interfaces between the spacecraft and launch vehicle. This comprehensive validation ensured that all systems communicated effectively and operated in unison.

On June 23, following a thorough review of data and system performance from the previous tests, the Joint Combined Systems Test was executed. This test aimed to simulate and validate the integrated functionality of all systems, including guidance, control, and communication interfaces crucial for mission success.

June 24, 1965, marked a pivotal day in the lead-up to the Gemini launch as crucial tests and preparations unfolded at the launch site.

The day commenced with the tanking test of Gemini launch vehicle (GLV) 10, a critical procedure to validate the integrity and readiness of its propellant systems. This test ensured that the vehicle could safely store and handle the necessary fuel for the upcoming mission.

Following the tanking test, meticulous post-tanking cleanup and systems testing procedures were conducted on the GLV. Simultaneously, spacecraft No. 10 underwent servicing of its hypergolic systems from June 27 to June 28, essential for its propulsion and maneuvering capabilities in space.

From June 28 to July 1, Final Systems Tests were meticulously executed on spacecraft No. 10, scrutinizing every aspect of its operational readiness. Crew stowage evaluations and checks of the extravehicular life support system further ensured that all onboard systems were primed for mission success by July 1.

On June 27, 1966, a significant milestone was achieved at Sunnyvale as the final acceptance test of Gemini Agena target vehicle 5006 was completed. This marked a crucial step in preparing the vehicle for its pivotal role in upcoming Gemini missions.

Following the completion of testing, meticulous preparations ensued. By July 6, the vehicle was disconnected from the test complex, paving the way for its formal acceptance by the Air Force on July 13—two days ahead of schedule. This timely acceptance underscored the vehicle's robust design and readiness for deployment.

However, adverse wind conditions delayed the planned shipment of the vehicle to the Eastern Test Range (ETR) until July 14. Despite this setback, the vehicle arrived at ETR in the early morning of July 15, ensuring it was positioned for its crucial role in supporting upcoming Gemini missions.

July 1, 1966, marked a pivotal moment in the Gemini program as Gemini Agena target vehicle 5005 underwent crucial integration and testing procedures at Complex 14.

On this day, Gemini Agena target vehicle 5005 was transferred to Complex 14 and meticulously mated to target launch vehicle 5305. This integration process was essential to ensure seamless compatibility and operational readiness for upcoming missions.

Over the following days, from July 1 to July 8, engineers and technicians conducted the Joint Flight Acceptance Composite Test. This comprehensive test validated the integrated functionality of the Agena target vehicle and the launch vehicle, ensuring they could work together effectively during mission operations.

Further testing and preparations continued at Complex 14, with systems tests completed by July 12. The culmination of these tests included the Simultaneous Launch Demonstration, a critical exercise to simulate launch conditions and verify the readiness of all systems for operational deployment.

On July 5, a significant milestone was achieved with the mechanical mating of the spacecraft and GLV, signifying their integration into a unified unit. The erector, responsible for positioning the spacecraft atop the launch vehicle, underwent cycling tests to confirm its operational reliability.

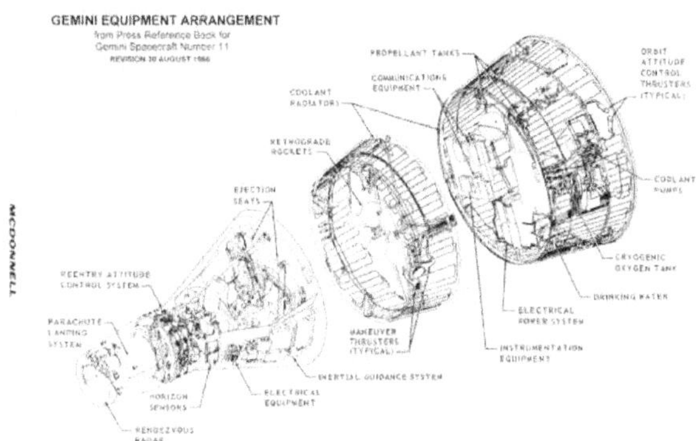

The Gemini Atlas-Agena target vehicle undergoing systems tests at complex 14 prior to the Gemini X mission. (NASA Photo No. 66-H-989, released July 18, 1966.)

The meticulous preparation continued on July 6, retesting the electrical interfaces between the spacecraft and GLV, crucial for seamless communication and control during the mission.

These rigorous preparations culminated with the Simultaneous Launch Demonstration on July 12 and the Simulated Flight Test on July 13. These final tests aimed to simulate launch conditions and validate the integrated functionality of all systems, ensuring readiness for the upcoming mission.

Delivery and Preparation of Gemini Spacecraft No. 11

On July 7, 1966, McDonnell Aircraft Corporation achieved a significant milestone in the Gemini program by delivering spacecraft No. 11 to Cape Kennedy. This event marked a crucial step in preparing for the upcoming mission, demonstrating the meticulous coordination and logistics involved in manned space missions.

Upon arrival at Cape Kennedy, spacecraft No. 11 underwent essential preparatory tasks. Fuel and pyrotechnics were installed, followed by preliminary checks to ensure all systems were functioning correctly and ready for the demanding mission ahead.

Subsequently, the spacecraft was transported to the Merritt Island Launch Area, where it would undergo Plan X integrated tests with the target vehicle on July 25. These integrated tests aimed to validate the spacecraft's seamless operation and compatibility with the target vehicle, Which was essential for mission success.

On July 15, 1966, a pivotal event took place at San Diego with the acceptance meeting for Atlas 5306, designated as the target launch vehicle for Gemini XI. This marked the commencement of critical preparations for an upcoming mission in NASA's Gemini program.

Following meticulous evaluations and testing, the final acceptance of Atlas 5306 was successfully concluded on July 18. This rigorous process ensured that the launch vehicle met stringent performance and safety criteria necessary for its vital role in supporting manned space missions.

With acceptance confirmed, Atlas 5306 was swiftly prepared for deployment. On the same day, July 18, the launch vehicle was expedited by air to Cape Kennedy, underscoring the urgency and precision of mission logistics. It arrived at its destination on July 19, ready to undergo final preparations and integration for the impending Gemini XI mission.

McDonnell personnel bolting the Gemini XI spacecraft to a support ring for boresighting in the Pyrotechnic Installation Building, Merritt Island. (NASA Photo S-66-47635, July 2, 1966.)

July 18, 1966, marked a significant milestone in the Gemini program as Gemini Agena target vehicle (GATV) 5006 was successfully mated to target docking adapter (TDA) 6 at Cape Kennedy. This crucial integration step paved the way for complex testing and preparations essential for upcoming manned space missions.

On July 7, McDonnell Aircraft Corporation delivered TDA-6 to Cape Kennedy, highlighting the coordinated efforts in logistics and preparation for space missions. The delivery ensured all necessary components were in place for the integration and testing phase.

Gemini X: Pioneering Rendezvous and Spacewalks

On July 18, the Gemini X mission commenced with the launch of the Gemini Atlas-Agena target vehicle from complex 14 at 3:40 p.m. EST. Shortly after, the Gemini

space vehicle, piloted by command astronaut John W. Young and astronaut Michael Collins, launched from Complex 19 at 5:20 p.m. The Gemini Agena target vehicle (GATV) achieved a near-circular orbit while Spacecraft No. 10 entered an elliptical orbit, maintaining a slant range close to the intended 1000 miles.

A pivotal moment came during the fourth revolution when the spacecraft successfully rendezvoused with the GATV at 5 hours 23 minutes into the mission. Docking followed approximately 30 minutes later, although it required more propellant than anticipated, necessitating an altered flight plan.

The spacecraft undocked successfully at 44 hours 40 minutes ground elapsed time, completing its second rendezvous using onboard thrusters three hours later. At 48 hours 42 minutes into the mission, a 39-minute umbilical EVA commenced, during which a micrometeorite collection package from the Gemini VIII Agena was retrieved. Following this, extraneous equipment was jettisoned before preparations for reentry.

After three hours of stationkeeping, the spacecraft separated from the GATV, followed by a true anomaly-adjust maneuver at 51 hours 39 minutes ground elapsed time to fine-tune reentry parameters. Retrofire occurred during the 43rd revolution at 70 hours 10 minutes after liftoff. The spacecraft landed just three miles from its planned point within view of the prime recovery ship, the aircraft carrier Guadalcanal, at 4:07 p.m. on July 21.

Astronauts John Young and Michael Collins launched on the Gemini X mission on July 18, 1966, landing nearly 3 days later on July 21.

Despite challenges with fuel consumption altering the mission timeline, Gemini X successfully achieved critical rendezvous and EVA objectives, advancing NASA's capabilities in space exploration and paving the way for future lunar missions.

On July 21st, immediately following the reentry of spacecraft No. 10, attention turned to the Gemini Agena target vehicle (GATV) 5005, which continued to demonstrate its versatility and capability in space. Under meticulous ground control, GATV 5005 executed three precise orbital maneuvers, showcasing its agility and the precision of ground-based operations.

Initially, the primary propulsion system (PPS) was engaged to adjust GATV 5005's orbit, transitioning it to a 750.5 by 208.6 nautical mile path. This maneuver aimed to study the thermal effects on the vehicle at higher altitudes, crucial data for future space missions. Interestingly, the temperature readings gathered showed no significant variance compared to those recorded at lower orbits, affirming the vehicle's robust design and adaptability.

Following this phase, another firing of the PPS circularized the GATV's orbit, ensuring stability and optimal operational conditions. A subsequent maneuver utilizing the secondary propulsion system Unit II then positioned the GATV into a 190-navigational-mile circular orbit. This strategic placement positioned GATV 5005 as a potential rendezvous target for the upcoming Gemini XI mission, highlighting its role in ongoing mission planning and preparation.

Throughout its orbital tenure, GATV 5005 received and executed an impressive array of commands—1700 in total. Ground controllers issued 1350 commands, while spacecraft No. 10 contributed 350 commands, underscoring the collaborative effort and meticulous coordination essential for successful space operations.

On July 21st, Gemini Agena target vehicle 5001 marked a significant milestone in its operational journey as it transitioned to systems test complex C-10 in Sunnyvale. This relocation followed an extensive refurbishment to ensure the vehicle's readiness for future missions. Despite these efforts, the vehicle remained incomplete, lacking several crucial pieces of equipment essential for its full operational capability.

The transfer to complex C-10 underscored ongoing efforts to optimize and prepare the vehicle for its next testing and deployment phase. This strategic move highlighted the dedication to maintaining and enhancing space exploration technologies and emphasized the meticulous attention to detail required in aerospace engineering and operations.

As preparations continued, the focus remained on completing the necessary equipment installations to bolster Gemini Agena target vehicle 5001's functionality and reliability. This ongoing refurbishment ensured the vehicle met stringent operational standards and contributed effectively to future space missions.

Following successful testing, GATV-5006 was relocated to the Merritt Island Launch Area on July 22. This move facilitated integrated tests involving spacecraft No. 11 and extravehicular equipment, crucial for validating all systems' operational readiness and functionality under simulated mission conditions.

On July 22nd, preparations for Gemini launch vehicle 11 began with its removal from storage and careful erection at complex 19. This marked a crucial step in readying the vehicle for its upcoming mission, highlighting the meticulous planning and logistical coordination inherent in space launch operations.

Following its erection, the vehicle underwent thorough inspections to ensure all components were in optimal condition. Umbilicals were meticulously connected, establishing vital communication and power connections for the vehicle's functionality and support systems.

On July 27th, power was applied to the vehicle, initiating the next phase of preparations. The Subsystems Reverification Tests (SSRT) commenced, focusing on verifying the functionality and performance of various subsystems critical to mission success. From July 27th to August 4th, these tests involved rigorous evaluations to validate the vehicle's readiness for spaceflight.

Concluding the SSRT phase on August 4th, the Prespacecraft Mate Verification Combined Systems Test was conducted the following day. This comprehensive test integrated and verified the operation of all systems in conjunction with the spacecraft, ensuring seamless compatibility and operational readiness ahead of launch.

On July 26th, following the completion of Plan X tests at Merritt Island Launch Area, Gemini target vehicle (GATV) 5006 embarked on the next phase of its journey. It returned to Hanger E, where it commenced systems verification tests, marking a pivotal stage in its preparation for an upcoming mission.

The rigorous testing continued with Combined Interface Tests commencing on August 4th and concluding on August 12th. These tests focused on ensuring seamless integration and communication between GATV 5006 and associated systems, critical for operational cohesion during the mission.

Subsequently, on August 13th, functional tests of the Primary Propulsion System (PPS) and Secondary Propulsion System (SPS) were initiated. The meticulous evaluation of the SPS functional tests concluded on August 18th, followed by the installation of SPS modules on August 19th. The PPS functional tests were completed on August 21st, further validating the operational readiness and performance capabilities of GATV 5006.

With these comprehensive tests successfully concluded, GATV 5006 transitioned to complex 14 for the final preparation phase—mating with the Atlas launch vehicle. This critical step represented the culmination of extensive preparations and testing efforts, underscoring the dedication to ensuring mission success and advancing capabilities in space exploration.

Atlas 5306: Preparations for Gemini XI

On July 28th, Atlas 5306, designated as the target launch vehicle (TLV) for Gemini XI, commenced its preparatory phase by being erected at launch complex 14. This marked the beginning of meticulous preparations essential for a successful mission, emphasizing the precision and coordination required in aerospace operations.

Electrical power was applied to Atlas 5306 the following day, initiating critical systems checks and preparatory activities. These initial steps set the stage for subsequent testing and readiness assessments leading up to launch.

As preparations continued, a significant milestone was reached with the dual propellant loading (DPL) scheduled for August 18th. This process encountered challenges, including resolving liquid oxygen leaks that necessitated careful troubleshooting and adjustments. A discrepancy noted in the vernier engine liquid oxygen bleed system during the initial loading prompted a second DPL procedure, successfully completed on August 22nd, ensuring optimal fueling conditions for the mission.

Concurrently, the Booster Flight Acceptance Composite Test, crucial for verifying the readiness and operational integrity of the entire launch system, was successfully conducted on August 19th. This comprehensive test validated the performance capabilities and readiness of Atlas 5306 under simulated flight conditions.

On August 22nd, another pivotal milestone was achieved as Atlas 5306, and the Gemini Agena target vehicle were successfully mated. This critical step in the pre-launch sequence ensured the seamless integration of the launch vehicle and its payload, setting the stage for final checks and preparations leading to launch day.

On July 28th, preparations for Gemini spacecraft No. 11 reached a critical phase as it was transported to complex 19 and meticulously hoisted atop its designated launch vehicle. This marked the beginning of intensive preparations to ensure the spacecraft's readiness for its upcoming mission, highlighting the meticulous planning and coordination essential in space exploration.

Over the following days, essential cabling work was completed by August 1st, establishing crucial connections for communication and operational support systems onboard the spacecraft. Concurrently, from August 1st to August 3rd, the Premate Systems Test was conducted. This comprehensive test validated the integration and functionality of all systems between the spacecraft and its launch vehicle, ensuring seamless operation during the mission.

However, as meticulous checks continued, issues with fuel cell sections were identified on August 4th, revealing high leakage rates. Prompt actions were taken to replace the affected sections, underscoring the proactive approach and stringent quality assurance measures applied to maintain operational integrity and safety.

Following the replacement, rigorous fuel cell activation and deactivation procedures were meticulously executed, culminating in their completion on August 6th. These procedures ensured that the fuel cells were ready to support the spacecraft's power requirements throughout the mission, essential for sustaining critical systems and scientific operations in space.

On July 29th, the launch vehicle acceptance test achieved a critical milestone in preparing the Gemini launch vehicle (GLV) 12. This test marked a pivotal step in verifying GLV-12's readiness and operational capability for its upcoming mission, highlighting the meticulous testing and evaluation processes essential in aerospace engineering.

Following the acceptance test, the launch vehicle acceptance team convened on August 9th to review the test results and assess the vehicle's readiness. After thorough deliberation and validation, GLV-12 was formally accepted on August 12th, affirming its readiness for deployment.

Subsequently, on August 17th, GLV-12 was carefully disassembled and prepared for transport, with Stage I airlifted to Cape Kennedy on the same day. Stage II followed, arriving on September 3rd. Both stages were meticulously placed in controlled access storage within Hanger T, ensuring optimal conditions and security pending the launch of Gemini XI.

During this period, additional preparations included revalidating the launch pad, a critical step completed on September 16th. This process ensured that all systems and infrastructure were aligned and ready to support the upcoming mission, reinforcing operational readiness and safety protocols.

On August 8th, a significant milestone was reached in the preparations for the Gemini mission as Gemini launch vehicle 11 and spacecraft No. 11 were electrically mated at complex 19. This critical step involved establishing essential electrical connections between the launch vehicle and spacecraft, laying the groundwork for integrated operations and mission readiness.

Immediately following the electrical mating, from August 8th to August 9th, the Electrical Interface Integrated Validation and Joint Guidance and Control Test were conducted. These rigorous tests aimed to validate and synchronize the launch vehicle's and spacecraft's electrical interfaces. They also included comprehensive checks of guidance and control systems, ensuring seamless communication and coordination essential for mission success.

Continuing the sequence of tests, the Joint Combined Systems Test took place from August 11th to August 12th. This integrated test session comprehensively evaluated the combined operational systems of both the launch vehicle and spacecraft. It encompassed a range of simulated scenarios and operational conditions to verify performance and readiness under various mission parameters.

On August 15th, preparations for the Gemini mission intensified with the tanking test conducted on the Gemini launch vehicle (GLV) 11. This critical test evaluated the launch vehicle's fueling operations, ensuring that all systems performed flawlessly under operational conditions, Which was essential for mission readiness.

Simultaneously, post-tanking operations for GLV 11 were meticulously executed to confirm the integrity and readiness of the vehicle following fueling procedures, emphasizing safety and operational protocols.

Meanwhile, from August 22nd to August 23rd, Final Systems Tests of Spacecraft No. 11 were conducted. These comprehensive tests thoroughly examined and validated all onboard systems and subsystems of the spacecraft, ensuring optimal performance and functionality during the mission.

On August 24th, another significant milestone was achieved as spacecraft No. 11 and GLV 11 were mechanically mated. This crucial step involved physically integrating the spacecraft with the launch vehicle, setting the stage for subsequent operational checks and validations.

Following the mechanical mating, erector cycling tests were conducted to ensure the proper functioning and reliability of the launch vehicle's erector mechanisms, crucial for the deployment process during launch.

Subsequently, from August 25th to August 29th, the electrical interface between spacecraft No. 11 and GLV 11 was revalidated. This phase focused on verifying and synchronizing electrical connections and interfaces, ensuring seamless communication and operational readiness between the spacecraft and launch vehicle.

As preparations neared completion, the simultaneous launch demonstration was conducted on August 31st, followed by the Simulated Flight Test on September 1st. These final tests simulated critical launch and flight scenarios, validating mission procedures and readiness under realistic conditions and ensuring all systems performed as expected.

On August 16th, Gemini Agena target vehicle 5001 completed its final acceptance testing, marking a significant milestone in its preparation for upcoming missions. This comprehensive testing phase rigorously evaluated all vehicle systems and subsystems to ensure readiness and reliability for its intended operational tasks.

Following the completion of testing, a thorough analysis of test data was undertaken, a process that concluded by August 24th. This analysis provided critical insights into the performance and operational capabilities of the vehicle, confirming its adherence to stringent quality and safety standards.

Subsequently, with the analysis completed and all test objectives met satisfactorily, Gemini Agena target vehicle 5001 was disconnected from the test complex. This step signified the conclusion of testing activities and affirmed the vehicle's readiness for deployment to support upcoming Gemini missions.

On August 22nd, Gemini Agena target vehicle 5006 reached a crucial milestone as it was successfully mated to the target launch vehicle 5306. This pivotal step involved physically integrating the target vehicle with its designated launch vehicle, marking a significant advancement in mission preparations.

Following the integration, on August 26th, the Joint Flight Acceptance Composite Test was conducted. This comprehensive test session aimed to validate the integrated performance and readiness of both the target and the launch vehicles under simulated flight conditions. The test encompassed various operational scenarios to ensure that all systems and subsystems operated harmoniously and reliably.

Continuing with the preparations, the simultaneous launch demonstration was carried out on August 31st. This critical demonstration simulated key launch procedures and sequences, validating mission procedures and readiness under realistic conditions. It provided essential insights into the coordination and integration of systems required for a successful launch.

On September 2nd, Gemini Agena target vehicle 5001 achieved a significant milestone. Following inspection by the vehicle acceptance team, the Air Force formally accepted it. This formal acceptance underscored the successful completion of rigorous testing and evaluation, affirming the vehicle's readiness for operational deployment.

Immediately after acceptance, on September 3rd, Gemini Agena target vehicle 5001 was prepared for transport from Sunnyvale, marking the beginning of its journey to the Eastern Test Range. The meticulous preparations ensured the vehicle was securely packaged and ready for transit.

By September 4th, Gemini Agena target vehicle 5001 arrived at the Eastern Test Range, where it would be positioned and readied for its upcoming missions. This deployment culminated in extensive preparations and testing efforts, highlighting the dedication to ensuring operational readiness and mission success.

Delivery and Integration of Gemini Spacecraft No. 12

On September 6th, McDonnell delivered Gemini spacecraft No. 12 to Cape Kennedy, marking a significant step in preparations for an upcoming mission. This delivery underscored the culmination of manufacturing and assembly processes, ensuring the spacecraft's readiness for integration and testing phases.

Upon arrival at Cape Kennedy, preliminary installations and preparations commenced to set up the spacecraft for integrated operations. These initial tasks focused on ensuring all necessary components and systems were in place and operational, laying the groundwork for subsequent testing and validation activities.

From September 19th to September 20th, the spacecraft was relocated to the Merritt Island Launch Area for integrated tests with the target vehicle. This testing phase aimed to verify the spacecraft's seamless integration and operational compatibility with its designated target vehicle, crucial for mission readiness.

On September 9th, the scheduled launch of Gemini XI faced an unexpected delay when a pinhole leak was detected in the stage I oxidizer tank of the launch vehicle shortly after propellants had been loaded. This critical discovery prompted immediate action to assess and address the issue to ensure the safety and integrity of the mission.

In response to the leak, the launch was postponed, and repair efforts were initiated. Propellants were safely unloaded from the launch vehicle to facilitate access to the affected area. The leak was repaired by plugging it with a sodium silicate solution and applying an aluminum patch to reinforce the repair.

These repair procedures were meticulously executed to restore the launch vehicle's structural integrity and operational readiness. The thoroughness of these efforts underscored the commitment to mission success and safety, ensuring that all systems operated flawlessly during the upcoming launch attempt.

Following the successful repair, the rescheduled launch of Gemini XI was planned for September 10th, allowing sufficient time to verify the integrity of the repair and conduct necessary checks to ensure readiness for launch. This postponement and subsequent repair process exemplified the meticulous planning and preparedness essential in space exploration missions, demonstrating resilience in overcoming challenges to achieve mission objectives.

On September 10th, the scheduled launch of the Atlas-Agena mission was delayed due to apparent issues with the target launch vehicle autopilot system. Initial assessments indicated anomalies likely caused by factors such as propellant sloshing, wind loading, and the sensitivity of the autopilot recorder.

Subsequent analysis determined that these conditions were within the expected parameters and did not necessitate hardware replacement. This conclusion reassured mission planners that the system's functionality remained intact despite the initial concerns.

As a result, the launch was rescheduled for September 12th, allowing sufficient time for final checks and adjustments to ensure optimal conditions for a successful launch attempt. This postponement underscored the meticulous attention to detail and commitment to safety inherent in space missions, ensuring that all systems operated effectively and reliably during critical phases.

Gemini XI: Pushing the Boundaries of Space Exploration

On September 12, the Gemini XI mission began with the launch of the Gemini Atlas-Agena target vehicle from complex 14 at 8:05 a.m., e.s.t. Subsequently, the Gemini space vehicle, piloted by command pilot Astronaut Charles Conrad, Jr., and pilot Astronaut Richard F. Gordon, Jr., was launched from complex 19 at 9:42 a.m. The mission's primary objective was to rendezvous with the Gemini Agena target vehicle (GATV) and dock during the first revolution. This objective was achieved with five maneuvers, completing the rendezvous at 1 hour 25 minutes ground elapsed time, followed by docking nine minutes later.

Secondary objectives included docking practice, extravehicular activity (EVA), conducting 11 experiments, performing docked maneuvers, conducting a tethered vehicle test, demonstrating automatic reentry, and parking the GATV.

All objectives were successfully accomplished except for one experiment - the evaluation of the minimum reaction power tool - which was not performed due to the premature termination of the umbilical EVA. The umbilical EVA began at 24 hours, 2 minutes of ground elapsed time. It ended 33 minutes later when Gordon became fatigued while attaching the tether from the GATV to the spacecraft docking bar.

At 40 hours 30 minutes after liftoff, the GATV primary propulsion system (PPS) was fired to raise the apogee of the docked vehicles to 741 nautical miles for two revolutions. Later, the PPS was fired again, 3 hours 23 minutes later, to reduce the apogee to 164 nautical miles. At 47 hours 7 minutes into the flight, a standup EVA commenced and lasted 2 hours 8 minutes. Following this, the spacecraft was undocked to initiate the tether evaluation.

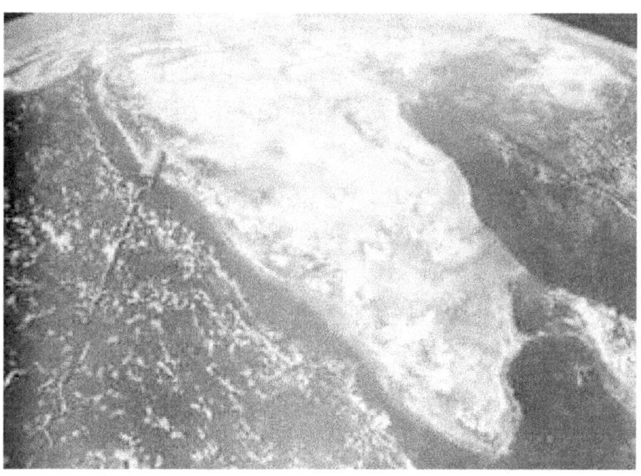

View of India and Ceylon from Gemini XI at 540 nautical miles looking north, with the Bay of Bengal to the right and the Arabian Sea to the left. (NASA Photo No. 66-H-1246 [66-HC-1608], released Sept. 17, 1966.)

Astronaut Richard F. Gordon, Jr., returning to the hatch of Gemini XI after extravehicular activity. (NASA Photo No. 66-H-1249, released Sept. 13, 1966.)

he Gemini XI spacecraft landing approach in the western Atlantic. (NASA Photo No. 66-H-1214, released Sept. 15, 1966.)

At 50 hours 13 minutes ground elapsed time, the crew began rotational maneuvers with initial oscillations damping out and stabilizing the combination after about 20 minutes. The tethered configuration became very stable, allowing the rotational rate to increase. At approximately 53 hours into the

277

mission, the crew released the tether, separated from the GATV, and maneuvered the spacecraft to an identical orbit with the target vehicle.

Despite a fuel cell stack failure at 54 hours 31 minutes, the remaining five stacks shared the load and operated satisfactorily. A rendezvous was completed at 66 hours 40 minutes ground elapsed time, after which the crew prepared for reentry. The spacecraft landed within three miles of the planned landing point at 71 hours 17 minutes after liftoff. The crew was retrieved by helicopter, and the spacecraft was brought aboard the prime recovery ship, the aircraft carrier Guam, about an hour after landing, marking the successful completion of the Gemini XI mission.

Masticating preparations were underway at Cape Kennedy in the lead-up to the Gemini XI mission. The Gemini Agena target vehicle 5001 was carefully mated to the target docking adapter (TDA) 7A. This critical step, completed on September 12th, ensured the seamless integration essential for the upcoming mission's success.

The Gemini-11 prime crew, astronauts Richard F. Gordon Jr. (left), pilot, and Charles Conrad Jr., command pilot, relax on deck of the NASA Motor Vessel Retriever after suiting up for water egress training in the Gulf of Mexico on July 13, 1966. NASA

In August, McDonnell delivered TDA 7A to Cape Kennedy, setting the stage for rigorous tests and verifications. From September 13th to 15th, functional verification tests were conducted to ensure all systems were operating at peak performance. Following these tests, the vehicle was relocated to the Merritt Island Launch Area on September 19th and 20th for Plan X integrated tests.

On September 16th, a pivotal acceptance meeting for the target launch vehicle (TLV) 5307 took place in San Diego, marking a significant milestone in the Gemini program's preparations. Following successful acceptance, the vehicle was promptly shipped to Cape Kennedy, arriving on September 20th, ready to fulfill its new role.

Initially earmarked for the Lunar Orbiter program, TLV 5307's journey to becoming a key component of the Gemini missions resulted from strategic adaptations and meticulous planning. The decision to repurpose TLV 5307 arose from setbacks encountered earlier in the Gemini program. The failure of Atlas 5305 on May 17th necessitated the use of Atlas 5304 for launching the augmented target docking adapter. This circumstance prompted the Gemini Program Office (GPO) to secure an additional TLV to meet program requirements.

In May, negotiations concluded successfully to acquire Atlas 7127 from Vandenberg Air Force Base in California. However, due to significant differences between Atlas 7127 and the standard Gemini TLV, the GPO opted to utilize Atlas 5803, initially designated for the Lunar Orbiter program. Despite minor engineering change proposals (ECPs), which were thoroughly analyzed and deemed acceptable, modifications to adapt TLV 5307 for the Gemini program were swiftly completed by August 22nd.

Factory testing on September 12th validated the vehicle's readiness, ensuring it met stringent operational standards for the upcoming Gemini missions. This adaptive approach underscored the program's flexibility and commitment to overcoming challenges, ultimately securing TLV 5307 as a crucial asset in successfully executing Gemini missions.

Preparing GLV 12 for Mission

On September 19th, Gemini launch vehicle (GLV) 12 emerged from storage to begin its journey towards readiness at complex 19. This pivotal step signaled the start of meticulous preparations for its upcoming mission.

Following its removal, the GLV underwent thorough inspection, ensuring all components were in optimal condition for the mission ahead. By September 21st, umbilicals essential for mission operations were meticulously connected, marking a crucial milestone in the preparation process.

Power was applied to the GLV the following day, initiating a series of Subsystems Reverification Tests (SSRT) starting September 23rd. These tests, spanning several days until October 2nd, rigorously verified the functionality and integration of each subsystem critical to mission success.

On October 4th, the Prespacecraft Mate Verification Combined Systems Test further validated GLV 12's readiness for its mission objectives. This comprehensive test ensured that all systems, from propulsion to communication, operated harmoniously as a cohesive unit.

On September 21st, after completing intensive Plan X tests at the Merritt Island Launch Area, the Gemini Agena target vehicle (GATV) 5001 was returned to Hangar E for critical systems testing. This phase, essential for verifying operational integrity, continued diligently until September 29th, ensuring that all systems performed reliably under simulated mission conditions.

Following successful systems testing from September 29th to October 13th, the GATV progressed to the Combined Interface Test. This comprehensive test phase meticulously evaluated the interaction between the GATV and associated systems, including functional tests of both primary and secondary propulsion systems. By October 22nd, all tests were satisfactorily completed, affirming the readiness of GATV 5001 for its upcoming mission.

With preparations finalized, GATV 5001 was subsequently relocated to complex 14, poised and prepared to play a pivotal role in supporting the Gemini program's objectives. This systematic approach to testing and readiness underscored the program's commitment to meticulous planning and thorough validation, ensuring mission success in the challenging space exploration environment.

On September 23rd, significant adjustments were made to the extravehicular activities (EVA) planned for Gemini spacecraft No. 12, highlighting NASA's ongoing efforts to refine astronaut maneuvering capabilities in preparation for future missions.

The astronaut maneuvering unit (AMU), initially installed in spacecraft No. 12 on September 17th, was removed as the spacecraft underwent final preparations for relocation to complex 19. This decision came amidst persistent challenges encountered during earlier EVA attempts, prompting a reevaluation of EVA protocols and objectives.

Under the guidance of George E. Mueller, Associate Administrator for Manned Space Flight, NASA Headquarters decided to delete the AMU experiment from the Gemini XII mission. Mueller emphasized the need for astronauts to master fundamental EVA skills, including precisely executing basic tasks such as removing, installing, and tightening bolts, operating connectors and hooks, manipulating Velcro strips, and cutting cables.

This strategic pivot aimed to consolidate the foundation of EVA proficiency, ensuring that astronauts could perform essential tasks reliably and precisely. As NASA prepared for the final Gemini mission, these adjustments underscored a commitment to safety and operational excellence in the challenging space exploration environment.

On September 23rd, Gemini spacecraft No. 12 underwent pivotal preparations as it was relocated to complex 19 and swiftly hoisted atop its designated launch vehicle. This maneuver marked a crucial step toward readiness for the upcoming mission.

Following its installation atop the launch vehicle, meticulous premate verification procedures commenced promptly. These rigorous checks and tests, essential for ensuring all spacecraft systems' seamless integration and functionality with the launch vehicle, were diligently conducted over the following days.

By October 3rd, all premate verification procedures were completed. This milestone affirmed the readiness of Gemini spacecraft No. 12 for its forthcoming mission objectives, underscoring NASA's commitment to meticulous planning and operational readiness in the pioneering era of manned space exploration.

On September 26th, target launch vehicle (TLV) 5307 was erected at complex 14, marking a significant step in preparing for its crucial role in the upcoming mission.

The following day, comprehensive systems tests commenced to ensure the TLV's readiness and reliability. These tests, conducted meticulously over nearly three weeks until October 18th, rigorously assessed all critical systems and subsystems.

On October 5th, Gemini launch vehicle 12 and spacecraft No. 12 reached a critical milestone as they were electrically mated at complex 19. This essential step ensured seamless communication and integration between the spacecraft and its launch vehicle, setting the stage for upcoming mission preparations.

Simultaneously, the Electrical Interface Integrated Validation and Joint Guidance and Control Test occurred on October 5th and 6th. This test rigorously evaluated the interface and communication systems between the spacecraft and launch vehicle, ensuring all systems functioned harmoniously. The subsequent data review on the following day validated the integrity and performance of these integrated systems.

On October 10th, the Joint Combined Systems Test further solidified the readiness of Gemini spacecraft No. 12 for its upcoming mission objectives. This comprehensive test verified the functionality of combined systems, including guidance and control mechanisms, essential for mission success.

October 11th marked a crucial phase in the preparations for Gemini launch vehicle (GLV) 12, as it underwent a tanking test to validate its fueling systems and readiness for mission deployment.

Following the tanking test, spacecraft No. 12 underwent the Final Systems Test from October 17th to 19th, ensuring all onboard systems were operating optimally and ready for integration with the GLV.

October 23rd marked a significant achievement at complex 14 with successfully mating the Gemini Agena target vehicle (GATV) 5001 to the target launch vehicle (TLV) 5307. This pivotal integration marked the consolidation of essential components in readiness for an imminent mission.

Following integration, rigorous testing protocols were implemented to validate the integrated system's readiness and performance. The Joint Flight Acceptance Composite Test, which lasted from October 23rd to October 28th, meticulously assessed the functionality and compatibility of GATV 5001 and TLV 5307 under conditions simulating actual flight scenarios.

On October 24th, the Booster Flight Acceptance Composite Test further validated the TLV's readiness for mission deployment.

This test, encompassing a comprehensive evaluation of the TLV's performance under simulated flight conditions, confirmed its capability to meet the mission's stringent requirements.

On October 25th, a significant milestone was achieved as spacecraft No. 12 and GLV 12 were mechanically mated, symbolizing the culmination of extensive testing and preparations.

Subsequent tests included cycling the erector on October 25th, retesting the spacecraft guidance system on October 26th and 27th, and revalidating the spacecraft/GLV electrical interface on October 28th. These tests aimed to verify the integrity and functionality of critical systems in preparation for launch.

On November 1st, the Simultaneous Launch Demonstration was successfully conducted, followed by the Simulated Flight Test on November 2nd. These final tests comprehensively validated the readiness of both spacecraft and launch vehicle for the upcoming mission, ensuring all systems operated seamlessly under simulated mission conditions.

Completing the Simultaneous Launch Demonstration further underscored the integrated system's operational readiness. This successful demonstration validated its capability to perform flawlessly, affirming its preparedness for the upcoming mission with confidence.

On November 8th, the scheduled launch of Gemini XII faced an unexpected setback due to a malfunction in the launch vehicle's secondary autopilot system, detected just before the countdown for the planned November 9th launch commenced.

Efforts were promptly initiated to address the issue: both the secondary autopilot package and the secondary stage I rate gyro package were replaced as part of troubleshooting procedures. Following these replacements, the mission was rescheduled for launch on November 10th.

However, during tests conducted on November 9th to verify the functionality of the replacement autopilot, another malfunction occurred. This new issue was swiftly addressed by replacing the secondary autopilot package.

With the necessary corrections, the launch was rescheduled for November 11th, ensuring thorough testing and readiness before proceeding with the mission. These meticulous efforts highlighted NASA's dedication to resolving technical challenges swiftly and ensuring the safety and success of manned space missions during the pioneering era of space exploration.

Gemini XII: Advancing Space Exploration

On November 11, the Gemini Atlas-Agena target vehicle for the Gemini XII mission was launched from complex 14 at 2:08 p.m., e.s.t. Following this, the Gemini space vehicle, crewed by command pilot Astronaut James A. Lovell, Jr., and pilot Astronaut Edwin E. Aldrin, Jr., launched from complex 19 at 3:47 p.m. The mission's significant objectives included achieving rendezvous, docking with the Gemini Agena target vehicle (GATV), and evaluating extravehicular activities (EVA). Secondary objectives encompassed tethered vehicle evaluation, experiments, rendezvous and docking during the third revolution, automatic reentry demonstration, docked maneuvering for a high-apogee excursion, docking practice, systems tests, and GATV parking.

September 1966) — The Gemini-12 prime crew (in front) is astronauts James A. Lovell Jr. (right), command pilot, and Edwin E. Aldrin Jr. (left), pilot. In the rear is the Gemini-12 backup crew, astronauts L. Gordon Cooper Jr., (right), command pilot, and Eugene A. Cernan, pilot. NASA

The high-apogee excursion was not attempted due to an anomaly detected in the GATV's primary propulsion system during insertion. Additionally, GATV parking was not executed because its attitude control gas was depleted. Nevertheless, all other objectives were successfully achieved. The spacecraft performed nine maneuvers to rendezvous with the GATV, overcoming an onboard radar malfunction during the terminal phase. The maneuver was initiated by using backup procedures for maneuver calculations.

Astronaut Edwin E. Aldrin, Jr., carrying a micrometeoroid package to the spacecraft from the adapter section during extravehicular activity on Gemini XII. (NASA Photo No. 66-H-753 [66-HC-1546], released Nov. 16, 1966.)

Rendezvous was achieved at 3 hours 46 minutes ground elapsed time, followed by docking 28 minutes later. Two phasing maneuvers were conducted using the GATV's secondary propulsion system, as the primary system was not utilized. The mission included two periods of standup EVA, with the first lasting 2 hours 29 minutes and starting at 19 hours 29 minutes into the flight. The second standup EVA lasted 55 minutes, ending at 67 hours 1 minute ground elapsed time. An umbilical EVA lasting more than two hours commenced at 42 hours 48 minutes, during which Aldrin attached a 100-foot tether from the GATV to the spacecraft docking bar.

The tether evaluation began at 47 hours 23 minutes after liftoff, during which the crew observed that the tether tended to remain slack. However, they believed the two vehicles slowly achieved gravity-gradient stabilization. The crew eventually jettisoned the docking bar and released the tether at 51 hours 51 minutes. Several spacecraft systems encountered issues throughout the mission, including two fuel cell stack failures and reduced power from two other stacks. At 39 hours 30 minutes ground elapsed time, two orbit attitude and maneuver thrusters provided little or no thrust.

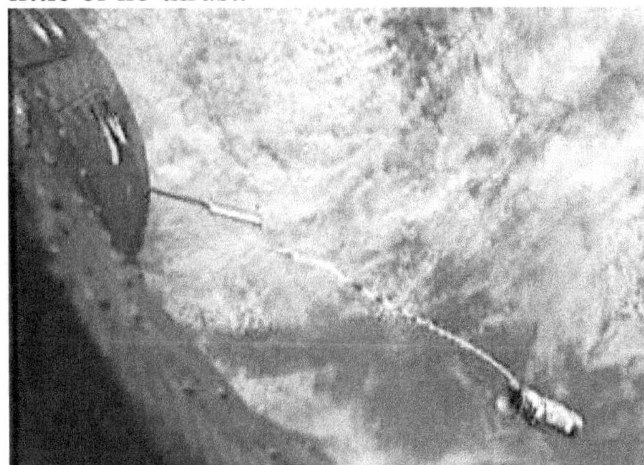

The Gemini Agena target vehicle tethered to the spacecraft during the Gemini XII mission. (NASA Photo No. 66-H-751, released Nov. 16, 1966.)

The retrofire sequence occurred 94 hours after liftoff, leading to an automatically controlled reentry. The spacecraft landed less than three miles from the planned landing point at 2:21 p.m. on November 15. The crew was swiftly retrieved by helicopter and brought to the prime recovery ship, the aircraft carrier Wasp, 28 minutes later. The spacecraft was recovered 67 minutes after landing, concluding the Gemini XII mission.

Conclusion of the Gemini Program

On February 1, 1967, a pivotal transition marked the conclusion of NASA's Gemini program with the abolition of the Manned Spacecraft Center's (MSC) Gemini Program Office. This office had been instrumental in overseeing the program's operations and missions, pushing the boundaries of human spaceflight.

Responsibility for final Gemini activities, including the disposal of equipment and resolution of contract matters, was entrusted to George F. MacDougall, Jr., newly appointed as Special Assistant for Gemini within MSC's Office of the Director of Administration. This administrative shift signified the beginning of a gradual winddown of the program, a process expected to span several years as efforts were gradually reduced.

Through its ten crewed missions, the Gemini program achieved numerous milestones critical to NASA's broader ambitions in space exploration. From perfecting orbital maneuvers and rendezvous techniques to pioneering extravehicular activities (EVAs), Gemini had laid the groundwork for subsequent missions, including the Apollo program's lunar landings.

On February 1-2, 1967, the Manned Spacecraft Center hosted a pivotal Gemini Summary Conference, marking a comprehensive review of the program's achievements and findings. Directed by Robert R. Gilruth, the conference brought together experts and officials to analyze the outcomes of the final Gemini missions through a series of 22 presentations.

The conference sessions were structured to highlight key aspects of the Gemini program's successes and lessons learned. Topics included in-depth discussions on orbital rendezvous and docking operations, which were critical for future space missions requiring multiple spacecraft interactions. Extravehicular activities (EVAs), another milestone of the Gemini program, were extensively reviewed to assess their operational feasibility and refine procedures for future missions.

Operational experience gleaned from the missions provided valuable insights into spacecraft systems and crew management in prolonged space missions. Additionally, the results of experiments conducted aboard Gemini flights were presented, offering scientific data crucial for understanding human adaptation to space environments and advancing technological capabilities.

The Gemini Summary Conference celebrated the program's accomplishments and disseminated knowledge to inform future endeavors in space exploration. It underscored NASA's commitment to continuous improvement and innovation in the pursuit of human spaceflight goals.

<u>Epilogue</u>

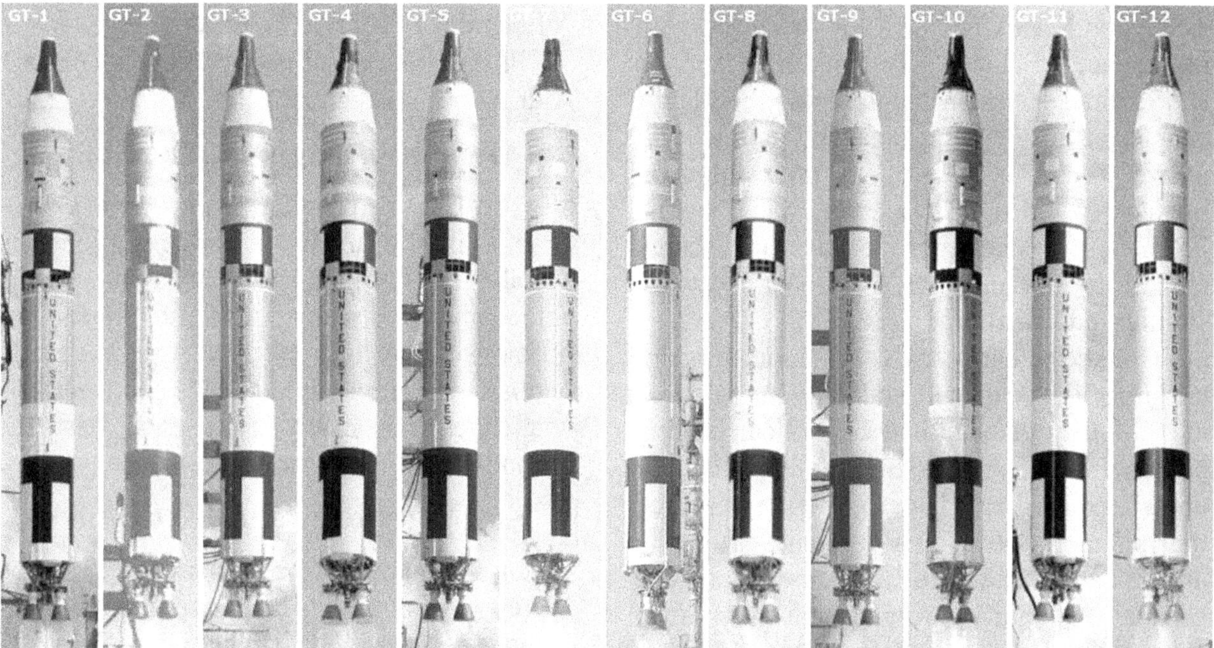

The Gemini program was designed as a bridge between the Mercury and Apollo programs, primarily to test equipment and mission procedures in Earth orbit and to train astronauts and ground crews for future Apollo missions. The general objectives of the program included long-duration flights over of the requirements of a lunar landing mission; rendezvous and docking of two vehicles in Earth orbit; the development of operational proficiency of both flight and ground crews; the conduct of experiments in space; extravehicular operations; active control of reentry flight path to achieve a precise landing point; and onboard orbital navigation. Each Gemini mission carried two astronauts into Earth orbit for 5 hours to 14 days. The program consisted of 10 crewed launches, two uncrewed launches, and seven target vehicles, costing approximately 1,280 million dollars.

Throughout the Gemini program, astronauts conducted 25 biomedical experiments, many repeated across multiple missions. These experiments, predominantly using astronauts as subjects, sought to explore how the human body responds to the unique challenges of space. While most experiments were meticulously controlled with specific dietary requirements, Gemini III was an exception, allowing for the novelty of a corned beef sandwich in space without disrupting scientific protocols.

Among the experiments flown was an ambitious study involving sea urchin eggs, intended to observe fertilization and cell division in microgravity. However, a mechanical mishap during Gemini III, when Grissom inadvertently broke the experiment's handling mechanism, thwarted its objectives. Undeterred, NASA redesigned the canister and substituted frog eggs, achieving their goals on subsequent missions like Gemini VIII and XII. These experiments ultimately demonstrated that crucial biological processes like cell division can occur in space, challenging prior assumptions about the effects of microgravity.

Since those pioneering experiments with sea urchin and frog eggs, NASA has continued advancing its space biology research. Notably, mice accompanied astronauts during the final Apollo mission to study the impacts of deep space radiation. Today, as NASA prepares for Artemis I with its Space Launch System rocket, the BioSentinel mission represents the latest chapter in this ongoing saga. BioSentinel will utilize yeast to investigate the biological impacts of deep space radiation beyond Earth's orbit, leveraging decades of research aboard the International Space Station, space shuttles, and small satellites.

Unlike the early days of Gemini, modern missions like Artemis I and BioSentinel were meticulously planned and controlled, leaving no room for clandestine sandwiches or unplanned mishaps. Instead, they represented the pinnacle of scientific achievement in understanding how life functions beyond the confines of Earth—a testament to human curiosity and the quest for knowledge in the vast expanse of space.

Project Gemini emerged as NASA's crucial bridge between the pioneering Mercury missions and the historic Apollo lunar landings. From 1965 to 1966, Gemini extended human spaceflight capabilities and paved the way for America's ambitious lunar exploration goals. Here are the missions and their pilots:

Gemini 3 (March 1965)
Pilots: Gus Grissom, John Young
Gemini 4 (June 1965)
Pilots: James McDivitt, Ed White (First American spacewalk)
Gemini 5 (August 1965)
Pilots: Gordon Cooper, Charles Conrad
Gemini 6A (December 1965)
Pilots: Wally Schirra, Thomas Stafford
Gemini 7 (December 1965)
Pilots: Frank Borman, James Lovell
Gemini 8 (March 1966)
Pilots: Neil Armstrong, David Scott
Gemini 9A (June 1966)
Pilots: Thomas Stafford, Eugene Cernan
Gemini 10 (July 1966)
Pilots: John Young, Michael Collins
Gemini 11 (September 1966)
Pilots: Charles Conrad, Richard Gordon
Gemini 12 (November 1966)
Pilots: James Lovell, Edwin "Buzz" Aldrin

Gemini: Bridging to the Stars

The Gemini program stands as a pivotal chapter in the annals of space exploration, seamlessly bridging Mercury's pioneering achievements with Apollo's audacious missions. Launched by NASA in the early 1960s, Gemini aimed to conquer the challenges of space travel and lay the groundwork for lunar missions.

Key Milestones and Innovations

In a departure from its predecessor, the Mercury program, Gemini spacecraft were launched exclusively by the Titan II rocket, save for the reflight of Gemini 2, which utilized the Titan IIIC for tests related to the Manned Orbiting Laboratory in 1966. This transition marked a technological leap, enhancing the capability and reliability needed for extended space missions.

Gemini 3, commanded by Gus Grissom and piloted by John Young, utilized the Mercury Control Center at Cape Kennedy due to the new Houston-based center still undergoing tests. Subsequent missions, starting from Gemini 4, were controlled from Houston, a shift that solidified the city's role as the heart of American space operations.

The program's initial ambition to implement a paraglider for land-based touchdown was abandoned in 1964, highlighting the adaptive nature of NASA's mission planning in response to technological challenges and operational considerations.

Recapping Gemini's Accomplishments

The Gemini program, conducted by NASA between 1964 and 1966, marked significant milestones in space exploration and paved the way for the Apollo missions to the Moon. Here are some of the key highlights:

NASA conducted two uncrewed Gemini missions in April 1964 and January 1965 to test spacecraft systems and the heat shield.

The first crewed Gemini mission was Gemini 3, launched on March 23, 1965, with astronauts Virgil "Gus" Grissom and John W. Young aboard. It was the first multi-crewed US mission and the first to use thrusters to change its orbit.

Gemini 4, launched on June 3, 1965, saw astronaut Ed White perform the first American spacewalk, or extravehicular activity (EVA).

Gemini 5, launched on August 21, 1965, demonstrated an 8-day mission duration, crucial for Apollo lunar missions. It was also the first use of fuel cells to generate electrical power.

Gemini 6A (originally intended as Gemini 6 but renamed after a launch attempt was aborted) accomplished the first space rendezvous with Gemini 7 in December 1965. Gemini 7 set a 14-day endurance record.

Gemini 8, launched on March 16, 1966, achieved the first space docking with an uncrewed Agena target vehicle.

Gemini 10, launched on July 18, 1966, demonstrated operations at high altitudes, performed another successful rendezvous with an Agena target, and utilized the Agena's rocket to change its orbit.

Gemini 11, launched on September 12, 1966, set a crewed Earth orbital altitude record of 739.2 nautical miles (1,369.0 km) using the Agena target vehicle's propulsion system.

Gemini 12, launched on November 11, 1966, featured astronaut Buzz Aldrin performing spacewalks with newly implemented tools like footholds and handholds, demonstrating that useful work could be conducted outside the spacecraft.

The Gemini program refined complex orbital maneuvers like rendezvous and docking, crucial for lunar missions. Astronauts trained extensively using simulators to practice these maneuvers.

The Gemini program, a pivotal phase in American space exploration during the 1960s, utilized the Gemini-Titan II launch vehicle, adapted from the U.S. Air Force's Titan II ICBM. This rocket, integral to all Gemini missions, featured Air Force serial numbers painted on it, reflecting its military origins and the collaboration between NASA and the Air Force.

Launch Complex and Operations:

Launch Complex 19 (LC-19): At Cape Kennedy Air Force Station in Florida, LC-19 was the launch site for all Gemini missions. It was specifically tailored for the Gemini-Titan II launches, providing a secure and controlled environment for mission preparations and liftoffs.

The newly constructed Mission Control Center at the Houston Manned Spacecraft Center was crucial in managing Gemini missions. This advanced facility monitored and controlled every aspect of the missions, ensuring astronaut safety and mission success from liftoff through to splashdown.

Astronaut Corps and Training:

The Gemini program included astronauts from multiple groups: the original "Mercury Seven," followed by "The New Nine," and later "The Fourteen." These groups collectively contributed their expertise and courage to advance American space capabilities.

Tragically, the Gemini program faced setbacks, including the loss of three astronauts in air crashes during training missions. This included both members of the prime crew designated for Gemini 9. The backup crews stepped forward despite these tragedies, demonstrating resilience and dedication to continue the mission objectives.

The Gemini program demonstrated crucial advancements in orbital rendezvous, spacewalking, and long-duration missions and laid the groundwork for the successful Apollo lunar landings. It exemplified the spirit of exploration and technical innovation that defined NASA's efforts during the space race era.

The Gemini program left a profound legacy on American space exploration, not only achieving its primary objectives but also influencing future space initiatives:

Military Applications and Manned Orbital Laboratory (MOL):

The U.S. Air Force explored adapting Gemini for military purposes, leading to projects like the Manned Orbital Laboratory (MOL). This involved reusing the Gemini 2 spacecraft atop a mockup of the MOL, marking the first instance of a spacecraft flying into space twice.

Blue Gemini was another Air Force initiative that modified Gemini for surveillance and satellite rendezvous tasks. However, this project was canceled in favor of utilizing uncrewed spy satellites more cost-effectively.

Extended Capabilities and Ambitious Proposals:

Jim Chamberlin, Gemini's chief designer, envisioned broader applications for the spacecraft, including lunar missions and cislunar operations before Apollo. NASA's administration did not pursue his proposals to utilize Gemini for lunar landings at a lower cost at the time.

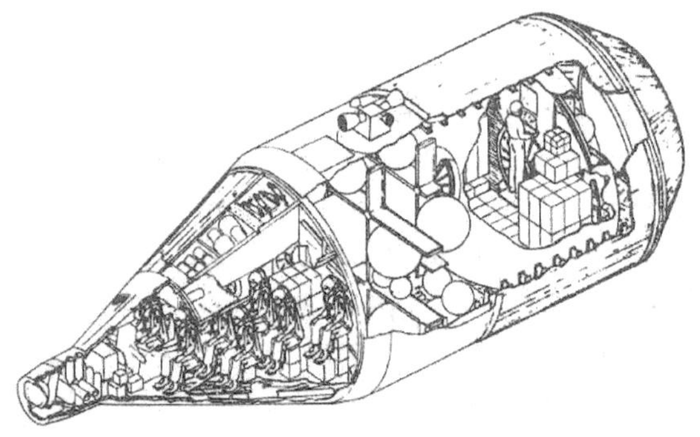

"Big Gemini" (Big G) was a conceptual enlargement of the Gemini spacecraft proposed to support missions like space station resupply and large-capacity crew transportation. Despite detailed studies and proposals, funding shifted towards other programs like Skylab, leading to the eventual shelving of Big Gemini.

Legacy and Influence:

Despite its relatively short operational span, Gemini significantly advanced American capabilities in orbital flight, astronaut training, rendezvous and docking procedures, and extravehicular activities (EVAs).

The program's success laid the groundwork for Apollo's lunar missions by demonstrating crucial capabilities such as long-duration spaceflight and precision orbital maneuvers.

Gemini's modular design and operational experience directly influenced subsequent NASA programs' development and operational concepts, including Skylab and the Space Shuttle.

Program Cost and Impact:

From 1962 to 1967, the Gemini program cost approximately $1.3 billion in 1967 dollars, equivalent to about $9.07 billion in 2023 dollars. This investment underscored its pivotal role in bridging the gap between Mercury and Apollo, ensuring NASA's readiness for ambitious lunar exploration goals.

International Context and Achievements

During its ten crewed flights, the Gemini program achieved several notable firsts in space exploration. Despite the Soviets' pioneering of the first Extravehicular Activity (EVA), commonly known as a spacewalk, in 1965, subsequent Soviet EVAs did not occur until January 1969, a testament to Gemini's pioneering role in advancing human space activities.

Gemini's final mission, highlighted by the rendezvous and manual docking with the Agena target vehicle, showcased astronaut Buzz Aldrin's record-breaking 5 hours and 30 minutes EVA. Aldrin's accomplishments during this mission set new benchmarked for spacewalk duration and demonstrated crucial advancements in resolving previous EVA challenges.

Astronaut Contributions and Personnel

The Gemini program also marked significant milestones in astronaut recruitment and training. Already an accomplished test pilot with NASA, Neil Armstrong became one of the first civilian astronauts alongside Elliot See in Astronaut Group 2. Armstrong's career trajectory within NASA, including his notable achievements and awards, underscores the program's role in shaping astronautical history.

The crew assignments for Gemini missions were not without changes. Thomas Stafford, initially designated as Pilot for Gemini 3 alongside Alan Shepard, swapped roles with John Young after Shepard's grounding due to Ménière's disease, highlighting the program's flexibility in adapting to medical and operational challenges.

Cultural and Media Impact

Beyond its scientific and technical achievements, Gemini captivated the public imagination and left an indelible mark on popular culture. The program featured prominently in films such as "You Only Live Twice" and "Countdown," where modified Gemini capsules embarked on fictional lunar missions, reflecting the era's fascination with space exploration.

Television series like HBO's "From the Earth to the Moon" and documentaries such as PBS's "Spaceflight" further cemented Gemini's legacy, immortalizing its contributions to humanity's quest for the stars.

Even decades later, the program inspires, as evidenced by its portrayal in recent films like "First Man" and occasional nostalgic references in television classics like "I Dream of Jeannie."

Apollo 1 Fire: Tragedy and Aftermath

On January 27, 1967, tragedy struck the American space program with a devastating fire aboard the Apollo 1 spacecraft during a ground test at Cape Kennedy's Launch Complex 34. The loss of astronauts Virgil I. "Gus" Grissom, Edward H. White, and Roger B. Chaffee reverberated across the nation and around the world, marking a profound setback for NASA's ambitious lunar exploration goals.

The Incident

Inside the Command Module (CM), designed for what was to be the first piloted Apollo flight, the astronauts were engaged in a routine countdown simulation. Sealed within the capsule, the crew and ground personnel pressurized the atmosphere with pure oxygen, preparing for the upcoming mission. Unexpectedly, at 6:31 p.m., a flash fire erupted within the cabin. Rapidly escalating temperatures and pressures caused structural damage to the CM, leading to the tragic loss of the astronauts.

Dense smoke and intense heat hindered efforts to rescue Grissom, White, and Chaffee. Despite valiant attempts by ground crews to open the hatches, the astronauts succumbed to asphyxiation from carbon monoxide and other toxic gases. Recovery efforts later that night confirmed the heartbreaking outcome, documenting the scene and conducting examinations that revealed the astronauts' cause of death.

National Mourning and Response

The deaths of these three pioneering astronauts prompted a national outpouring of grief and solemn remembrance. President Lyndon B. Johnson and Vice President

Hubert H. Humphrey expressed heartfelt condolences, acknowledging the profound loss to the nation and the space exploration community. Former President Dwight D. Eisenhower and even Radio Moscow offered condolences, underscoring the tragedy's international impact.

In the aftermath, NASA launched a comprehensive investigation to determine the cause of the fire. The resulting investigative board identified critical technical and management shortcomings that contributed to the accident. Concurrent congressional hearings scrutinized NASA's protocols and spurred reforms, including redesigning the Apollo spacecraft and significant changes to safety protocols and organizational culture.

Memorial and Legacy

Memorial services held in Houston honored each astronaut individually, attended by NASA leadership and their fellow astronauts. The funerals, conducted with full military honors, underscored their service and sacrifice. Grissom and Chaffee were laid to rest at Arlington National Cemetery, with President Johnson in attendance. At the same time, White was interred at West Point Military Academy, honored by First Lady "Lady Bird" Johnson and Vice President Humphrey.

The legacy of Grissom, White, and Chaffee endures as a reminder of the risks inherent in space exploration and the unwavering commitment of those who venture beyond Earth's boundaries. Their tragic deaths galvanized NASA's resolve to forge ahead, leading to safer spacecraft designs and procedures that would ultimately propel humanity to the lunar surface and beyond.

The Apollo 1 fire was a poignant chapter in the history of space exploration. It memorializes the bravery and sacrifice of these courageous astronauts while shaping the future of manned spaceflight.

Investigation and Reform: Lessons from the Apollo 1 Tragedy

Following the catastrophic fire on January 27, 1967, that claimed the lives of astronauts Virgil I. "Gus" Grissom, Edward H. White, and Roger B. Chaffee during a ground test at Cape Kennedy's Launch Complex 34, NASA swiftly mobilized to investigate the incident and implement critical reforms.

Establishing the Apollo 204 Review Board

In response to the Apollo 1 accident, NASA's Kennedy Space Center (KSC) imposed stringent security measures at LC-34 and sequestered all related materials. On January 28, Deputy NASA Administrator Robert C. Seamans appointed the Apollo 204 Review Board to conduct a thorough investigation. The board, led by Floyd L. Thompson of NASA's Langley Research Center, included key figures like Maxime A. Faget and astronaut Frank Borman. They commenced their inquiry immediately, visiting the accident site and initiating interviews with witnesses.

Investigative Procedures and Findings

NASA transported a nearly identical Command Module (CM-014) from North American Aviation (NAA) in California to KSC on February 1 to facilitate the investigation. Engineers practiced component removal on CM-014 before tackling the damaged CM-012, which still housed the remnants of the tragic event. Hardware removal began after disassembling the Launch Escape System tower and relocating components to the Pyrotechnics Installation Building (PIB) for detailed examination.

By March 7-9, workers had dismantled CM-012, meticulously documenting 1,261 individual items, focusing on the crew compartment and heat shields. The investigation culminated in a comprehensive 3,000-page report released on April 9. While the exact ignition source remained elusive, investigators identified arcing in wiring near

Grissom's position, exacerbated by the spacecraft's oxygen-rich environment pressurized to 16.7 psi.

Congressional Hearings and Reforms

Both chambers of Congress convened hearings to review the board's findings and NASA's proposed reforms. The House Committee on Science and Astronautics and the Senate Committee on Aeronautical and Space Sciences scrutinized NASA's safety protocols and oversight. Senior NASA officials and astronauts testified extensively, outlining responses to the board's recommendations and highlighting the need for organizational and procedural changes.

In the aftermath, NASA enacted significant reforms. Associate Administrator George E. Mueller announced structural changes within NASA, appointing George M. Low to manage the Apollo Spacecraft Program Office (ASPO). Joseph F. Shea assumed a pivotal role in technical oversight, marking a shift towards enhanced safety and reliability practices. At the agency level, NASA established the Aerospace Safety Advisory Panel (ASAP) to provide independent oversight—a crucial step towards preventing future tragedies.

Legacy and Impact

The Apollo 1 fire remains pivotal in space exploration history, galvanizing NASA's commitment to safety and resilience. The reforms spurred by the tragedy reshaped NASA's spacecraft design, operational protocols, and organizational culture. The establishment of the ASAP continues to safeguard manned spaceflight endeavors, ensuring that the sacrifices of Grissom, White, and Chaffee serve as enduring lessons in pursuing human exploration beyond Earth's bounds.

Post-Accident Reforms and Legacy

In the wake of the Apollo 1 tragedy, NASA undertook sweeping reforms to enhance spacecraft safety and crew survivability, setting a new standard for space exploration.

Hardware and Design Modifications

NASA transitioned all subsequent Apollo missions to the more advanced Block II version of the Command Module (CM), abandoning the Block I configuration used by Apollo 1. The Block II CM incorporated significant safety improvements, including:

The cumbersome three-piece hatch that impeded the escape of Grissom, White, and Chaffee was replaced with a unified hatch. This new design allowed for rapid opening from both inside and outside the spacecraft, reducing egress time to three seconds.

Block II CMs featured a docking probe and transfer tunnel, facilitating docking maneuvers with the Lunar Module (LM). This capability enabled crew transfers between docked vehicles, enhancing mission flexibility and safety.

Material and Safety Enhancements

To mitigate fire risks, NASA implemented stringent controls over materials used in spacecraft construction:

NASA significantly restricted and monitored the presence of combustible materials within spacecraft interiors, not only in the CM but also extending these measures to the Lunar Module.

A new generation of spacesuits was developed using fire-resistant materials, providing added protection to astronauts during emergencies.

Crew Assignments and Mission Planning for Project Apollo

In May 1967, NASA announced the crew assignments for the first piloted Apollo mission post-Apollo 1. Walter M. Schirra, Donn F. Eisele, and R. Walter Cunningham were selected as the prime crew, with Thomas P. Stafford, John W. Young, and Eugene A. Cernan designated as backups. This decision underscored NASA's commitment to resuming manned spaceflight operations while prioritizing crew safety.

Preservation and Commemoration

Following the investigation's conclusion, the Apollo 1 Command Module (CM-012) was relocated to a secure facility at Langley as directed by the Review Board. In 2007, recognizing the historical significance of the accident, NASA moved the capsule to a modern, environmentally-controlled warehouse. To honor the 50th anniversary of the tragedy in 2017, NASA displayed the original three-piece hatch at the Apollo/Saturn V Center at Kennedy Space Center, serving as a poignant reminder of the sacrifices made in pursuing space exploration.

Enduring Impact

The Apollo 1 fire prompted a fundamental reassessment of NASA's safety protocols and operational practices. The reforms implemented in the aftermath of the tragedy enhanced spacecraft design and crew training and fostered a culture of rigorous safety standards that continues to guide human spaceflight endeavors to this day. The legacy of Gus Grissom, Ed White, and Roger Chaffee remains indelible in NASA's history, inspiring future generations of astronauts and engineers to push the boundaries of exploration while prioritizing safety above all else.

About the Author

Thornton D. "TD" Barnes, author and entrepreneur, grew up on a ranch at Dalhart, Texas. He graduated from Mountain View High School in Oklahoma and embarked on a ten-year military career. Following a tour as an Army intelligence specialist in Korea, he continued his education, attending two and a half years of missile and radar electronics by day and college courses at night. Midway through the first combat Hawk missile battalion deployment, Barnes attended the Artillery Officer Candidate School, where an injury ended his military career.

Barnes first became involved with US intelligence agencies while in the Army during the Agency's Project Palladium. An ongoing relationship with the agency continues today. Barnes's career includes serving as a field engineer at the NASA High Range in Nevada for the X-15, XB-70, lifting bodies and lunar landing vehicles; working on the Nuclear Engine for Rocket Vehicle Application, NERVA project at Jackass Flats, Nevada for crewed flight to Mars; and serving in Special Projects for the CIA at Area 51. Barnes later formed a family oil and gas exploration company, drilling and producing oil and gas and mining uranium and gold.

Barnes currently serves as the president of Roadrunners Internationale, an association of Area 51 CIA, Air Force, and contractor veterans. He is the executive director of the Nevada Aerospace Hall of Fame.

Two National Geographic Channel documentaries feature Barnes: Area 51 Declassified and CIA—Secrets of Area 51.

Numerous documentaries on the History Channel, the Discovery Channel, the Travel Channel, and others also feature him. Barnes authored CIA Station D Area 51 and several books approved by the CIA PRB, including the CIA Area 51 Chronicles, a three-book series about the declassified CIA U-2, A-12, MiG, and stealth projects at Area 51. Barnes remains active with oral history projects at the University of Nevada Las Vegas, the Central Intelligence Agency, the Defense Intelligence Agency, and the National Reconnaissance Office. Barnes lives in Henderson, Nevada.

References

National Aeronautics and Space Administration Project Gemini, Technology and Operations, A Chronology, Published as NASA Special Publication-4002., Prepared by, James M. Grimwood and Barton C. Hacker with Peter J. Vorzimmer - https://www.nasa.gov/history/SP-4002/contents.htm - Published June 1968

https://www.nasa.gov/gemini/